Emerging Challenges to Food Production and Security in Asia, Middle East, and Africa

Mohamed Behnassi · Mirza Barjees Baig ·
Mahjoub El Haiba · Michael R. Reed
Editors

Emerging Challenges to Food Production and Security in Asia, Middle East, and Africa

Climate Risks and Resource Scarcity

 Springer

Editors
Mohamed Behnassi
College of Law, Economics and
Social Science
Ibn Zohr University and CERES
Agadir, Morocco

Mahjoub El Haiba
College of Law, Economics and
Social Science
Hassan II University
Casablanca, Morocco

Mirza Barjees Baig
Prince Sultan Institute for Environmental
Water and Desert Research
King Saud University
Riyadh, Saudi Arabia

Michael R. Reed
University of Kentucky
Lexington, KY, USA

ISBN 978-3-030-72989-9   ISBN 978-3-030-72987-5 (eBook)
https://doi.org/10.1007/978-3-030-72987-5

This Springer imprint is published by the registered company Springer Nature Switzerland AG
The registered company address is: Gewerbestrasse 11, 6330 Cham, Switzerland

Preface

Although it was the food crisis in 2008 and its various implications that initially stimulated renewed focus on the food security issue, assessments of the underlying dynamics have recognized that what is at stake exceeds the sole objective of simply meeting growing food demands and that food security is not by any measure a trivial challenge. For instance, most people are increasingly hungry not because of insufficient food, but because they cannot afford it. Therefore, it is the effective demand for food and how food is distributed and accessed across and within countries that ultimately matter. The governance of food security is thus becoming of paramount importance and this shift in perception of the food security challenge has to shape future response mechanisms at all levels. In terms of food availability, some emerging risks, especially climate change and resource scarcity, have the potential to undermine the capacity of food systems to go on producing food according to current or expected levels. Indeed, the scientific and empirical evidence increasingly demonstrates that environmental and climatic changes pose major challenges to food and water security, especially that food production systems are still highly dependent on natural resources—such as water, land, and biodiversity—and that unsustainable agricultural and food systems are negatively impacting, in turn, their natural base.

More specifically, the current global food system—as a social-ecological system—is highly impacted by environmental and climatic dynamics through resource scarcity, desertification, changes in temperature and precipitation patterns, extreme events (such as flooding, drought, heat, and cold waves), ocean acidification, and sea-level rise. The considerable impact of climate change on the quality and availability of water and soil resources in many regions, especially Asia, Middle East, and Africa, is increasingly worrying. Global climate model (GCM) ensemble projections predict that by the 2050s, the increased crop water demand and intensified evapotranspiration resulting from global warming will substantially reduce water resources surplus and increase significantly the irrigation water demand in crop growth periods. Therefore, the potential demand for crop irrigation water will increase under climate change projections; however, the ability to expand irrigation will be constrained by the availability of water resources. Irrigation supplies are particularly vulnerable to shifts in precipitation, water cycling, and the demand of water from non-agricultural sectors. All these dynamics are disturbing the stability and sustainability of food production systems in concerned regions, thus threatening their food security.

In the meantime, the current global food system is a major driver of climate change, biodiversity loss, and health insecurity given its higher carbon footprint, its negative impacts on water and soil resources, its contribution to the degradation of aquatic and terrestrial ecosystems, its excessive use of chemical inputs, and its responsibility for increased mortality and animal suffering.

In addition to these environmental and climate-induced dynamics, among the factors strongly influencing food systems, and yet difficult to foresee, are regulatory frameworks. Laws and policies, particularly in the realm of agricultural development, food and water security, and land management, are generally made at the national level in response to domestic dynamics. Therefore, the challenges we are now facing in reorienting food systems for improved sustainability, climate resilience, and food, water, and health security are in part an outcome of past trajectories, which carried huge productivity and production achievements, but also introduced new negative environmental, social, and cultural externalities. The intensification of food production has progressively weighed on the world's natural resources and ecosystems, with critical concerns around water availability and quality, soil degradation, air quality, and carbon emissions. Less easy to measure and reverse, but now known to be vitally important, are the economic, social, environmental, and cultural implications of such political and technological choices. This shift from the traditional smaller mixed crop-livestock agricultural production units to the development of specialized large production units and homogenized agricultural landscapes, a trend which is massively underway in many developing countries as well, is associated to many problems affecting the sustainability and resilience of social-ecological systems. Debates continue on the question of biophysical limits in these critical areas and the extent, but not the direction, of change required, especially in light of projected food and water demand increases from a larger and richer population in many countries in addition to pressing related health issues.

In this perspective, it seems that improving long-term resilience of food and water security in the context of a changing climate, biodiversity loss, pressure on scarce resources, especially land and water, increasing global population, and changing dietary preferences will be very costly to achieve under the persistence of current trends. Actually, the way we produce, distribute, and consume food, especially in Northern and the so-called emerging countries, is increasingly considered as a human and environmental disaster, not to mention the ethical aspects. Scientists predict that within the coming decades, the ecological, climatic, and health challenges of the global food system could dramatically increase in the absence of appropriate response mechanisms, reaching levels beyond the planetary boundaries that define a safe operating space for humanity. It was also shown that options for reducing the negative environmental and health effects of the food system, while meeting increasing food demands sustainably and inclusively, need to be implemented simultaneously and in an integrated way according to innovative conceptual and policy frameworks.

While adaptation is more the norm than the exception when it comes to facing climate change impacts on food, water, and soil resources, the key challenge in the coming decades will be the scale and nature of environmental and climatic changes compared with the coping capacity of concerned systems. If future changes are incrementally small, these systems may adapt to emerging risks by adjusting structurally. However, it is unlikely that incremental change alone will be sufficient to enable, for instance, some food production systems and rural regions to function within the limits of adaptive social and ecological systems which often have significant but finite capacities to adapt. Exceeding adaptation limits will result in multiplying losses or require a transformational change capacity that many countries are lacking.

Technically, the potential exists to narrow, for instance, the food productivity or yield gap between the highest and lowest producers in a region and thus increases food production, including livestock and animal feed production, to meet expected food demands in the future and reduce losses and wastes along the food chain. A key question here is how to incorporate well-understood technologies and approaches to different scales and types of producers considering, at the same time, farmer innovation and knowledge, improving yields and diversifying production systems, and allowing poor producers to have access to land, water, and biodiversity resources in order to improve the design and management of their agricultural systems. The potential also exists to improve the diets, nutritional status, and health of poor and vulnerable people but there are at least three key questions to address in actually realizing that potential. *First*, how can food and water security be achieved in ways and in systems that conserve natural resources, reduce pollution, adjust to climate change, and respect established social and cultural values? *Second*, how can effective policies be designed and implemented to achieve food and water security inclusively across and within different countries and societal groups? *Third*, what are the ways and means to address the trade-offs involved in moving toward more sustainable and climate-friendly food systems?

This contributed volume attempts to provide answers to many of these questions. Chapters are mostly empirically based and providing context-specific analyses and recommendations based on a variety of case studies from Africa (Morocco, Tunisia, Egypt, Nigeria, and sub-Saharan countries), Middle East (Gulf countries), and Asia (Pakistan and Thailand). The main topics and conclusions of these chapters are presented below.

In Chap. 1, "Enhancing Resilience for Food and Nutrition Security Within a Changing Climate," Behnassi et al. acknowledge that climate change, even with a 1.5 °C scenario, is a major impediment to the performance of food production systems since it is adversely impacting the systems' processes and their natural base while increasing the vulnerability of human societies—especially resource-poor small producers—and diminishing their resilience to food and nutrition insecurity. Accordingly, the authors attempt to analyze the dynamics through which climate change affects food and nutrition security and drawing the pathways toward resilience building in this area. They start by analyzing the predominant impacts of climate change on food security and resilience, then assess these dynamics from international and regional perspectives, and finally explore some best pathways and approaches toward building resilience for food and nutrition security in a changing climate. The authors conclude that the resilience to food and nutrition insecurity should be based on deep changes of food and farming systems and methods that support small-holder farmers in particular. Compassion for farmed animals, enhancement of soils, the protection of biodiversity and wildlife could meanwhile massively reduce negative impacts on the climate. In addition, climate-resilient and sustainable agricultural livelihoods are possible and can yield mitigation, adaptation, and resilience co-benefits. Therefore, it is essential and promising to build climate resilience by working with nature and reshape investments across terrestrial and marine systems.

In Chap. 2, "Impacts of Climate Change on Agricultural Sector of Pakistan: Status, Consequences, and Adoption Options," Usman et al. claim that Pakistan is an agrarian economy and agriculture still serves as the backstone of the economic and social development. However, this sector is vulnerable to the impacts of climate change, and it is increasingly categorized as a risky occupation. Furthermore, Pakistan has an insufficient coping capacity to adjust to these adverse impacts. The country has the world's largest irrigation system which is mainly fed by snow and glacial melt, but this system is threatened due to fast melting glaciers, risk of floods, droughts, landslides, power shortages, and avalanches. It is a continually mounting concern with unlimited importance owing to its pronounced, comprehensive socioeconomic effects. Based on this context, the authors analyze these dynamics, demonstrate the role of agriculture under climate change and sustainable development goals (SDGs), and formulate policy-oriented recommendations with the potential to help address the impacts of climate change on agriculture in the country.

In Chap. 3, "Impacts of Climate Change on Agriculture and Food Security in Tunisia: Challenges, Existing Policies, and Way Forward," Ouessar et al. start with the fact that climate change is expected to negatively affect many sectors, but the heaviest impacts will affect the agricultural sector as it depends largely on natural resources and meteorological conditions. In the Tunisian context, where agriculture plays a key economic and social role, the main impacts of climate change on this sector would be a disruption of the cropping cycle for the main agricultural products. As a result, the country's food security will be severely threatened. To face such a challenge, Tunisia's food security strategy tries to improve the living standard of the population—where about 83% of the severely food-insecure households belong to the poorest (50%) and poor (33%) sections of the population—while taking full advantage of the contributions of the agricultural sector to economic and rural development, poverty reduction, nutrition and food security, and employment generation. Also, to cope with the above-cited challenging issues, Tunisia has been engaged during the last three decades in developing various climate mitigation and adaptation strategies. In such strategies, a special focus has been devoted to the agriculture sector as it is the main pillar of food security through implementing institutional and technical measures. Nevertheless, the authors believe that the integration with other sectors and the mainstreaming of climate imperative in economic development plans are still insufficient, hence the need to bridge such a gap.

In Chap. 4, "Food Security in the MENA Region: Does Agriculture Performance Matter?" El Mahmah and Amar start with the evidence that food security is still a growing challenge for the Middle East and North-African (MENA) countries, despite their continued efforts in developing agricultural production and improving access to food. To understand such a situation, the authors investigate the potential of agriculture sector in affecting food security in the MENA region by providing empirical evidence on the macroeconomic pillars of food security, namely availability, accessibility, utilization, and stability. While doing so, the authors set out to construct a panel data model for each pillar over the period 1990–2017 using the Generalized Method of Moments (GMM) approach to estimate the relationship between all variables. The authors found that the determinants of food security in the MENA vary according to the adopted pillars. Also, food security in this region depends not only on agricultural factors but essentially on various macroeconomic factors, such as trade openness and international food prices, which may affect directly or indirectly the country's food security status, regardless of its dependency on agriculture or oil sector.

In Chap. 5, "Emerging Threats to Food Security in Nigeria: Way Forward," Ozkan and Fawole focus on Nigerian context where agriculture is still playing an important role in maintaining food security and employment. However, the country is increasingly facing many challenges negatively affecting its food production potential, including: the abandonment of agricultural sector because of oil boom in the late 1960s; the lack of modern farming implements; the low agricultural value addition; and recent political conflicts in the North-Eastern part of the country. The emerging impacts of climate change are also drivers of vulnerability which significantly affect both crop production and livestock. Against this background, the authors examine the impacts of these risk dynamics on the Nigerian agricultural sector, review the potential of agricultural sector and rural development public policies in making the sector attractive to the youth, and make some policy-oriented recommendations within the perspective of revamping the agricultural sector for a better agricultural delivery and achievement of SDGs by 2030.

In Chap. 6, "Food Security in Morocco: Risk Factors and Governance," Zahour focuses on the Moroccan context where several sectoral policies have been implemented to modernize the agricultural sector, to sustainably manage domestic natural resources, and to ensure a social and ecological sustainable agricultural transition. The author believes that the potential of such policies in ensuring and improving the national food security is still low and ineffective, as the degradation of biodiversity and unsustainable use of resources is increasing in addition to a noticeable rise of social-ecological inequalities. Drawing on this, the author attempts to assess the food security situation in Morocco by identifying and evaluating the various challenges facing food security and relevant response mechanisms adopted by policymakers. The results reveal the lack of appropriate and robust planning and assessment tools of policies and projects' impacts, a weakness which is increasingly hindering the capacity to meet social demands, particularly in rural areas. The authors recommend the adoption of decisive actions to support income-generating initiatives, to adapt to the impacts of climate change, and to strengthen knowledge and social-ecological systems. The success of the national food security policy remains as well predicated on the inclusive mobilization of all stakeholders.

In the same context, Saidi et al. recall in Chap. 7, "Climate Change, Agricultural Policy and Food Security in Morocco," the fact that the agricultural sector is a key factor in ensuring food security and meeting dietary and nutrition needs in Morocco in addition to its important potential in terms of economic and social development. However, Moroccan agriculture is still vulnerable to climate risks (such as recurrent droughts and floods) and water scarcity which are expected to intensify in the coming decades. This, combined with water-intensive practices in already environmentally stressed territories (like the Souss-Massa region), in particular within the framework of the Green Morocco Plan (2009–2020), increases the vulnerability of the agricultural sector to existing risks. These challenges will have as well negative impacts on potential agricultural yields, employment opportunities, and purchasing power of rural people. In such a scenario, the authors believe that it is imperative to develop sustainable agricultural practices and to foster the coping capacity of the country face to climate change.

In Chap. 8, "Impacts of Climate Change on Livestock and Related Food Security Implications—Overview of the Situation in Pakistan and Policy Recommendations," Hashmi et al. focus on the livestock sector in Pakistan where it is playing a major economic and social role. However, given the severe impacts of climate change on this sector, the authors believe that a detailed understanding of these dynamics is necessary, especially with regard to the future impacts of climate change on the productivity of ruminants. Indeed, the performance of small ruminants—a primary source of animal protein—is adversely affected by the rise in temperature. Moreover, the authors claim that the importance of climate change and the value of food security are expected to increase even more in the context of COVID-19 Pandemic. Within such a perspective, the authors suggest a myriad of mitigation measures to help address productivity losses in the livestock food supply in a context marked by a rising food demand.

In Chap. 9, "Water Scarcity Threats to National Food Security of Pakistan—Issues, Implications, and Way Forward," Munir et al. start with the fact that Pakistan is primarily an agrarian country with an agriculture sector that is a major source of economic activity, foreign exchange earnings, and the livelihood of the majority of population, caretaker of food and nutritional security, a means to combat rural poverty, and a supplier of raw material for the industries. Almost 80% of the cultivated area is irrigated and supported with the world's largest contiguous canal irrigation system—Indus Basin Irrigation System (IBIS)—that is highly dependent on transboundary sources—mainly snow and glacier melt—which are as well vulnerable to the impacts of climate change. This situation, in addition to the pressure exerted by demographic growth and urbanization, is worsening the water scarcity in Pakistan, therefore generating serious implications for the agriculture sector with the potential to compromise the national food security of the country. Against this background, the authors attempt to investigate the linkages between water scarcity, climate change, and food security with the objective to develop policy guidelines which may help different stakeholders better understand these challenges within the perspective of adopting efficient resource management measures.

In Chap. 10, "Climate Change Impact on Water Resources and Food Security in Egypt and Possible Adaptive Measures," Abdelfattah focuses on Egypt as a country that is highly vulnerable to climate-induced impacts and found paradoxical the fact that adaptation to such impacts—given their worrying implications especially for water and food security—is still a low priority in domestic policies. For the author, Egypt has reached a unique juncture: its traditional water resources, mainly surface and groundwater, are fully utilized, while its water demands grow due to demographic growth and rising living standards; on the other hand, climate change will affect agriculture and food production in complex ways, directly through changes in agroecological conditions, and indirectly by affecting growth and income distribution, and thus demand for agricultural produce. Therefore, the author believes that improving the efficiency of the country's limited water resources, that is producing more food with less water, is currently an imperative. Within such a perspective, the authors provide a comprehensive review on the impacts of climate change on water resources, agriculture, and food security before highlighting various issues related to climate change and water resources in Egypt, including water supply (conventional and non-conventional), water demands, water quality, challenges facing water sector, construction of the Grand Ethiopian Renaissance Dam (GERD), and adaption and mitigation measures.

In Chap. 11, "The Combined Impact of Climate Change and the Use of Solar Energy on the Water Consumption in Agriculture: A Case Study from Souss Massa Region," Elame et al. simulate the impact of climate change and the use of renewable energy on water resources in the study area using a dynamic management model for irrigation water allocation. The simulation was conducted and based on a comparison of water pumping costs, which are introduced in a dynamic model and compared to solar energy pumping cost (Scenario B). The results showed that the average shadow price of water remains below MAD $3/m^3$ in the first five years, knowing that it has already exceeded MAD $5/m^3$ compared to scenario A assessing only the climate change impact. The authors found that the economic price directly affects the amount of water consumed for irrigation, and this change has the potential to adversely affect the availability of groundwater resources if no measure limiting water pumping is applied. Therefore, the authors recommend the implementation of an immediate strategy for an appropriate combination of technology and regulation by taxing groundwater at the farm level within the perspective of sustainably managing the aquifer depletion and environmental scarcities from which the study area is highly suffering.

In the same vein, Hajiyev et al. in Chap. 12, "A Mathematical Model for Control of Drainage in an Irrigation System," note that drainage systems are widely used in irrigation systems to achieve multiple objectives; thus, it is necessary to consider the choice of mathematical models that are adequately describing the processes of drainage and control of groundwater levels given their important role. For this purpose, the authors propose an algorithm for the regulation and prevention of salinization of arable land and swampy areas and estimate the amount of drainage water. Based on this model, the behavior of the groundwater (temporal variation) was analyzed. Under the given conditions, it was derived that, for an isotropic and homogeneous case, the level of the priming relative to axis x is approximately stable. However, a potential change in other parameters leads to the reduction of the effectiveness of drainage system and, consequently, to decreasing the level of water.

Hence, the authors highlight the need to increase the velocity of water through the drainage system using appropriate engineering processes.

In Chap. 13, "Readiness of Entrepreneurs Towards Group Performance Development of OTOP Product: A Case Study in Northeastern Thailand," Chouichom take us to another area of investigation by focusing on the One Tambon (Village) One Product (OTOP) model, which is a public project intended to boost the household income for villagers in Thailand. In this empirical research, the authors recruited 35 OTOP entrepreneurs—who engaged in rice production and processing—as participants. The collection of data was based on an interview schedule with open-ended and close-ended questions and the interpretation of such data had been done through descriptive statistics. The results show that the respondents strongly agreed on both production and buying–selling issues and relatively agreed on supporting and competition issues. However, a large proportion of OTOP entrepreneurs felt that the marketing for OTOP products—which encompass general consumable goods and other non-consumable products linked to traditional and indigenous knowledge—is the most troubling issue. Therefore, the main obstacles for the OTOP entrepreneurs observed in this survey included the lack of marketing knowledge for OTOP products, difficulty in improving packaging design, and inferior production system.

Agadir, Morocco   Mohamed Behnassi
Riyadh, Saudi Arabia   Mirza Barjees Baig
Casablanca, Morocco   Mahjoub El Haiba
Lexington, KY, USA   Michael R. Reed

Acknowledgements

This contributed volume is partly based on the outcome of the International Conference on *Social-Ecological Systems—From Risks and Insecurity to Viability and Resilience (SES2019)*, organized on October 24–25, 2019, in Marrakech, Morocco, by the Research Center for Environment, Human Security, and Governance (CERES). More than 50 researchers and experts from different parts of the world, and representing a myriad of disciplines and institutions, have participated in this conference. The event was an opportunity to undertake a multidisciplinary debate about social-ecological systems (SES) in a changing and uncertain context marked by many risk dynamics. The approach consists of: assessing the structural drivers of vulnerability, crises, and insecurity of SES; identifying areas of integration and synergy given the interdependence of SES; assessing the ability and inability of existing response mechanisms (governance frameworks, conceptual referential, cultural patterns, and values…) to foster the resilience and viability of SES; and promoting approaches and ways of action to reverse undesirable trends. The key topics of the conference were multidisciplinary to facilitate the fruitful interaction between numerous scientific disciplines. They were also relevant to experts, practitioners, and decision-makers from different scales and spheres.

I have been honored to chair the SES2019 and to share the editorship of this volume with my colleagues: Dr. Mirza Barjees Baig, Professor at the Prince Sultan Institute for Environmental, Water and Desert Research, King Saud University, Saudi Arabia; Dr. Mahjoub El Haiba, Full Professor at the Faculty of Law, Economics, and Social Sciences, Hassan II University of Casablanca, Morocco; and Dr. Michael R. Reed, Emeritus Professor of Agricultural Economics at the University of Kentucky, USA. I seize this opportunity to warmly thank all of them for their collaboration and support during the publishing process. Their professionalism, expertise, and intellectual capacity made the editing process an exciting and instructive experience and definitely contributed to the quality of this publication.

I would also like to seize this opportunity to pay tribute to all chapters' authors without whom this valuable and original publication could not have been produced. Their collaboration, reactivity, and openness during the process were very remarkable and impressive.

The chapters in this volume are also the result of the invaluable contribution made by peer reviewers, who generously gave their time and energy to provide insight and expertise to the selection and editing process. On behalf of my co-editors, who actively participated in the peer-reviewed process, I would specifically like to acknowledge, with sincere and deepest thanks, the following peer reviewers:

- Dr. Ashrafuzzaman Chowdhury, Professor, Department of Economics, J. B. College, Jorhat, Assam, India
- Dr. Boubaker Dhehibi, Senior Agricultural Resource Economist, Sustainable Intensification, and Resilient Production Systems Program (SIRPS), International Center for Agricultural Research in the Dry Areas (ICARDA), Amman, Jordan
- Dr. Guillaume Baggio, Research Associate, UNU-INWEH, Hamilton, Canada
- Dr. Jan W. Hopmans, Distinguished Professor Emeritus, Soil Science and Irrigation Water Management, University of California Davis, USA
- Dr. M. Kamal Sheikh, Former Chief Scientific Officer/Technical Staff Officer to Chairman Pakistan Agricultural Research Council (PARC) and Professor PIASA, Islamabad, Pakistan
- Dr. M. Brodsky, President and Rector, Lincoln University, Oakland, California, USA
- Prof. Dr. Hossein Dehghanisanij, Agricultural Research Education and Extension Organization (AREEO), Agricultural Engineering Research Institute (AERI), Karaj, Iran
- Dr. Manzoor Ahmad Malik, Director (R), Pakistan Council of Research in Water Resources, Khayaban-e-Johar, Islamabad, Pakistan
- Dr. Michael R. Reed, Emeritus Professor, Agricultural Economics, University of Kentucky, Lexington, KY, USA
- Dr. Mohamed Mujithaba Mohamed Najim, Vice-Chancellor, South Eastern University of Sri Lanka and Professor of Environmental Conservation and Management, Department of Zoology and Environmental Management, Faculty of Science, University of Kelaniya, Sri Lanka
- Dr. Mohamed Taher Sraïri, Department of Animal Production and Biotechnology, Head of the School of Agricultural Sciences, Hassan II Agronomy and Veterinary Medicine Institute, Rabat, Morocco
- Dr. Monica Butnariu, Professor, Department of Chemistry and Biochemistry, Banat's University of Agricultural Sciences and Veterinary Medicine, Timis, Romania
- Dr. Muhammad Ashfaq, Professor, Institute of Agricultural and Resource Economics, Faculty of Social Sciences, University of Agriculture, Faisalabad, Pakistan
- Dr. Muhammad Qaiser Alam, Associate Professor, Department of Economics, D.S. (Postgraduate) College, Aligarh (Uttar Pradesh), India
- Dr. Muhammad Younas, Dean, Faculty of Animal Husbandry, University of Agriculture, Faisalabad, Pakistan

- Dr. Peyman Falsafi, Professor of Agricultural Extension Education, Head of Halal Food Commission, Ministry of Jihad-e-Agriculture, Tehran-Islamic Republic of Iran
- Dr. Sadraddin Gurbanov, Institute of Genetic Resources of Azerbaijan, National Academy of Sciences, Azerbaijan
- Dr. Sawsan Hassan, International Center for Agricultural Research in Dry Areas (ICARDA), Amman, Jordan
- Dr. Sonia Maciel, Deputy Vice-Chancellor, University Lurio (UNILURIO), Consultant on Livestock, Environment and Community Development, Mozambique
- Dr. Syed Ghazanfar Abbas, National Consultant (Innovation, E-Agriculture and E-Learning) of the Food and Agriculture Organization (FAO), Islamabad, Pakistan
- Dr. Taye Tolemariam Ejeta, College of Agriculture and Vet Medicine, Jimma University, Jimma, Ethiopia
- Dr. Usman Haruna, Department of Agricultural Economics and Extension, Faculty of Agriculture, Federal University Dutse, Jigawa, Nigeria
- Dr. Usman Mustafa, Senior Consultant, Pakhtunkhwa Economic Policy Research Institute (PEPRI), AWK University, Mardan, Khyber Pakhtunkhwa, Pakistan
- Dr. W. M. Wishwajith W. Kandegama, Senior Lecturer, Department of Horticulture and Landscape Gardening, Faculty of Agriculture and Plantation Management, Wayamba University of Sri Lanka, Makandura, Sri Lanka
- Ms. Ezzahra Mengoub, Agro-Economist, Policy Center for the New South, Morocco

Agadir, Morocco   Mohamed Behnassi

About the Research Institution

The Center for Research on Environment, Human Security, and Governance (CERES)

The CERES, previously the North-South Center for Social Sciences (NRCS), 2008–2015, is an independent and not-for-profit research institute founded by a group of researchers and experts from Morocco and other countries. The CERES aims to develop research and expertise relevant to environment and human security and their governance from a multidimensional and interdisciplinary perspective. As a think tank, CERES aspires to serve as a reference point, both locally and globally through rigorous research and active engagement with policy-making processes. Currently, CERES is a member of MedThink 5+5 which aims at shaping relevant research and decision agendas. Through its research and expertise program, the CERES aims to investigate the links between environmental/climate change, their implications for human security and the needed shifts to be undertaken in both research and policy. The CERES, led by Dr. Mohamed Behnassi and mobilizing a large international network of researchers and experts, aims to undertake original research, provide expertise and contribute to effective science and policy interactions through its publications, events, capacity building, and expertise.

Contents

Editors and Contributors

About the Editors

Dr. Mohamed Behnassi is a professor at the Faculty of Law, Economics, and Social Sciences, Ibn Zohr University of Agadir, Morocco. He is as well a senior researcher of international law and politics of environment and human security focusing on some specific regions such as the MENA and the Mediterranean. He has a Ph.D. in International Environmental Law and Governance (Hassan II University of Casablanca, 2003) and a Diploma in International Environmental Law and Diplomacy (University of Eastern Finland and UNEP, 2015). He is currently the founding director of the Center for Environment, Human Security, and Governance (CERES)—Former North-South Center for Social Sciences (NRCS)—which is a member of MedThink 5+5 aiming at shaping relevant research and decision agendas. From 2015 to 2018, he was the director of the Research Laboratory for Territorial Governance, Human Security and Sustainability (LAGOS) in the same university. Recently, he was appointed as an expert for the Intergovernmental Science-Policy Platform on Biodiversity and Ecosystem Services (IPBES), the National Center for Scientific and Technical Research (CNRST/Morocco), and Mediterranean Experts on Climate and Environmental Change (MEDECC). He is among the lead authors who elaborate the 1st Assessment Report (MAR1): *Climate and Environmental Change in the Mediterranean Basin— Current Situation and Risks for the Future* (MEDECC, 2021). He has published 15 books, including *Human*

and Environmental Security in the Era of Global Risks (Springer, 2019); *Climate Change, Food Security and Natural Resource Management: Perspectives from Africa, Asia and the Pacific Islands* (Springer, 2019); *Environmental Change and Human Security in Africa and the Middle East* (Springer, 2017); *Sustainable Food Security in the Era of Local and Global Environmental Change* (Springer 2013). In addition, he has organized many international conferences covering the above research areas and managed many research and expertise projects on behalf of various national and international organizations. He is regularly requested to provide scientific expertise nationally and internationally. Other professional activities include social compliance auditing and consultancy by monitoring human rights at work and the sustainability of the global supply chain.

Dr. Mirza Barjees Baig is a professor at the Prince Sultan Institute for Environmental, Water, and Desert Research, King Saud University, Saudi Arabia. He earned his MS degree in International Agricultural Extension in 1992 from the Utah State University, Logan, Utah, USA, and was placed on the "Roll of Honor." He completed his Ph.D. in Extension for Natural Resource Management from the University of Idaho, USA, and was honored with the "1995 outstanding graduate student award." He has published extensively on the issues associated with natural resources in the national and international journals. He has also made oral presentations about agriculture and natural resources and role of extension education at various international conferences. Food waste, water management, degradation of natural resources, deteriorating environment and their relationship with society/community are his areas of interest. He has attempted to develop strategies to conserve natural resources, promote environment, and develop sustainable communities. He started his scientific career in 1983 as a researcher at the Pakistan Agricultural Research Council, Islamabad, Pakistan. He served at the University of Guelph, Ontario, Canada, as the Special Graduate Faculty from 2000 to 2005. He served as a foreign professor at the Allama Iqbal Open University (AIOU), Pakistan, through the Higher

Education Commission from 2005 to 2009. He served as a professor of Agricultural Extension and Rural Society at the King Saud University, Saudi Arabia, from 2009 to 2020. He serves as well on the editorial boards of many international journals and the member of many international professional organizations.

Dr. Mahjoub El Haiba is a professor at the Faculty of Law, Economics, and Social Sciences, Hassan II University of Casablanca, Morocco. He graduated from the Hague Academy of International Law, the Center for Studies and Research in International Law and International Relations in 1984, and in 1994, he had a State Doctorate in Public Law and Political Science from the same University. He held many academic and official positions, including being: Former Member of the National Consultative Human Rights Council (2002–2010), and his Secretary General (2005–2010); first Inter-ministerial Delegate for Human Rights (March 2011–December 2018); former Member ex officio of the Higher Council for Education, Training, and Scientific Research, and of its Executive Office; nominated Member of the Equity and Reconciliation Instance (IER) (2004–2005); former and current Member of the United Nations' Human Rights Committee (2008–2011/2020–at present); former Member and President of the Jury to the League of Arab States' Environment Award, Cairo; former Member and President of the Jury to the Hassan II Environment Award; former Member of the Jury to the Kingdom of Saudi Arabia's Award on Environmental Administration; Founding Member of the Moroccan Human Rights Organization (OMDH); Founding Member of the Arab Network for Environment and Development (RAED), Cairo; Founding Member of the Center for Studies on Migration, Faculty of Law, Economics, and Social Sciences, Hassan II University of Casablanca; and former Deputy Dean of the same Faculty. His areas of teaching, research, and expertise include human rights, transitional justice, and environmental law. He has a publication record in these areas, supervised numerous doctorate and master theses, and made many presentations in national and international scientific meetings.

 **Dr. Michael R. Reed** is an emeritus professor of Agricultural Economics at the University of Kentucky, USA. He holds a Ph.D. in economics from Iowa State University (1979); a Doctor Honoris Causa (Honorary Ph.D.) from Bucharest University of Agricultural Sciences and Veterinary Medicine (Romania); and an Honorary Ph.D. from the Faculty of Business Administration, Maejo University (Thailand). His principal area of research is international trade in agricultural products, including the effects of macroeconomic policies and exchange rates on US food exports, international commodity price dynamics, consumer demand in various countries, and the effects of competition patterns on world agricultural trade patterns.

Contributors

Mahmoud A. Abdelfattah Soils and Water Department, Faculty of Agriculture, Fayoum University, Fayoum, Egypt;
Institute of Strategic Research and Studies for Nile Basin Countries, Fayoum University, Fayoum, Egypt;
Food and Agriculture Organization of the United Nations (FAO), Fayoum, Egypt

Saidi Abdelmajid Department of Economics, University Moulay Ismail, Meknes, Morocco

Anwar Ahmad Pakistan Council for Research on Water Resources, Islamabad, Pakistan

Amine Amar Moroccan Agency for Sustainable Energy (MASEN), Rabat, Morocco

Mirza Barjees Baig Water and Desert Research, Prince Sultan Institute for Environmental, King Saud University, Riyadh, Saudi Arabia

Mohamed Behnassi Global Environmental and Human Security Politics, College of Law, Economics and Social Sciences, Ibn Zohr University of Agadir, Agadir, Morocco;
Research Center for Environment, Human Security and Governance (CERES), Agadir, Morocco;
Faculty of Law, Economics and Social Sciences of Agadir, Ibn Zohr University, Agadir, Morocco

Azaiez Ouled Belgacem Sustainable Rangeland Management Expert, Food & Agriculture Organization of the United Nations (FAO), Riyadh, Saudi Arabia;

Arabian Peninsula Regional Program, International Center for Agricultural Research in the Dry Areas, Dubai, UAE

Seksak Chouichom Department of Agricultural Extension and Communication, Faculty of Agriculture, Kasetsart University, Chatuchak, Bangkok, Thailand

Joyce D'Silva Compassion in World Farming, Godalming, UK

Fouad Elame National Institute of Agronomic Research, Agadir Regional Center, Rabat, Morocco

Wasiu Olayinka Fawole Faculty of Agriculture, Department of Agricultural Economics, Akdeniz University, Antalya, Turkey

Aymen Frija North Africa Regional Program, International Center for Agricultural Research in the Dry Areas (ICARDA), Tunis, Tunisia

Himangana Gupta JSPS-UNU Postdoctoral Fellow, Institute for the Advanced Study of Sustainability, United Nations University (UNU), Tokyo, Japan

Asaf Hajiyev Azerbaijan National Academy of Sciences, Institute of Control Systems, Baku, Azerbaijan

Hammad Ahmed Hashmi Paws and Claws Animal Consultancy Pakistan (R&D), ADS and OIC DHA Lahore KC, Lahore, Pakistan

Jan W. Hopmans Soil Science and Irrigation Water Management, University of California Davis, Davis, USA

Khalid Javed Department of Livestock Production, University of Veterinary and Animal Sciences, Lahore, Pakistan

Hayat Lionboui National Institute of Agronomic Research, Tadla Regional Center, Rabat, Morocco

Assil El Mahmah National Bank of Kuwait, Kuwait, Kuwait

Ahmed Mukhtar Islamic Educational, Scientific and Cultural Organization (ISESCO), Rabat, Morocco

Muhammad Umar Munir Pakistan Council for Research on Water Resources, Islamabad, Pakistan

Usman Mustafa Pakhtunkhwa Economic Policy Research Institute (PEPRI), AWK University, Mardan, Khyber Pakhtunkhwa, Pakistan

Mohamed Mujithaba Mohamed Najim Department of Zoology and Environmental Management, Faculty of Science, University of Kelaniya, Kelaniya, Sri Lanka

Mohamed Ouessar Laboratory of Eremology and Combating Desertification, Institute of Arid Regions (IRA), University of Gabes, Medenine, Tunisia

Burhan Ozkan Faculty of Agriculture, Department of Agricultural Economics, Akdeniz University, Antalya, Turkey

Olaf Pollmann SCENSO—Scientific Environmental Solutions, Sankt Augustin, Germany

Doukkali Rachid Agronomy and Veterinary Hassan II Institute, Rabat, Morocco

Mohammed Ataur Rahman Centre for Global Environmental Culture (CGEC) and Program on Education for Sustainability, Dhaka, Bangladesh;
WWOOF Bangladesh & RCE Greater Dhaka, International University of Business Agriculture and Technology (IUBAT), Dhaka, Bangladesh

Gopichandran Ramachandran NTPC School of Business, Noida, U.P, India

Nira Ramachandran ICSSR Senior Fellow, Institute of Economic Growth, Delhi University, New Delhi, India

Michael R. Reed Agricultural Economics, University of Kentucky, Lexington, USA

Yasin Rustamov Azerbaijan National Academy of Sciences, Institute of Control Systems, Baku, Azerbaijan

Abderrahman Sghaier Laboratory of Eremology and Combating Desertification, Institute of Arid Regions (IRA), University of Gabes, Medenine, Tunisia

Mongi Sghaier Laboratory of Economics and Rural Societies, Institute of Arid Regions (IRA), University of Gabes, Medenine, Tunisia

Gary S. Straquadine Career and Technical Education, Utah State University, Logan, UT, USA

Mohamed Zahour Faculty of Law, Economics, and Social Sciences, Ibn Zohr University, Agadir, Morocco

Abbreviations and Acronyms

ADB	Asian Development Bank
AQD	Animal Quarantine Department (Pakistan)
ASEAN	Association of Southeast Asian Nations
AU	African Union
BDP	Bangladesh Delta Plan
BMI	Body Mass Index
CbA	Community-Based Adaptation
CBD	Convention on Biological Diversity
CBN	Central Bank of Nigeria
CCAFS	Climate Change, Agriculture, and Food Security
CCI	Council of Common Interest (Pakistan)
CERES	Center for Environment, Human Security, and Governance
CESE	Economic, Social, and Environmental Council (Morocco)
CGIAR	Consultative Group on International Agricultural Research
CGPM	Clean Green Pakistan Movement
CRI	Climate Risk Index
CSA	Climate-Smart Agriculture
CSCCC	Civil Society Coalition for Climate Change
CTA	Technical Center for Agricultural and Rural Cooperation
DMI	Dry Matter Intake
DNM	Department of National Meteorology (Morocco)
ESF	Energy Smart Food for People and Climate
FAO	Food and Agriculture Organization of the United Nations
FDA	Fund for Agricultural Development (Morocco)
FDA	Food and Drug Administration (Thailand)
FMD	Foot and Mouth Disease
GAMS	General Algebraic Modeling System
GCC	Gulf Council Countries
GCM	Global Climate Model
GCOS	Global Climate Observing System
GDP	Gross Domestic Product
GERD	Grand Ethiopian Renaissance Dam

GHG	Greenhouse gas
GHI	Global Hunger Index
GMM	Generalized Method of Moments
GMP	Green Morocco Plan
GMST	Global mean surface temperature
HAD	High Aswan Dam (Egypt)
HPLE	High-Level Panel of Experts on Food Security and Nutrition
HRC	Human Rights Council
IARC	International Agency for Research on Cancer
IBIS	Indus Basin Irrigation System
ICRISAT	International Crops Research Institute for the Semi-Arid Tropics
IFAD	International Fund for Agricultural Development
IPBES	Intergovernmental Science-Policy Platform on Biodiversity and Ecosystem Services
IPCC	Intergovernmental Panel on Climate Change
IRA	Arid Land Institute (Tunisia)
IWMI	International Water Management Institute
KJWA	Koronivia Joint Work on Agriculture
LCA	Life Cycle Assessments
LDDB	Livestock and Dairy Development Board (Pakistan)
MDGs	Millennium Development Goals
MENA	Middle East and North African
MGPO	Mountain and Glacier Protection Organization
NbS	Nature-Based Solutions
NDC	Nationally Determined Contributions
NEPAD	New Partnership for Africa's Development
NFP	National Food Policy (Pakistan)
NIAR	National Institute of Agronomic Research (Morocco)
NIDH	National Initiative for Human Development (Morocco)
NRLPD	National Reference Laboratory for Poultry Diseases (Pakistan)
NSCC	National Strategy for Adapting to Climate Change (Tunisia)
NVL	National Veterinary Laboratory (Pakistan)
NWP	National Water Plan (Morocco)
NWP	National Water Policy (Pakistan)
OECD	Organization for Economic Cooperation and Development
ONAGRI	National Observatory of Agriculture
OTOP	One Tambon (Village) One Product
PCRWR	Pakistan Council of Research in Water Resources
PMKSY	Pradhan Mantri Krishi Sinchayee Yojana
PMP	Positive Mathematical Programming
PPR	Peste des Petits Ruminants (Pakistan)
RCP	Representative Concentration Pathway
RDA	Recommended Daily Allowance
SDGs	Sustainable Development Goals
SFA	Sustainable Food and Agriculture

SMEs	Small and Medium-Sized Enterprises
SMOG	Simple Measure of Gobbledygook
SRCCL	Special Report on Climate Change and Land
SSA	Sub-Saharan Africa
TDN	Total Digestible Nutrients
TFSS	Tunisia's Food Security Strategy
THI	Temperature Humidity Index
TVET	Technical and Vocational Education and Training
UAA	Utilized Agricultural Land
UN	United Nations
UNFCCC	United Nations Framework Convention on Climate Change
UNICEF	United Nations Children's Fund
US-HHFSS	United States Households' Food Security Scale
VAR	Vector Auto-Regression
WFP	World Food Program
WHO	World Health Organization
WMO	World Meteorological Organization
WRI	World Resources Institute
WRVI	Water Resources Vulnerability Index

List of Figures

Chapter 12

List of Tables

Chapter 1
Enhancing Resilience for Food and Nutrition Security Within a Changing Climate

Mohamed Behnassi, Mohammed Ataur Rahman, Joyce D'Silva, Gopichandran Ramachandran, Himangana Gupta, Olaf Pollmann, and Nira Ramachandran

Abstract Climate change is adversely affecting food production systems while increasing the vulnerability of human societies—especially resource-poor small producers—and diminishing their resilience to food and nutrition insecurity. Even with a 1.5 °C scenario, climate change is believed to leave disadvantaged populations weakly resilient to food, health, and livelihood insecurity. Additionally, the

M. Behnassi (✉)
Global Environmental and Human Security Politics, College of Law, Economics and Social Sciences, Ibn Zohr University of Agadir, Agadir, Morocco
e-mail: m.behnassi@uiz.ac.ma

Research Center for Environment, Human Security and Governance (CERES), Agadir, Morocco

M. A. Rahman
Centre for Global Environmental Culture (CGEC) and Program on Education for Sustainability, Dhaka, Bangladesh
e-mail: marahman@iubat.edu

WWOOF Bangladesh & RCE Greater Dhaka, International University of Business Agriculture and Technology (IUBAT), Dhaka, Bangladesh

J. D'Silva
Compassion in World Farming, Godalming, UK
e-mail: joyce@ciwf.org

G. Ramachandran
NTPC School of Business, Noida, U.P, India
e-mail: gopichandran@nsb.ac.in

H. Gupta
JSPS-UNU Postdoctoral Fellow, Institute for the Advanced Study of Sustainability, United Nations University (UNU), Tokyo, Japan

O. Pollmann
SCENSO—Scientific Environmental Solutions, Sankt Augustin, Germany
e-mail: o.pollmann@scenso.de

N. Ramachandran
ICSSR Senior Fellow, Institute of Economic Growth, Delhi University, New Delhi, India

New Delhi, India

scale of change required to limit warming to 1.5 °C is historically unprecedented and can only be achieved through strategically important societal transformation and ambitious mitigation measures, a requirement still not efficiently met by the majority of countries, especially key carbon emitters. This chapter accordingly aims at analyzing the dynamics through which climate change affects food and nutrition insecurity and drawing the pathways towards resilience building in this area. The analysis starts with investigating the predominant impacts of climate change on food security and resilience; then assesses these dynamics from international and regional perspectives; and finally explores some best pathways and approaches toward building resilience for food and nutrition security in a changing climate.

Keywords Climate change · Food and nutrition insecurity · Food production systems · Resilience · Gender · Chemical ecology · Community engagement · Scientific uncertainty

1 Introduction

The impact of global warming is much worse than anticipated (Wallace-Wells, 2019). Scientists believe that climate risks could disrupt ecosystem services and lead to severe effects on livelihoods, which in turn would affect the human security—especially in the areas of food, health and water—of millions of people worldwide, migration dynamics and the potential for conflicts in many countries. For the Intergovernmental Science-Policy Platform on Biodiversity and Ecosystem Services (IPBES), climate change affects food production systems—including marine ecology which is conquered by rising ocean temperatures (Cheng et al., 2019)—and poses dire economic and social threats (IPBES, 2019). As a result, millions face malnutrition due to devastating droughts, and many more will have to choose between starvation and migration. Under these circumstances, business as usual is a response that increasingly invites disaster (HRC, 2019a).

According to the World Bank (2010: 5), at 2 °C of warming, 100–400 million more people could be at risk of hunger and 1–2 billion more people may no longer have adequate water. Even with adaptation measures, climate change could result in global crop yield losses of 30% by 2080 and exacerbate health shocks that already push 100 million into poverty every year (Hallegatte et al., 2016). Between 2030 and 2050, climate change is expected to cause approximately 250,000 additional deaths per year due to malnutrition, malaria, diarrhea, and heat stress (WHO, 2014).

Even with a 1.5 °C scenario, rather than 2 °C as targeted by the Paris Agreement, this could mean: reducing the number of people vulnerable to climate-related risks by up to 457 million; 10 million fewer people will be exposed to the risk of sea level rise; reducing exposure to floods, droughts, and forest fires. This will also limit damage to ecosystems and reduction in food and livestock; reduce the number of people exposed to water scarcity by half; and up to 190 million fewer premature deaths over the century (IPCC, 2018). On the other hand, catastrophes on account of even 1.5 °C of warming will prevail, albeit at a lesser scale. This is expected to be at around 500 million people exposed and vulnerable to water stress, 36 million people with lower crop yields, and up to 4.5 billion people exposed to heat waves (IPCC, 2018).

In all these scenarios, the worst affected are the least well-off members of human society across the world (HRC, 2019a). As per the IPCC (2018), the scale of change required to limit warming to 1.5 °C is historically unprecedented and can only be achieved through 'societal transformation' and ambitious emissions reduction measures. This applies as well to building resilience for food and nutrition security which will remain one of the ultimate challenges to be dealt with, especially in a changing climate.

Against this background, this chapter aims at analyzing the dynamics through which climate change affects food and nutrition security and drawing the path towards resilience building in this area. It also draws attention to the dynamics of science-based policy making to target integrated mitigation and adaptation outcomes. To this end, the chapter investigates the predominant impacts of climate change on food security and resilience; then assesses these dynamics from international and regional perspectives; and finally explores some best pathways and approaches toward building resilience for food and nutrition security in a changing climate.

2 Food Security, Climate Change, and Resilience: Impacts and Projected Changes

2.1 Climate Change, Food Security and Resilience: A Literature Review

Government commitments and sustained country efforts towards attaining the Millennium Development Goals (MDG 1C), and later, the Sustainable Development Goals (SDGs) of freedom from hunger and malnutrition (SDG 2. 1 and 2.2) resulted in slow, yet steady progress from the early 2000s. The percentage of undernourished people globally decreased from 14.5 to 10.6 (2001–2015). However, from 2016 onwards, a reversal in this positive trend has emerged. In fact, the absolute number of people in the world affected by undernourishment, or chronic food deprivation, is now estimated to have increased by 17 million over just a year; from 804 million in 2016 to nearly 821 million in 2017 (FAO, 2018b). Not only are Africa and South America emerging as currently hunger hotspots, and trends in undernourishment in Asia are quite discouraging as well.

According to the UN's latest *World Population Prospects* released in 2017, global population has increased from 2.536 billion to 7.383 billion over the period 1950 to 2019. This is expected to grow to 9.314 and 10.301 billion in 2050 and 2100, respectively. This increase in population will exert significant problems on food supply and distribution. Food production has to accordingly increase by nearly 70%

to cover worldwide demand. The UN Children's Fund, indicates that nearly half of all deaths of children under the age of five are due to under-nutrition. This translates into a loss of about 03 million young lives a year. FAO et al. (2018) in *The State of Food Insecurity and Nutrition in the World 2018: Building Climate Resilience for Food Security and Nutrition*, indicates new evidence about growing numbers of the hungry; despite the possibility that our planet could produce enough food to feed everyone. In addition, more than two billion people are affected by hidden hunger (see Fig. 1), meaning, they suffer from micronutrient deficiencies (WHO, 2006). In 2017, almost 124 million people across 51 countries and territories faced 'crisis' levels of acute food insecurity or worse, calling for immediate emergency action to safeguard their lives and livelihoods (FSIN, 2018).

The re-emergence of hunger can be largely attributed to two main causes: climate change and conflict. Of the two, the effects of climate change are more pervasive. Climate-related extreme events—particularly drought, floods and tropical storms—decimate crops, livestock and fisheries, food stocks and alternative means of livelihoods. The emergent food and nutrition insecurity and concomitant conflicts further compound deficits (Godfray & Garnett, 2014).

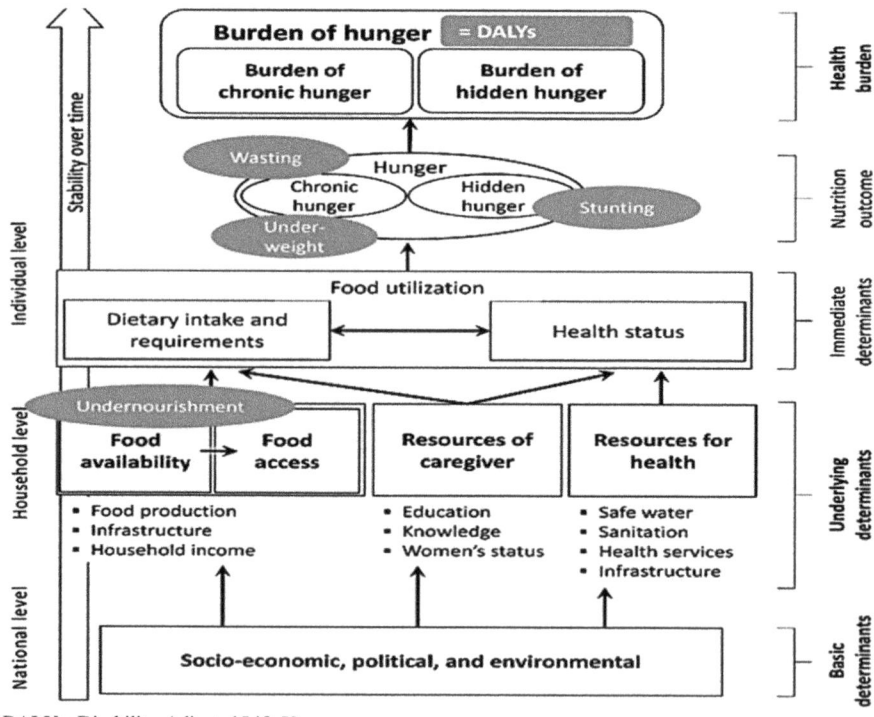

DALYs: Disability-Adjusted Life Years

Fig. 1 Determinants of hunger including related proxy measures. *Source* Gödecke (2017)

The FAO defines food security as "a situation that exists when all people, at all times, have physical, social and economic access to sufficient, safe and nutritious food that meets their dietary needs and food preferences for an active and healthy life" (FAO, 2001). The Human Rights Council (HRC) in its *Resolution on the Right to Food* adopted in 2019 also reaffirms everyone's right to safe, sufficient and nutritious food. This is aligned with the fundamental right of freedom from hunger, to fully develop and maintain one's physical and mental capacities (HRC, 2019b).

Hunger is seen as an outrage and violation of human dignity (HRC, 2019b). It is, therefore, an absolute imperative to tackle this on priority, as rightly addressed by the SDG framework (FAO, 2017, 2018; Barrett, 2010; Allen & de Brauw, 2018). According to Godfray et al. (2010), despite major advances in food production and distribution, knowledge about nutrition and health, and substantial reductions of hunger and malnutrition in developing countries, food security—as defined above—remains a formidable global challenge. The HRC (2019b) recognizes the complex character of food insecurity and its likely recurrence due to several major factors. These include global financial and economic crisis, volatility in commodity prices, environmental degradation, poverty, armed conflicts, and the lack of appropriate technologies. Related investment and capacity-building to confront emergent impacts, particularly in developing countries, least developed countries and small island developing States have to be addressed systematically.

FAO et al. (2018) urge a special focus on understanding the impacts of climate variability and extremes on food security and nutrition. They highlight the nexus between climate variability and exposure to more complex, frequent and intense climate extremes that erode and even reverse gains in ending hunger and malnutrition. Such impacts could be significantly worse in countries with agricultural systems sensitive to rainfall and temperature variability. Livelihoods too are hit hard.

As indicated above, SDG1 focuses on ending poverty, SDG2 on ending hunger and SDG13 on combating climate change with implications for resilience. Resilience is defined by the IPCC (2012) as "the ability of a system and its component parts to anticipate, absorb, accommodate, or recover from the effects of a hazardous event in a timely and efficient manner, including through ensuring the preservation, restoration, or improvement of its essential basic structures and functions". Miller et al. (2010), indicate the importance of institutions, social capital, leadership, and learning in this context. A wider resilience lens is also being applied to institutional responses to climate change effects and to climate solutions in livelihood transformations. Moreover, the resilience approach tends to prefer a systemic perspective, whereas climate change adaptation and vulnerability literature tend to take an actor-oriented approach. The complex systems approach adopted by many resilience theorists emphasizes the complexity of social, ecological, and geophysical systems, and their interactions across various levels. Hence, resilience has advanced our understanding of system dynamics and interconnections, ecological thresholds, social-ecological relations, and feedback loops (Miller et al., 2010), as an integral element of decision support systems.

2.2   Impacts of Climate Change on Major Food Crop Yields and Food Production Systems

As discussed in the previous section, climate change has already affected food security through warming, changing precipitation patterns, and greater frequency of some extreme events (IPCC, 2019; Rosegrant et al., 2017; FAO, 2017; Vermulen, 2010). However, in addition to climatic drivers, the number and strength of non-climate drivers, such as cultivar improvement or increased use of irrigation and fertilizers in the case of crops, make it difficult to define a clear baseline (IPCC, 2014).

Studies using different crop models and input data produce different results. Therefore, there is a lack of clear understanding of regional impacts. In many lower-latitude regions, yields of some crops (e.g., maize and wheat) have declined, while in many higher-latitude regions, yields of some crops (e.g., maize, wheat and sugar beets) have increased over recent decades (IPCC, 2019). Yields under the climate change scenario reportedly show slower growth compared to the no-climate change scenario resulting in reduced aggregate yields by 2050 (Rosegrant et al., 2017). The negative impacts of climate trends on crop and food production have been more common than the positive ones (IPCC, 2014). Impacts are mostly negative in Europe, Southern Africa and Australia but generally positive in Latin America, while impacts in Asia and Northern and Central America are mixed (Ray et al,. 2019). Wang et al. (2018) consider severity of droughts and suggest that developing countries and regions (e.g., Southern Africa, Middle Africa, Northern Africa, Central Asia, Southern Asia, South-Eastern Asia, Southern America, and Central America) are generally more susceptible to extreme droughts. They suffer more crop losses than developed countries and regions (Europe, Oceania, Eastern Asia, and Northern America). The trend of yields of main food crops viz. maize, rice, wheat, soybeans, sorghum and barley from 1961 to 2017 is shown in Fig. 2.

Yields of most of these food crops have significantly increased since 1961. On the contrary, Lobell and Field (2007) on the other hand, indicate that recent climate trends have suppressed global yield progress for wheat, maize and barley, with significant yield suppression for soybean and sorghum since 1990. Averaged globally, yields changed between −2551 (oil palm) and +982 (sugarcane) kg/ha/year, and the percentage change in recent yield over all harvested croplands ranged from −13.4% (oil palm) to +3.5% (soybean). Yields have decreased for rice (−0.3% or ~ −1.6 million tons (MT) annually) and wheat (−0.9% or ~ −5.0 MT annually) and increased negligibly for maize (0% or ~0.2 MT annually) (Ray et al., 2019).

Current food production trends are not compatible with the rising demand; as approximately 795 million people continue to suffer from hunger, and their food security is increasingly threatened by climate change (FAO, 2017). Yields of maize, rice, wheat, and soybean are increasing at 1.6%, 1.0%, 0.9%, and 1.3% per year, non-compounding rates, respectively. This is less than the 2.4% per year required to double global production and meet projected demands in 2050 (Ray et al., 2013). A study of 23 major crop producing countries, showed that low and mid-latitudes in the Southern Hemisphere (0–40 °S) experienced significantly increased year-to-year variation in

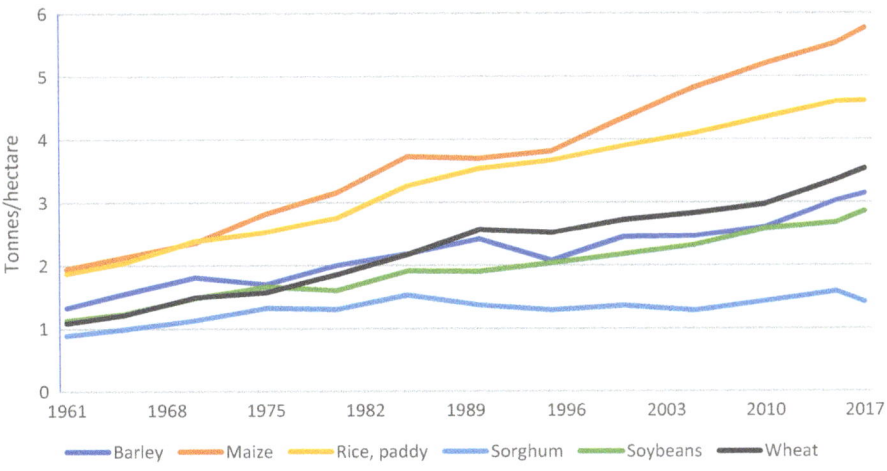

Fig. 2 Trends of yields of major food crops globally in Tons/hectare. *Source* Data extracted from FAOSTAT (2019) (http://www.fao.org/faostat/en/#data/QC)

maize, rice and wheat yields in 1994–2006 compared with 1982–93 context (Iizumi et al., 2014). Another study indicates that global average yield decreases for all maize simulations under Representative Concentration Pathway (RCP) scenario 8.5 by the 2080s, whereas corresponding yields of spring wheat and soybean increase throughout the twenty-first century owing to large positive responses in C3 crops due to CO_2 fertilization effects (Deryng et al., 2014). This is an important finding as the cereal grains including wheat, rice, barley, oats, and other crops like cotton, sugar beet, tobacco and soybean are C3 crops. A study indicates that yield growth of maize and soybean crops in high-income countries located at mid and high latitudes will stagnate, though that of rice and wheat will not (Iizumi et al., 2017).

While global production appears stable, regional differences in crop production are likely to grow stronger through time (Parry et al., 2004), as low-income countries located at low latitudes will benefit from limited warming trends (1.8 °C). On the contrary, with 2 °C of local warming, losses in aggregate production are expected for wheat, rice and maize in both temperate and tropical regions, while adaptations can help wheat and rice production more than maize (Challinor et al., 2014). Yield losses are expected to be greater in magnitude for the second half of the century. Global wheat production is estimated to fall by 6% for each 1 °C rise in temperature (Asseng et al., 2015).

Soybean yields in USA, China and Brazil have reportedly significantly increased due to the CO_2 fertilization effect (Sakurai et al., 2015). On the contrary, research in the Pacific islands shows that climate change will adversely affect the supply of food from agriculture and fisheries (Barnett, 2011). According to Soora et al. (2013), climate change is likely to exhibit three types of impacts on rice crop: gain in net productivity with adaptation; vulnerability despite adaptation gains; and increased production due to increased rainfall in low rainfall but rain-fed regions. Both rain-fed and irrigated agriculture face different climate risks (Calzadilla et al., 2013).

It is also projected that by 2030, the impact of climate change on food consumption will be moderate but in the long run, the impact could be much stronger, with global average calorie losses (Kcal/capita/day) of 6% by 2050 and 14% by 2080 (Havlík et al., 2015). In a temperature scenario (for increases of 1.3 °C and 2.6 °C) based study in the Peruvian Andes, where several traditionally grown varieties of potato and maize were planted at different elevations, maize production declined by 21%–29% in response to new soil conditions. Production of maize and potatoes declined by >87% when plants were grown under warmer temperatures (Tito et al., 2018). In a global study, the impact of climate change on agricultural production by 2050 will reportedly be negative with a range of 0.5%–2.5%, but relatively small at the aggregated global level under RCP6.0 and RCP2.6 scenarios (Meijl et al., 2018). End-of-century (2070–2099) maize yield changes with CO_2 effects for RCP8.5 show substantial impacts, and models indicate high-latitude increases and low-latitude decreases for maize and wheat, while rice and soybean are consistent in the mid- and high-latitude regions showing yield increases (Rosenzweig et al., 2014). Among major crops, the drought-affected sown areas for maize more than doubled from 8.51 to 18.63%, and drought became more widespread for the sown areas of sorghum from 16.01 to 28.12% between 1961 and 2006 (Li et al., 2009). It is predicted that an increase in mean air temperature by 1, 2 and 4 °C relative to the baseline climate could result in a yield reduction of 1–21%, 3–34% and 17–67%, respectively (Tesfaye et al., 2018). For sugarcane, climate change impacts depend on geographic location and on degree of adaptation, while cane yields in most developing countries still tend to increase by improved cultivars and management practices (Zhao & Li, 2015).

In addition to climate change, urban expansion is also enhancing the risks to food production systems (IPCC, 2019).

2.3   Impacts of Industrial Animal Farming on Climate Change, Biodiversity, and the Achievement of Food Security and Resilience

The impacts of industrial animal farming/'factory farming' demand attention with respect to climate change. Industrial animal farming involves captive breeding of animals to maximize growth/productivity. This involves, keeping large numbers of them permanently indoors, either isolated in cages or stalls or crowded together in pens or large buildings. Although it has huge impacts beyond the farm, it also impacts the health and welfare of the animals being farmed.

Broiler (meat) chickens are usually crammed into barren sheds, each containing 20,000 or more birds. As they are bred for unnatural, super-fast growth, over a quarter of them are likely to suffer from lameness, as their skeletons can no longer bear the weight of their fast-growing bodies (EFSA, 2010). They are slaughtered at only 5 or 6 weeks old.

The milk yield of many dairy cows has doubled over the last 40 years, partly due to genetic selection. This is the major factor causing poor health and welfare in cows. Some of these include increased risk of lameness, mastitis (painful infection of the teat/udder), infertility and metabolic disorders such as fatty liver syndrome and acidosis, and resultant laminitis and immune-suppression (EFSA, 2010). This leads to high use of antibiotics, at a time when global antibiotic resistance is forecast to become a major health problem for humans and animals alike. Many dairy cows are now kept permanently indoors in 'zero-grazing' farms.

Piglets would naturally wean themselves from their mother from 11 weeks of age. In factory farms they are usually taken away from her at 3–4 weeks age and reared indoors in barren conditions. The breeding sows may be confined in narrow stalls in which they cannot turn around. In natural conditions, pigs would spend up to 50% of their time using their snouts to root in the soil, seeking tubers and grubs to eat and another 23% in foraging behavior (Stolba & Wood-Gush, 1989). Such behavior is totally impossible in industrial farms.

Industrially reared laying hens are kept in cages. In the barren battery cage, they spend their productive lives (from about 18 weeks of age) until slaughter, about a year later, standing on a sloping wire mesh floor, unable to stretch their wings, scratch in the soil or dust-bathe and are unable to lay their eggs in a nest—a behavior of great importance for the hen (Cooper & Albentosa, 2003). Since 2012, the barren cage has been banned throughout the European Union. Most hens are however caged in slightly larger so-called 'modified' cages (European Commission, 1999). The barren cage is still widely used in many countries.

Because the stressful, overcrowded conditions of intensive livestock production cause distress to the animals, their immune systems can be impaired, making them even more susceptible to disease. This leads to high use of antibiotics, at a time when global antibiotic resistance is forecast to become a major health problem for humans and animals alike.

Factory farming is mainly profit-oriented. It may seem economically viable, but if you count its negative impacts on the climate, biodiversity, ecosystem resilience and food security, as explained in the sections below, you soon realize that the costs of industrial animal farming systems are very high compared to its supposed benefits.

2.3.1 Impact of Industrial Animal Farming on Climate

The links between large amounts of greenhouse gas emissions (GHG) and livestock production are well known. IPCC's Special Report on Climate Change and Land (SRCCL) stresses the importance of a reduction in food waste and meat consumption to tackle methane production (IPCC, 2019). It indicates balanced diets featuring plant-based foods, such as coarse grains, legumes, fruits and vegetables, and animal-sourced food produced sustainably in low greenhouse gas emission systems, which present major opportunities for adaptation (IPCC, 2019). Indeed, UNEP's Global Environment Outlook (2019) emphasizes that even with a move in the global energy supply from fossil fuel-based to renewables, it will not be possible to stay below 1.5 °C above pre-industrialized levels, without addressing agriculture, food production and food waste (UNEA, 2019).

FAO's most recent estimate implicates livestock production for 14.5% of anthropogenic (human-caused) GHG (Gerber et al., 2013). That is roughly the same amount as all kinds of transport put together. Gases associated with livestock farming are mainly methane emissions from ruminants and nitrous oxide from growing animal feed and from animal waste. Although ruminants do emit methane, they can graze on land unsuitable for crop production. Pasture land itself sequesters carbon in the soil. It is unwise to call for ruminant beef and sheep to be replaced by monogastric pigs and poultry, as these animals are usually reared in industrial farms. As these indoor-raised animals are unable to seek their own food, it has to be bought in and fed to them. This has led to a huge demand for cereals and soy for animal feed, with resulting detrimental impacts on soils, water and biodiversity and further deforestation in places like South America (Willett et al., 2019).

In a recent blog, leading experts Bojana Bajželj and Tim Benton argued "Our demand for food alone could virtually guarantee that the Paris aspirations are unachievable". They call for "cutting down on consumption of intensively-produced meat and dairy. Raising livestock is a much less efficient way of producing food than growing crops. Currently, a third of the crops we grow are fed to livestock to produce meat. If we used the land growing feed to grow food, and ate only meat from pasture-fed animals, there is scope for very significant reductions in emissions" (Carbon Brief, 2016). A Chatham House study stresses that it is unlikely that global temperature rises can be kept below 2 °C without a reduction in meat and dairy consumption (Bailey et al., 2014).

Furthermore, beef, pork and poultry production involve much higher GHG emissions than fruit, vegetables, cereals or pulses. The latter is a climate-friendly solution with global food-related GHG emissions lessened by 29–70% (Springmann, 2016). From a health perspective, the World Health Organization's specialist cancer branch, the International Agency for Research on Cancer (IARC), ruled that red meat (beef, pork, lamb and goat) was "probably carcinogenic to humans" and that processed meat (ham, bacon and salami) was definitely carcinogenic to humans and "causes colorectal cancer" (IARC, 2015).

2.3.2   Impact of Industrial Animal Farming on Biodiversity

The world is losing species at unprecedented rate, now between 100 and 1000 times faster than the background rate of species loss (Ceballos et al., 2015). There are multiple reasons for this accelerated rate, including impacts of population growth, urban expansion, deforestation to provide pasture for cattle grazing and soya production, and the widespread chemical spraying of croplands. This reinforces a direct link with industrial animal farming. Forests and savannahs are being destroyed, not only for cattle grazing, but also to grow soy and cereals for industrially farmed animals. In fact, 98% of global soya bean meal is used as animal feed (Soyatech, 2015). Industrial agriculture relies on the intensive use of fossil fuel and chemical inputs to grow these crops, often intended for animal feed. This has resulted in increased agricultural productivity at the expense of biodiversity and soil fertility, land degradation and chemical contamination of soil and water, with major consequences on human, animal and planetary health (Kremen & Miles, 2012).

Excessive fertilizer use on intensively cropped land not only harms the soil, but frequently causes pollution of rivers. Indeed, the United Nations (UN) has concluded that "intensive livestock production is probably the largest sector-specific source of water pollution" (United Nations, 2011). The 2019 report by the High-Level Panel of Experts on Food Security and Nutrition (HPLE) of the Committee on World Food Security (HPLE, 2019), concludes as well that "it has become evident that the widespread use of synthetic fertilizers entailed very high environmental costs, including air, water and soil pollution. Because of their high solubility, synthetic fertilizers pollute surface and groundwater, including coastal and marine watersheds, and provoke toxic algal blooms and aquatic dead zones (Campbell et al., 2017; Howarth et al., 2012; Kirchmann & Bergström, 2007; Swaney et al., 2012). Indeed, the FAO has predicted that if current intensive farming methods are not stopped, we may have only 60 harvests left (FAO, 2015).

Therefore, a move to more agro-ecological and regenerative farming coupled with a reduction in the consumption of livestock products in high-consuming populations would ease the pressure on wildlife and biodiversity. With these methods, soil quality can be rebuilt through rotations, legumes, green manure and animal manure. Soils with high levels of organic matter can store carbon and improve water retention, thus reducing flooding risks and enhancing plants' ability to withstand drought. This approach to farming could restore biodiversity, enabling farmland birds, pollinators and other wildlife to flourish. This reinforces the importance of agro-biodiversity in guaranteeing food security and fulfill rights to food for all (HRC, 2019b).

2.3.3 Impact of Industrial Animal Farming on Resilience and Food Security

Industrial farming completely contradicts the ancient concept of the farmer or shepherd as a person who cares for their flock and the land. It has no place for the small-scale farmer who may have a 'house cow' and some free-roaming hens which provide eggs and, on occasion, meat. Ninety percent of the breeding stock for the 66 billion broiler chickens slaughtered globally per annum (FAOStat) comes from just two breeder companies. Most chicken farmers are tied in by contracts to companies which specify the feed, medication and conditions in which the birds must be grown. Farmers have been turned into contract 'growers'. In such situations, the profits of industrial animal farming do not 'trickle down' to local communities; instead, they are concentrated in the hands of a small number of agribusiness companies. Moreover, conflicts with industrialized animal operations over land and forest resources threaten the ability of smallholders and indigenous peoples to overcome poverty. Local people are vulnerable to 'land grabbing' by powerful companies who wish to use the land to grow soy and grain for animal feed (Pretty, 2006).

Within the same perspective, the High-Level Panel of Experts on Food Security and Nutrition of the Committee on World Food Security has declared: "The social benefits of agriculture can be eroded as production becomes more concentrated and intensive. Intensive agricultural systems are associated with negative effects on employment, wealth distribution, ancillary economic activity in rural areas [and] service provision in rural areas (such as schools and health facilities)" (HLPE, 2016). The World Bank has recognized as well that intensification of livestock production carries "a significant danger that the poor are being crowded out" (HLPE, 2016).

Additionally, industrially reared pigs, poultry and cattle undermine food security as they consume human-edible crops and convert them very inefficiently into meat and milk, whereas pasture-based ruminants can boost food security as they convert materials we cannot eat, such as grass, into food we can eat. For every 100 cal fed to animals as cereals, just 17–30 cal enter the human food chain as meat. Many more people can be fed if crops are used for direct human consumption rather than as animal feed.

With mixed rotational farming, animals feed on grass supplemented with crop residues and forage crops where necessary. Their manure, rather than being a pollutant, enhances soil quality. With outdoor rearing, the potential for high animal welfare is increased.

2.4 Linkages Between Ecosystem Management and Resilience to Food Insecurity in a Changing Climate

Ecosystems serve as platforms of interaction between biotic and abiotic components (Acquaah, 2002). In an ecosystem, energy transfers into the living systems, convert biomass utilizing nutrients from the soil, water and air; thus, maintaining the overall food system (Ricard, 2014). The productivity of an ecosystem depends on geographical position, landscape, climatic factors and availability of soil, water, and biotic components (Astanin & Blagosklonov, 1978). With advances in agriculture, urbanization and industrialization, humans have increasingly regulated/managed production system to harness benefits. Application of modern technologies has greatly enhanced profitability of cultivation in many areas of the world, which were hitherto, not suitable for cultivation (Jones et al., 2017).

Although much progress has been made in reducing hunger and poverty and improving food security and nutrition, millions of people still suffer from hunger and malnutrition. Global food security is in jeopardy due, among others, to mounting pressures on natural resources and climate change (FAO, 2017). The well-being of people depends on the management of healthy ecosystems for goods—like food and water—and services—like climate regulation and protection from natural hazards—in addition to the mitigation of the negative effects of climate change (TEEB, 2010; Meckey, 2009; WHO, 2019), through human adaptation and behavioral change (Munang et al., 2011). Regarding food security, climate change affects people's ability to access and use food efficiently by altering food safety conditions, and thereby increasing the risks of vector-, water-and food-borne diseases (Githeko et al., 2000; Patz et al., 2005). In addition, nearly US\$577 billion in annual global crops are at risk from pollinator loss and 100–300 million people are at increased risk of floods and hurricanes because of loss of coastal habitats and protection (IPBES, 2019). Land degradation has reduced as well the productivity of 23% of the global land surface.

Forests play a crucial role in climate change adaptation efforts. They act as a food safety net during climate shocks, help reduce risks of disasters—like coastal flooding—and regulate water flows and microclimates. Improving the health of these forest ecosystems and introducing sustainable management practices increase the climate resilience of social-ecological systems (IUCN, 2017). Deforestation, however, is a great concern; the tropics lost 12 million hectares in 2018, the fourth-highest annual loss since records began in 2001, according to Global Forest Watch (WRI, 2019). For safeguarding wildlife, curbing and adapting to climate change, and boosting water security, some countries, including China, India, Malawi, Cameroon, Brazil, Indonesia, Madagascar, Colombia, the Philippines, Vietnam, Myanmar, Thailand, Ivory Coast, Rwanda, Uganda, Burundi, Togo, South Sudan and Madagascar have already launched large-scale tree planting. Under the Bonn Challenge 2011, it was set to restore 350 million hectares of degraded land worldwide by 2030. Now more than 70% of hotspots are found in countries with reforestation commitments (Taylor, 2019). The European Commission promotes Nature-Based-Solutions (NbS) to sequester local carbon, reduce pollution, lower temperature, increase biodiversity and provide a pleasant environment (European Commission, 2015; Calfapietra & Cherubini, 2019). The Government of France will support a series of urban forests to transform its cities as carbon neutral entities by 2050 (Stinson, 2019). It should be noted that for successful establishment of secondary forests, trees should be chosen according to their ecological characteristics and habitats i.e. the right plant at the right place (Basak et al., 2015; Kallio, 2013).

In the Ganges and Brahmaputra basins of South Asian region, traditional floodplain management rely on digging ponds and raising land for homes. Indeed, vegetation with forest-groves are unique adaptation practices against floods, which help protect houses from cyclones and tornados and regulate water uses and fish practices. A cross-sectional view of this is presented in Fig. 3 (Dewan, 2015; Rahman & Rahman, 2015; Rahman et al., 2018).

Growing crops in an adverse environment, climatic manipulation practices, i.e., adaptation with climatic and environmental changes, are becoming effective and popular. Common practices are greenhouse agriculture, hydroponics and aquaponics, and floating agriculture. Floating agriculture for growing vegetables and spices has existed in the wetlands of the south-central coastal districts of Bangladesh for a long time. Using water hyacinth and other aquatic weeds, local communities have

Fig. 3 Cross-sectional view of traditional flood-plain management. *Source* Rahman and Rahman (2015)

developed a technique to construct reasonably-sized floating platforms or rafts on which vegetables and other crops can be cultivated (APEIS & RIPSO, 2004). It is also practiced in India (Chatterjee, 2016), Myanmar (NASA, 2015), Mexico, China, and Thailand (Pantanella et al., 2011).

Climate variability is well buffered by agroforestry with permanent tree cover and varied ecological niches. Agroforestry provides resilience through soil moisture conservation, wind protection and maintaining water cycle, enables recovery after a disturbance via diversified temporal and spatial management options, improves the quality of soil, and offers a win–win opportunity due to its carbon sequestration potential and its crucial role in ensuring food security, increasing farm income, and discouraging deforestation (Rao et al., 2007; FAO, 2013b).

Urban agriculture is growing as an option to secure fresh food, employment and recycle wastes to strengthen food security (FAO, 2019b). This complements climate-smart agriculture, designed to deliver multiple benefits (FAO, 2019a; World Bank, 2018).

Therefore, the issue of ecosystem management should be integrated with other measures to enhance the resilience for food security in the context of climate change. It is recommended that the ecosystem approach should be mainstreamed into relevant policy-making processes at all levels to ensure both a healthy ecosystem and a food-secure future.

3 Food Security, Climate Change, and Resilience: International and Regional Perspectives

3.1 The International Action Supporting the Achievement of Climate-Resilient Agriculture

According to the FAO et al. (2018), it is an imperative to enhance efforts to address challenges faced especially by the hungry. There is a risk of falling far short of achieving the target of the SDGs on ending hunger by 2030, should such efforts not materialize. To reverse this trend, many multilateral instruments set binding commitments for governments and international organizations especially to foster resilience to food and nutrition insecurity. In this regard, the Human Right Council has recently stressed the importance of outcomes of the second International Conference on Nutrition, hosted by the WHO and the FAO in Rome from 19 to 21 November 2014 (HRC, 2019b). These are the Rome Declaration on Nutrition and the Framework for Action. The same Resolution also affirms that each State must adopt a strategy consistent with its resources and capacities to achieve its individual goals in implementing the recommendations contained in the Rome Declaration on World Food Security and the World Food Summit Plan of Action. Indeed, it is recognized that food security is a national responsibility, and that any plan for addressing food security challenges must be nationally articulated, designed, owned and led, and built on consultation with all key stakeholders. In the meantime, each country should commit to strengthening the multilateral system about channeling resources and dedicate policies to fight hunger and malnutrition. It should cooperate regionally and internationally to organize collective solutions to tackle global issues of food security in a world of increasingly interlinked institutions, societies and economies.

Another landmark framework is the UN General Assembly resolution 70/1 of 25 September 2015, titled "Transforming our world: the 2030 Agenda for Sustainable Development". The Assembly adopted a comprehensive far-reaching and people-centered universal and transformative SDGs 2030 Agenda (HRC, 2019b).

The UN Decade of Action on Nutrition 2016—2025 provides an operational framework to strengthen efforts to end hunger and eradicate all forms of malnutrition worldwide, including through nutrition-sensitive disaster risk reduction and climate adaptation policies and programs for holistic outcomes relating livelihoods and food systems for healthy diets.

The 23rd Conference of the Parties to the United Nations Framework Convention on Climate Change (UNFCCC) adopted the Koronivia Joint Work on Agriculture (KJWA). This addresses vulnerability of agriculture to climate change and approaches to address food security (UNFCCC, 2017). This was considered a momentous decision on agriculture for transformative action to make farmers' livelihoods and food supply more resilient (Dinesh et al., 2017). In response to the decision, most Party submissions (AGN, Argentina, Benin, Brazil, Burundi, China, EU, India, Japan, Kenya, Malawi, New Zealand, Norway, Switzerland, Viet Nam) considered

it paramount to address vulnerability under the KJWA aligned with national priorities (Chiriaco et al., 2018). The FAO in its submission to the UNFCCC suggests
ensuring that climate finance reflects the vital importance of agriculture, e.g. by
unlocking private sector investment providing access to innovation and technologies
for adaptation, particularly in the least developed countries (FAO, 2018b).

3.2   Regional Perspectives

3.2.1   South and South-East Asia

Agriculture in South Asia is quite vulnerable to climate change (Aryal et al., 2019),
and will negatively impact agricultural yields as well in South-east Asia (ADB, 2010).
This is due to the high rate of population growth, and significant impacts are likely to
be borne by small-hold rain-fed farmers (Hossain et al., 2016). According to the IPCC
Special Report on global warming of 1.5 °C, exposure to multiple and compound
climate-related risks increases between 1.5 °C and 2 °C of global warming, with
greater proportions of people affected and susceptible to poverty, in addition to food
security and health risks (IPCC, 2018). Climate change has reduced consumable food
calories in numerous Asian countries including Bangladesh, Nepal and India (Ray
et al., 2019). Some trends specific to South and South East Asia are highlighted in
the following paragraph.

By 2100, higher temperatures are likely to reduce rice yield potential by up to 50%
on average compared to 1990 levels. This could cascade through, prompt conversion
of even more land to agriculture in the Southeast Asian Region (Gonsalves et al.,
2015). However, maize yields are almost doubling as in India, Pakistan, Indonesia,
Bangladesh, Laos, Cambodia and Vietnam (Ray et al., 2013). In addition, per capita
maize harvests are projected to further increase in Indonesia, Vietnam, Thailand,
Myanmar, Bangladesh, Cambodia, and Bhutan by 2050 (Ray et al., 2013). The case
is different for wheat and rice. There could be no significant change/increase in per
capita rice harvests in Pakistan, Nepal, Malaysia, Myanmar, Sri Lanka, and Bhutan,
and the wheat harvests are also showing a decreasing trend (Ray et al., 2013). Wheat
yields for 31% of the global harvested area, including South Asia, are anticipated
to stagnate with less significant temperature increases (Iizumi et al., 2017). Climate
change would reduce rain-fed maize yield by an average of 3.3–6.4% in 2030 and
5.2–12.2% in 2050 and irrigated yield by 3–8% in 2030 and 5–14% in 2050 if current
varieties were grown under the future climate in South Asia (Tesfaye et al, 2017). A
study on eight major crops in Africa and South Asia shows that the projected mean
change in yield of all crops is −16% (maize) and −11% (sorghum) in South Asia
(Knox et al., 2012).

Consumable food calories reduced in India ~0.8% overall on average annually,
and losses in rice production have also occurred in Vietnam (~−1.0 MT) and in
the Laguna province of Philippines (~11 kg /ha/year or −0.2% in yields), although
overall rice production increased in the Philippines (Ray et al., 2019). A study in

the Vietnam concludes that climate change will negatively impact food security and make the country more dependent on and vulnerable to changes in the world market, whereas strong, agricultural-based growth has the opposite effect (Rutten et al., 2014). Ricardian studies in India show that agriculture would be sensitive to even modest warming and residual harm would remain even with adaptation. Additionally, the wet eastern region of India would mildly benefit from warming whereas the dry western region would suffer large damages (Mendelsohn, 2008).

As a result of increasing climate change stressors, farmers are trying to find adaptation solutions like changing their agricultural and farming practices for the current and unforeseen events (Tripathi & Mishra, 2017). Over half of the farmers surveyed in Bangladesh, Indonesia, Sri Lanka, Thailand, and Vietnam claimed to have changed their irrigation, timing, or crop choices because of climate change (Abidoye et al., 2017). A study in Myanmar verified farmers' perception by historical meteorological data analysis and observed that farmers' perceptions of the amount of rainfall across monsoon periods were highly accurate. They were however less accurate with respect to changes in winter and monsoon seasons (Hein et al., 2019). Climate change risk perception plays a significant role in deciding adaptation strategies. In this scenario, farmers' choice of adaptation strategies varies with their perception of climate risks, assets, education level, income, irrigation access, and recent climate related shocks (Hein et al., 2019; Jain et al., 2015; Waibel et al., 2018). In a survey based study in Bangladesh, locals acknowledged climate change and perceived it as having negative impacts on agriculture and livelihoods (Islam & Nursey-Bray, 2017). A study in Nepal shows that farmers perceive climate change in the form of increased evapotranspiration, more infestations, new weeds and reduction in crop yield (Khanal et al., 2018b). Another study in the Sundarbans region found that most farmers were aware of climate variability (Dubey et al., 2017). Tripathi and Mishra (2017) found that farmers in India are unable to identify changes in extreme climatic events and variability as climate change. In a study in Indonesia, people from lower organizational level believed that climate change was happening and thought of it as a natural phenomenon, while those from higher organizational level attributed it to human activities (Bohensky et al., 2016). Respondents to a study in two provinces of Thailand observed the length of dry season to have increased while in Vietnam, lower total rainfall variability was noted (Waibel et al., 2018).

Adaptation strategies in different regions of South and Southeast Asia include change in cropping patterns and sowing dates, crop and income diversification, assets selling and migration, use of water efficient technologies, tolerant crop varieties (heat, salt, flood, drought), soil and water conservation, inter- and mixed cropping and crop diversification, climate-smart agriculture, chemical input management and planting trees (Ahmed & Suphachalasai, 2014; Alauddin & Sarker, 2014; Gonsalves et al. 2015; Bahinipati & Venkatachalam, 2015; Menike & Arachchi, 2016; Khatri-Chhetri & Aggarwal, 2017; Tripathi & Mishra, 2017; Tesfaye et al., 2017; Waibel et al., 2018; Khanal et al., 2018b; Thi Nhung et al., 2019; Aryal et al., 2019).

Bangladesh, for instance is among countries, internationally recognized for its cutting-edge achievements in addressing climate change (World Bank, 2016). The Bangladesh Delta Plan (BDP) 2100 is a combination of long-term strategies and subsequent interventions for ensuring long-term water and food security, economic growth and environmental sustainability while effectively reducing vulnerability to natural disasters and building resilience to climate change and other delta challenges through robust, adaptive and integrated strategies, and equitable water governance (Alam, 2019). Bangladesh has also developed salinity tolerant rice varieties viz. BRRI dhan55, BRRI dhan61 and BINA dhan8, BRRI dhan47, 48 and BRRI dhan28 (Islam et al., 2016).

Community-based adaptation (CbA) for developing adaptation strategies (Karim & Thiel, 2017), participatory approach for mapping climate risks (Yen et al., 2019), and strengthening capacity of local institutions (Singh et al., 2019) could play enhance agriculture adaptation. Even small-scale adjustments made by farmers at the local level are effective in improving farmers' efficiency in agriculture production (Khanal et al., 2018a). Risk pooling insurance, resource conserving agricultural technologies, mobile-based climate services and innovative collaboration are initial steps taken to assure adaptation in South and Southeast Asian countries (Enenkel et al., 2015; Mittal & Hariharan, 2018; Pathak et al., 2012; Rosegrant et al., 2014).

3.2.2   Sub-Saharan African (SSA) Perspective

A continuously growing population, severe climate impacts, herewith linked the rising demand for food production and decreasing availability of farm land and water resources, are extreme challenges for the African continent. Indeed, most African countries face the triple challenge of raising a large portion of their populations out of poverty, ensuring a climate-compatible development in a sustainable manner, and safeguarding natural and human systems from climate change vulnerability and adverse impacts. Africa is one of the continents most affected by climate change impacts despite having contributed the least. The HRC in its *Resolution on the Right to Food* recognizes the need to urgently assist some African countries that are facing drought, starvation and famine threats that could affect millions of people, most of whom are women and children, at risk of losing their lives (HRC, 2019b).

Food price increases are not the only feature of food insecurity in sub-Saharan Africa (SSA) that is mostly linked to environmental change. Droughts are frequent in SSA and lead often to regional livelihood losses that severely aggravate the plight of the rural poor. Rural conditions make it difficult for households to cope with adverse environmental events. Population densities in rural communal areas tend to be high, which, over time, contribute to over-grazing and severe soil erosion. In the event of drought, the production of even a small amount of grain or other food is diminished and livestock is lost. Many case studies across SSA reveal environmental impacts influenced community livelihoods, particularly in the case of more intense and variable climatic events (Mubirua et al., 2016) e.g., heavy rain, flooding, drought and even snow, hail and frost. Long-term erosion of livelihoods and food insecurity among vulnerable communities, particularly in rural areas, has caused malnutrition, with severe impact on children's physiology.

The current rapid population growth rate in SSA, coupled with persistent high malnutrition rates caused by climate change, water and food insecurity, requires immediate action in the form of nutrition-focused food (Foglianoa, 2018). With the current change in climate, the pressure on water supply is also likely to increase in SSA. This calls for innovative location specific water management strategies and broader adaptation strategies. To support these adaptation strategies, specific adapted types and cultivars, heat protection measures, and water efficient technologies are needed (Archer, 2019; SmartAgri, 2016).

Modern technologies as well as organic and chemical fertilizers can enable optimized plant breeding, plant protection, and food production. This is specially the case of SSA, that needs better education, capacity development and infrastructure to achieve higher productivity.

Knowledge of optimized agriculture adapted to different soil characteristics can significantly reduce loss of yield, balance nutrient supply and enhance the quantity of food available for many communities in SSA. It is increasingly proven that SSA needs knowledge in climate-smart agriculture focusing: on early maturing, drought tolerant and water efficient crops and crop varieties; pastures and fodder varieties; soil and water management technologies; humane livestock management options; and strategies that can improve adaptive capacity and explore mitigation co-benefits, as well as sustainable productivity. The knowledge gained from research is primarily usable for schools, public lectures and graduate schools to transfer knowledge across generations and support knowledge sharing within local, regional and national academic institutions.

These facts and challenges are currently well known by the African Union (AU), and are specifically named in the current '*Agenda 2063 - The Africa We Want*' First Ten-Year Implementation Plan 2014–2023. Since Agenda 2063 has a 50-year horizon, the goals are broader. The priority areas and their associated targets define the goals. While the goals are fixed, the priority areas and their associated targets can change over the various ten-year plan cycles. In the case of the MDGs (with their 15-year horizon), the goals are very specific with clear-cut targets within the ten-year plan results framework. To underline this strategy, the New Partnership for Africa's Development (NEPAD) has already established and focused on strengthening agriculture, fostering food and nutrition security, promoting the blue economy, improving environmental governance, facilitating the adoption of climate change mitigation and adaptation strategies, and promoting sustainable mining at national, regional and continental levels since 2013 by publishing the 'Declaration of the first NEPAD Ministerial Conference on Science and Technology'. The program's objective is to provide the AU Member States with innovative development and implementation capacities for viable Natural Resources Management and placing them on the path of inclusive growth and sustainable development.

To meet the above challenges for the SSA, the AU plans to achieve transformation in five key areas at national, regional and continental levels by 2023. Among others, 'Improvements in Living Standards' is one of the main key areas to reduce hunger, especially amongst women and youth to only 20% of actual 2023 levels. Malnutrition, maternal, child and neo-natal deaths as of 2013 would be reduced by half. Nine out of ten persons will have access to safe drinking water and sanitation. Importantly, for SSA at least one out of every three children will be having access to kindergarten education. Currently every child of secondary school age at school and seven out of ten of its graduates don't have access to tertiary education, or enroll in Technical and Vocational Education and Training (TVET) programs. Lessons learnt from the implementation of past continental frameworks indicate the necessity for building capacities of stakeholders, politicians and youth at the continental, regional and national levels (AU 2015).

4 Pathways to Ensuring Resilience for Food and Nutrition Security in a Changing Climate

4.1 Empowering Women to Ensure Household Food Security in a Changing Climate

Women have long been sidelined in agriculture, as in other sectors. Not only is this a case of women's rights and gender justice, from an economic viewpoint, it is a question of willfully losing out on agricultural productivity, maintenance of farmlands, and sustainability of food supplies. Women form between 20 and 50% of the agricultural workforce in the developing world (FAO, 2011), but their labor, being unpaid is also unaccounted for. It is by now well known that women in many parts of the world lack legal rights to land and are consequently denied access to credit, technology, extension services, and are not even considered a factor in agricultural planning or policy (Deliver for Good Campaign, 2018). Only 5% women farmers spanning 97 countries have access to agricultural and other training activities (FAO, 2013c). Where women own land, it is usually of poor quality and small in size. For example, in Kenya, men's landholdings are on average, three times larger, and in Bangladesh, Ecuador and Pakistan they are twice the size of women's (FAO, 2011; Razavi, 2007). Though UNFCCC COPs include the agenda on gender and climate change every year now after COP-18, they have little say in decision making both at the local and national level making their representation at the international level, more difficult (Gupta, 2015). Additionally, policy makers design programs in a gender-blind fashion, across marketing, feeding the family and/or tending land.

Women and men have always played complimentary roles in food production be it cultivation, animal husbandry or fishing. Women are however more severely affected in case of any disaster (Fordham, 2003). Of course, women, in addition bear the burden of feeding and caring for the household. Today, this disproportionate burden on women has almost grown into a total burden. As the vagaries of climate have taken control of agriculture and farming is increasingly becoming unviable, men have begun to migrate to cities to seek new ways to earn their livelihoods. Women remain behind to look after their land, animals and families and eke out an existence without male support. In this context, it is no longer enough to focus on women farmers as those most vulnerable to climate shocks, disasters or change and in need of protection; they are now, more importantly, the key players in the challenge of maintaining agricultural lands and waters, and ensuring sustainable food security within the changing and unpredictable climatic scenario.

Case studies from across the world present a similar pattern. In Bangladesh, a country particularly vulnerable to tropical storms and floods, increasing salinity of the soils in the coastal areas has led to large-scale male migration. Women are turning to their home gardens to grow cash crops like chillies, which require little water, or small-scale chicken farming to make ends meet. Increasing salinity also affects drinking water supplies and makes life more difficult (ICI, 2012). In West Africa, where women have begun to gain land rights to dry wastelands, an ICRISAT initiative has trained women in Nigeria to grow drought tolerant rain-fed fruit trees and vegetables on degraded land. This enables them to earn an income and grow nutritious food (CGIAR, 2011). Across the Pacific, in Vanuatu, devastated by Cyclone Pam in 2015 followed by an El Nino created drought, women are engaged in an endless struggle to regain food security and water supplies. Assisted by the UN Markets for Change program, it is once again women who meet the challenge of dealing with shriveled crops, severely limited water supplies and hungry families (UN Women, 2016).

With awareness creation, training and access to small grants, women can exert more positively on household and community food security, as well as the regeneration of the environment. In India, women are working as knowledge transmitters, assisting International Crops Research Institute for the Semi-Arid Tropics (ICRISAT) scientists by measuring rainfall, thus enabling the creation of drought prediction maps. This helps plan for drought spells and sustain harvests by switching from rice to drought resistant millet, when appropriate (CGIAR, 2011). Women shellfish producers in Galicia, Spain were once uneducated and engaged in a low status extractive operation under harsh conditions. An equal opportunities program instituted by the EU and the Galicean government led them to organize themselves, access training, and cultivate shellfish on beaches, increasing production and improving their economic status. They now plan the entire exploitation process paying due attention to long-term sustainability of coastal resources. This model could be scaled up to enable sustainable use of marine resources (Williams et al., 2005).

With gender specific economic and social constraints to overcome, it is difficult for individual women to break out of the norms imposed on them and take their rightful role in agriculture and environmental protection. On the other hand, women in close knit community groups or cooperatives, acquire the strength of numbers and are more than equal to face the climate challenge.

4.2 Science-Based Public Policy Thrusts to Strengthen Mitigation and Adaptation Approaches and Outcomes: The Case of Climate-Resilient Agriculture

The preceding sections substantiated some important recent insights also from the IPCC (2019), community perspectives and institutional mechanisms to enhance resilience. The FAO (2013a, 2017, 2018) and the UNDP have highlighted benefits of system—specific governance. The impacts of climate finance (Nicholson et al., 2017; UNDP, 2019, undated) and multilateral initiatives across regions with special reference to community-based successes are also well known (FAO, 2018c). The UN Environment's Low Carbon Technology Partnership Initiative and life cycle assessments (LCA) of agricultural production enrich the landscape of management decisions further. Community-led crop rotation, tillage, crop intensification, integrated pest/nutrient/fish/livestock management, crop-friendly landscaping, agroforestry, soil and water conservation have been at the center of such initiatives, including systems perspectives (OECD, 2019). The World Bank has assisted local level resilience initiatives as in Odisha India (World Bank, 2019a) and Nepal for instance (Garcia, 2018) and across regions of the world (World Bank, 2019b) and presents lessons learnt thereof (Multilateral Development Banks et al., 2018). These also reveal a consensus across regions of the world that human activities influence qualitative and quantitative profiles of land, water and bio resources through agriculture and closely aligned forestry. This implies mutually reinforcing impacts on mitigation and adaptation strategies and outcomes.

A logical way ahead of the above stated, is to relate to the Sendai Framework on Disaster Risk Reduction and the Global Stocktake process of Paris Agreement. Nationally Determined Contributions (NDC) and initiatives of several multilateral institutions also highlight the much-needed convergence. These elements of public policy can be infused with overarching deeper ecological principles of systems management including chemical ecology and the growing link between climate and ozone protection. This perspective is meant to strengthen mitigation and adaptation outcomes across three levels of aggregation. The first is about soils, water and bio resources. The second is about the need to eliminate uncertainties in the public domain, especially of the farming community, and to strengthen scientific temper for well-informed collective action. The third is about some elements of energy use beyond the farm, with implications of integrated troposphere stratosphere benefits. The following section accordingly establishes the need for immediate attention on three facets of science-based public policy to strengthen the cause-effect relationship vis-à-vis climate change.

4.3 Overcoming Gaps in Knowledge About Responses and Adaptation Patterns of Crops and Related Biodiversity to Changes in Micro-Climate

Chemical and microbial ecology should be given greater attention to understand ecosystem dynamics better. Gaps in knowledge about responses and adaptation patterns of crops and related biodiversity to changes in micro-climate should be addressed on priority. A stronger chemical ecological perspective will probably enrich our insights about appropriate safeguards.

The IPCC fifth report (AR5) (2014) substantiates concerns and creates a call for concerted action. Is it therefore possible to strengthen our adaptation and mitigation strategies by applying principles of ecology and ecosystem services in decision support systems? Four important facets of chemical ecology therefore, rise to the forefront. They relate to delayed induced resistance, rapid induced resistance and related biochemical profiles of plants and the consequences of such changes on the occurrence and distribution of pests and their natural enemies. Deeper insights about changes in the root zone and beyond in the soil, of preponderance and activity of microbes that mediate mineralization and naturally and artificially induced age-correlated biochemical profiles of all related bio-resources are central to holistic perspectives. It will be useful to enhance the importance of these parameters in decision-making systems. They will help minimize noise in decision-support systems that relate observed and projected yields of crops, water demand and pest dynamics including microbial invasions. The FAO (2018a) went a step further, calling for an integrated approach to assess systemic impacts across nutrition and agricultural livelihoods, including horticulture, livestock, fisheries and aquaculture.

A recent analysis of trends in agriculture productivity in north-east India revealed cognizable impacts. These were related to changes in temperature, resultant water dynamics, adaptive abilities of plants and a variety of such bio resources that can be optimized for resilience benefits (Roy et al., 2019). The logical way ahead will be to delve deeper into chemical ecological profiles of such bioresources, especially through the framework of tritrophic interactions. The need for such investigations is captured succinctly by the Entomological Society of America (2019) this year. This is true also of system-wide relationships between plants and arbuscular-mycorrhizal fungal elements that determine plant nutritional and defense traits (Frew & Price, 2019). Our knowledge is not adequate even in these cases with reference to GHG-driven warming. Decision makers have to go beyond their immediate and medium-term gain horizons, to long-term perspectives. This is especially important when the levels of uncertainty in decision making are quite high. This uncertainty is reflected in the growing spread and depth of vulnerability of systems in all regions of the world.

While these are essentials of science for resilience, it is important to also relate to other facets of public policy for integrated outcomes. These relate to institutional mechanisms that facilitate action at the grassroots and public engagement for timely buy-in of mitigation and adaptation measures. A recent treatise on the science of science communication (National Academy of Sciences, 2014; Baruch & Davis, 2015) elaborated on the dynamics of public understanding of science and the need for focused engagement strategies. Communication strategies (Green Climate Fund, 2019; UNFCCC Adaptation Committee, 2019) pertaining to climate change and related impacts should be revisited based on three important elements of the above cited framework to secure much needed significantly enhanced groundswell:

- The *first* is about controversies in science that should be addressed on priority. This is signified by the fact that diametrically opposite view-points prevail in the applications of science in public policy. These could arise on account of incomplete understanding of the phenomenon and consequences of inaction.
- *Secondly*, citizens, and decision-makers, should be assisted to imbibe values of preventive management and application of the precautionary principle (Patterson & Mclean, 2019; Winterfeldt, 2013). Empirical evidences relevant at the local level, should be adequately highlighted without losing sight of the global scale and parameters that determine projections. Limits and limitations of such analyses should be duly highlighted up-front. Decision-support systems for public policy should be enriched with evidence about barriers in public engagement that prevail, despite periodic information support. Resistance to engage could be caused by incongruent market and institutional mechanisms and not necessarily on account of attitudes.
- The *third* aspect is about the performance of social media tools of mass engagement, which are predominantly meant to only deliver information. Indeed, these tools at best deliver appropriate information in a timely manner and to many people. However, this outreach should not be mistaken for outcomes of communication. Detailed assessments of target specific information needs should guide communication strategies. They should also define opportunities for technical-capacity building to overcome challenges. The Capacity Building Initiative for Transparency of the Global Environmental Facility (GEF, 2018) substantiates this call, based on dozens of projects involving a wide variety of stakeholders from several countries. The specific emphasis is on transparency and track fulfillment of commitments.

An interesting body of literature (Pouyat, 1999; Partey Samuel et al., 2018) delves into the dynamics of scientific uncertainty with implications for public engagement. This continually evolving challenge tends to influence policy making and confounds solutions and highlights the framework of limits and limitations of knowledge for truth-based engagement with citizens to tackle uncertainties. The challenge is exacerbated when political processes ignore science-based solutions and the public is not able to visualize the positive outcomes of options. Importantly, legal and regulatory remedies may dominate if science-based solutions are not adequately embedded in decision making (Rare and The Behavioral Insights Team, 2019). Decision makers look for signals of change and links between causes and effects and could tend to choose from among fixed options. It is therefore, imperative to characterize and assess uncertainties and rationalize options for long-term benefits. Unknown-unknowns too tend to dominate this landscape of decision making. The meta-precautionary principle therefore, seems to provide some hope in this context since it says uncertainty should not be a cause for inaction. The triple win paradigm (Lewis & Rudnick, 2019) is too precious to be lost in the quagmire of uncertainties.

Citizen engagement as a public policy tool should be taken seriously. A special emphasis on scientific uncertainties and the need to build back-casting abilities of stakeholders[1] is essential to translate intents of public engagement into reality. Schemes like India's Pradhan Mantri Krishi Sinchayee Yojana (PMKSY) are designed to enhance the area of land under micro-irrigation, with a special focus on benefits for small and marginal farmers. Water stresses due to inclement weather call for location-specific water management strategies and appropriate capacity building to sustain transitions. Early warning systems that relate water quality and quantity could highlight consequences of impending droughts on the qualitative and quantitative profiles of crop yields under stress. The Accelerated Pulse Production Program (Crop Division, 2018) under the National Food Security Mission is another case of preventive management through government initiatives. These relate also to challenges in public understanding of mitigation benefits, calling for more concerted efforts to establish as many correlates as possible to reinforce cause-effect relationships.

An equally interesting window of opportunity is to enhance energy efficiencies along the entire chain of agricultural production and consumption to optimize on climate gains. This presents a typical case of policy convergence to tackle two disparate yet closely linked globally significant climate challenges. India, for instance, emphasizes increasingly robust post-harvest handling and storage, processing and logistics duly recognizing the shelf life of produce. This has enormous implications for cold storages in particular, related to India's Cooling Action Plan. It is now well-known that countries, as part of the Kigali Amendment to the Montreal Protocol on Substances that Deplete the Ozone Layer which entered into force on 1 January 2019, look for integrated climate and ozone protection outcomes. This is true also of the OECD (2016). The importance of cold chain cannot be overemphasized, especially because of the shelf life of agro-products. Such aspects as cost of electricity, operational efficiencies, and logistics of access attract preventive management approaches, as referred above. Energy efficiency enhancement of cooling operations and services are targeted through alternative technologies that could reduce emissions and increase affordability. This includes: phase change materials, use of renewable energy sources, and waste heat recovery systems for storage and refrigerated reefers.

It is imperative to focus on small and medium enterprises in the production sector and a special focus on emerging cold chain management. The FAO (2013a) has elaborated on the emission dynamics of various segments of the agri-food system. It's Energy Smart Food for People and Climate (ESF) program promotes energy-smart

---

[1] Back casting is a planning method the first step of which is to define a desirable future. This is followed by working backwards on the timeline on policies, plans and programmes to fulfil aspirations at each stage. This exercise is done while seamlessly connecting that specified future to the present.

agri-food systems that relate appropriate energy, water, food security and climate-smart strategies. This recognizes the seamless influence of energy efficiency on all aspects of agri-productivity that is however, determined by access to and use of locally adapted energy sources. This converges on energy-smart food chains and climate-smart agriculture (CSA) including the scope to grow trees on farms for energy purposes. Management interventions have to also take note of the dynamics of technology choices exerted by farmers (Khatri-Chhetri et al., 2017).

The World Bank (2018) highlighted the scope for emissions reduction through CSA practices. The case of more than one thousand small and medium agribusiness that have adopted environmentally sustainable energy technologies, to reduce more than three million tons of CO_2 in Mexico inspires confidence. Its recent interventions in Maharashtra in India cover a larger portfolio of crosscutting initiatives to deliver significant mitigation and adaptation benefits.

Recent inputs by Odhong et al. (2019) about the Kenya context, the IFAD (2019) about Egypt and MANAGE (Bhardwaj et al., 2018) in India have highlighted the importance of an integrated approach including financing for large-scale mitigation, especially by small holders and the need to minimize counter impacts of policies that promote extraction of resources and do not incentivize conservation.

4.4 Global Community-Based Adaptation Initiatives to Climate Proof Agriculture

This creates the context for concerted efforts to upscale community successes. A recent compilation by the CGIAR Research Program on Climate Change, Agriculture and Food Security (CCAFS) and the Technical Centre for Agricultural and Rural Cooperation (CTA) (2013) highlights benefits of integrated approaches across Africa and Asia to go beyond short-term weather events into sustained resilience. The spread and depth of benefits include sequestration and productivity benefits of natural regeneration. Such benefits are reportedly tangible over millions of hectares of degraded land. They have also reduced water, fuel, seed and fertilizer demand. Small holders in particular could benefit from the use of locally adapted cooking and milling practices and disease/pest resistance, crop insurance, advisories, safety nets and related benefits. Efforts to transfer knowledge on best practices pertaining to the above referred and wider crop-specific insights (World Bank, 2019c) elicited keen interest from farmers across India and Africa. The FAO has already presented evidences of comparable initiatives even in mountain and river basin areas from around the world covering such aspects as agro-forestry and sustainable grazing (FAO, 2013d). These too have delivered significant mitigation benefits. The Ecosystems Services for Poverty Alleviation initiative (Schaafsma & Bell, 2018) focused on the value of commodities derived through adapted agriculture, and the need for a pro-poor focus. The role of women-led institutions in bridging communities has also been highlighted recently (Chanana et al., 2018), and therefore, the opportunity to strengthen such initiatives further. These are quite encouraging and yet raise an important question about the ecological foundations of sustainability over longer periods of time.

5 Conclusion

Currently, mounting evidence points to the fact that climate change is already affecting agriculture and food security significantly. This further enhances the challenge of ending hunger, achieving food security, improving nutrition and promoting sustainable agriculture. While hunger is on the rise, it is equally alarming that the number of people facing crisis-level food insecurity continues to increase. Therefore, ending hunger is still a challenge that the international community and national governments find difficult to meet.

Additionally, changes in climate undermine production of major crops in tropical and temperate regions and, without adaptation, the situation is expected to worsen as temperatures increase and become more extreme. Climate-related disasters have come to dominate the risk landscape to the point where they now account for more than 80% of all major internationally reported disasters. Of all the natural hazards, floods, droughts and tropical storms affect food production the most. Drought causes more than 80% of the total damage and losses in agriculture, especially for the livestock and crop production subsectors. In relation to extreme events, the fisheries subsector is most affected by tsunamis and storms, while most of the economic impact on forestry is caused by floods and storms (FAO et al., 2018).

Resilience to food and nutrition insecurity should be based on deep changes of our food and farming systems and methods that support smallholder farmers in particular. Compassion for farmed animals, enhancement of soils, biodiversity and wildlife could meanwhile massively reduce negative impacts on the climate. Climate-resilient and sustainable agricultural livelihoods are possible and can yield mitigation, adaptation and resilience co-benefits (FAO et al., 2018, 2019a). It is essential to build climate resilience by working with nature and reshape investments across terrestrial and marine systems.

We should as well move to regenerative, agro-ecological farming to enhance productivity, simultaneously safeguard other life forms and eco system functions too. Agro-ecology reduces farmers' reliance on costly external inputs, thus improving livelihoods of the poorest farming households (HLPE, 2019). This move to a more sustainable and environmentally-friendly farming should be accompanied by advice and action on healthy diets, and management of food loss and waste. Moreover, food security research must go beyond its focus on production to also examine food access and utilization. Scientific investigations should help better understand trade-offs and synergies, uncertainties regarding the magnitude and direction of climate change, particularly at the downscaled, local level to help design mutually reinforcing and co-evolving preventive strategies (Jarvis et al., 2011).

Finally, among the predominant policy prescriptions for CSA, we can focus here on three. The *first* is centered on community initiatives that have delivered mitigation and adaptation benefits and call for further strengthening. The specific policy prescription is about infusing chemical ecological foundations in decision making about management of bio-resources to sustain productivity. The *second* relates to public understanding of science, especially of the farming community and the mechanics of science communication for optimal engagement. It is therefore, essential to systematically tackle contours of uncertainty in science as a public policy focus. This comes also against the backdrop of calls for enhanced action on nature-based solutions at the latest G20 meeting too. This wades further into the area of energy-efficient production and consumption systems and especially with reference to cooling-based shelf life of agro-produce. The *third* is to take cognizance of the convergence of climate and ozone layer protection, with implications for CSA.

References

Abid, M., Schneider, U. A., & Scheffran, J. (2016). Adaptation to climate change and its impacts on food productivity and crop income: Perspectives of farmers in rural Pakistan. *Journal of Rural Studies, 47*, 254–266. https://doi.org/10.1016/j.jrurstud.2016.08.005

Abidoye, B., Kurukulasuriya, P., & Mendelsohn, R. (2017). South-East Asian farmer perceptions of climate change. *Climate Change Economics, 8*, 1740006. https://doi.org/10.1142/S20100078 17400061

Acquaah, G. (2002). *Principles of crop production: Theory, techniques and technology, eastern economy edition* (pp. 285–287). Prentice-Hall of India Private Limited. ISBN-81-203-2161-8.

African Union. (2015a). *Agenda 2063—A shared strategic framework for inclusive growth and sustainable development.* https://www.un.org/en/africa/osaa/pdf/au/agenda2063-first10yearimpl ementation.pdf

African Union. (2015b). *Agenda 2063—A shared strategic framework for inclusive growth and sustainable development.* https://www.un.org/en/africa/osaa/pdf/au/agenda2063-first10yearimpl ementation.pdf

Ahmed, M., & Suphachalasai, S. (2014). *Assessing the costs of climate change and adaptation in South Asia.* Asian Development Bank, Mandaluyong City, Philippines

Alam, S. (2019). Bangladesh Delta Plan 2100: Implementation challenges and way forward. *The Financial Express*, March 23, 2019

Alauddin, M., & Sarker, M. A. R. (2014). Climate change and farm-level adaptation decisions and strategies in drought-prone and groundwater-depleted areas of Bangladesh: An empirical investigation. *Ecological Economics, 106*, 204–213. https://doi.org/10.1016/j.ecolecon.2014. 07.025

Ali, A., & Erenstein, O. (2017). Assessing farmer use of climate change adaptation practices and impacts on food security and poverty in Pakistan. *Climate Risk Management, 16*, 183–194. https:// doi.org/10.1016/j.crm.2016.12.001

Allen, S., & de Brauw, A. (2018). Nutrition sensitive value chain: Theory, practice, and open questions, global. *Food Security, 16*, 22–28. https://doi.org/10.1016/j.gfs.2017.07.002

APEIS and RIPSO. (2004). *Floating Agriculture in the flood-prone or submerged areas in Bangladesh (Southern regions of Bangladesh).* Bangladesh: APEIS and RIPSO. https://enviro scope.iges.or.jp/contents/APEIS/RISPO/inventory/db/pdf/0146.pdf

Archer, L., et al. (2019). South Africa's winter rainfall region drought: A region in transition? *Climate Risk Management.* https://doi.org/10.1016/j.crm.2019.100188

Aryal, J. P., Sapkota, T. B., Khurana, R., et al. (2019). Climate change and agriculture in South Asia: Adaptation options in smallholder production systems. *Environment, Development and Sustainability.* https://doi.org/10.1007/s10668-019-00414-4

Asian Development Bank (ADB). (2010). *Climate Change in Southeast Asia: Focused actions on the Frontlines of Climate Change.* ADB.

Asseng, S., Ewert, F., Martre, P., et al. (2015). Rising temperatures reduce global wheat production. *Natural Climate Change, 5,* 143–147. https://doi.org/10.1038/nclimate2470

Astanin, L. P., & Blagosklonov, K. N. (1978). *Conservation of nature.* (pp. 20–30). Progress Publishers.

Bahinipati, C. S., & Venkatachalam, L. (2015). What drives farmers to adopt farm-level adaptation practices to climate extremes: Empirical evidence from Odisha, India. *International Journal of Disaster Risk Reduction, 14,* 347–356. https://doi.org/10.1016/j.ijdrr.2015.08.010

Bailey, R., Froggatt, A., & Wellesley, L. (2014). *Livestock—Climate change's forgotten sector.* Chatham House, London, https://www.chathamhouse.org/sites/default/files/field/field_doc ument/20141203LivestockClimateChangeForgottenSectorBaileyFroggattWellesleyFinal.pdf

Barnett, J. (2011). Dangerous climate change in the Pacific Islands: Food production and food security. *Regional Environmental Change, 11,* 229–237. https://doi.org/10.1007/s10113-010-0160-2

Barrett, C. B. (2010). Measuring food insecurity. *Science, 327,* 825–828

Baruch, F., & Davis Alex, L. (2015, September 16). Communicating scientific uncertainty. *PNAS, 111*(4), 13664–13671.

Basak, S. R., Basak, A. C., & Rahman, M. A. (2015). Impacts of floods on forest trees and their coping strategies in Bangladesh. *Weather and Climate Extreme Journal, 07.*

Bhardwaj, D., Raj, S., & Bhattacharjee, S. (2018). *Climate smart agriculture towards triple win: Adaptation, Mitigation and Food Security.* Discussion Paper 5, MANAGE-Centre for Agricultural Extension Innovations, Reforms, and Agripreneurship (CAEIRA). https://www.manage.gov.in/ publications/discussion%20papers/MANAGE-Discussion%20Paper-5.pdf

Bohensky, E. L., Kirono, D. G. C., Butler, J. R. A., et al. (2016). Climate knowledge cultures: Stakeholder perspectives on change and adaptation in Nusa Tenggara Barat, Indonesia. *Climate Risk Management, 12,* 17–31. https://doi.org/10.1016/j.crm.2015.11.004

Broom, D. (2014). *Sentience and animal welfare.* Wallingford CABI, UK.

Calfapietra, C., & Cherubini, L. (2019, January). Green infrastructure: Nature-Based solutions for sustainable and resilient cities. *Urban Forestry & Urban Greening, 37,* 1–2.

Calzadilla, A., Rehdanz, K., Betts, R., et al. (2013). Climate change impacts on global agriculture. *Climate Change, 120,* 357–374. https://doi.org/10.1007/s10584-013-0822-4

Campbell, B. M., Beare, D. J., Bennett, E. M., Hall-Spencer, J. M., Ingram, J. S. I., Jaramillo, F., Ortiz, R., Ramankutty, N., Sayer, J. A., & Shindell, D. (2017). Agriculture production as a major driver of the earth system exceeding planetary boundaries. *Ecology and Society, 22*(4), 8. https:// doi.org/10.5751/ES-09595-220408

Carbon Brief. (2016). http://www.carbonbrief.org/guest-post-failure-to-tackle-food-demand-could-make-1-point-5-c-limit-unachievable

Ceballos, G., Erlich, P., Barnosky, A., García, A., Pringle, R., & Palmer, T. (2015). Accelerated modern human–induced species losses: Entering the sixth mass extinction. *Science Advances, 1*(5), e1400253.

Consultative Group on International Agricultural Research (CGIAR). (2011). *Women fighting the effects of climate change.* Accessed September 23, 2019. https://ccafs.cgiar.org/fr/women-fig hting-effects-climate-change

Challinor, A. J., Watson, J., Lobell, D. B., et al. (2014). A meta-analysis of crop yield under climate change and adaptation. *Natural Climate Change, 4,* 287–291. https://doi.org/10.1038/ncl imate2153

Chanana, N., Khatri-Chhetri, A., Pande, K., & Joshi, R. (2018). *Integrating gender into the climate smart village approach of scaling out adaptation options in agriculture.* CCAFS Info

Note. CGIAR Research Program on Climate Change, Agriculture and Food Security (CCAFS). Copenhagen, Denmark.

Charles, O., Andreas, W., Suzanne, van D., Miriam, V., Samuel, N., Brian, S., & Lucy, K. (2019, 19 March). Financing large-scale mitigation by smallholder farmers: What roles for public climate finance? *Frontiers in Sustainable Food Systems.* https://doi.org/10.3389/fsufs.2019.00003/full

Chatterjee, J. (2016). India water portal. Floating gardens for the landless.

Cheng, L., et al. (2019, January 11). How fast are the oceans warming? *Science.* http://science.sci encemag.org/content/363/6423/128.full

Chiriaco, M. V., Perugini, L., Bernoux, M., et al. (2018). Koronivia joint work on agriculture: Analysis of submissions: Submissions under UNFCCC decision 4/CP.23 provided by Parties and observers as at 20 May 2018. Food and Agriculture Organization of the United Nations.

Climate Investment Fund. (2018). *Promoting climate resilient agriculture in Nepal—Building climate change resilient communities through private sector participation.* https://www.climat einvestmentfunds.org/sites/cif_enc/files/knowledge-documents/cif_case_study_nepal_1.pdf

Combined Project Information Documents/Integrated Safeguards Datasheet (PID/ISDS) Appraisal Stage | Date Prepared/Updated: 14-May-2019 | Report No: PIDISDSA25858The World Bank Odisha Integrated Irrigation Project for Climate Resilient Agriculture (P163533). http://docume nts.worldbank.org/curated/en/202281558084012152/pdf/Project-Information-Document-Integr ated-Safeguards-Data-Sheet-Odisha-Integrated-Irrigation-Project-for-Climate-Resilient-Agricu lture-P163533.pdf

Cooper, J. J., & Albentosa, M. J. (2003). Behavioural priorities of laying hens. *Avian and Poultry Biology Reviews, 14*, 127–149

Crop Division. (2018). *Pulses revolution, from food to nutritional security.* Government of India, Ministry of Agriculture & Farmers Welfare. https://farmer.gov.in/SucessReport2018-19.pdf

CTA. (2013). *Climate-smart agriculture success stories.* 44p. https://cgspace.cgiar.org/bitstream/ handle/10568/34042/Climate_smart_farming_successesWEB.pdf

Deliver for Good Campaign. (2018). *Women deliver.* Policy Brief. Accessed September 24, 19. https://womendeliver.org/wp-content/uploads/2017/09/Deliver_For_Good_Brief_10_09.17. 17.pdf

Deryng, D., Conway, D., Ramankutty, N., et al. (2014). Global crop yield response to extreme heat stress under multiple climate change futures. *Environmental Research Letters, 9*, 34011. https:// doi.org/10.1088/1748-9326/9/3/034011

Detlof, von Winterfeldt. (2013, August 20). Bridging the gap between science and decision making. *PNAS, 10*(3), 14055–14061.

Dewan, T. H. (2015, March). Societal impacts and vulnerability to floods in Bangladesh and Nepal. *Weather and Climate Extremes, 7*, 36–42.

Dinesh, D., Campbell, B., Wollenberg, L., et al. (2017). A step forward for agriculture at the UN climate talks—Koronivia Joint Work on Agriculture.

Dubey, S. K., Trivedi, R. K., Chand, B. K., et al. (2017). Farmers' perceptions of climate change, impacts on freshwater aquaculture and adaptation strategies in climatic change hotspots: A case of the Indian Sundarban delta. *Environmental Development, 21*, 38–51. https://doi.org/10.1016/ j.envdev.2016.12.002

EFSA. (2009). Scientific opinion of the panel on animal health and Welfare on a request from European Commission on the overall effects of farming systems on dairy cow welfare and disease. *The EFSA Journal, 1143*, 1–38. September 25, 2014. http://www.efsa.europa.eu/en/efsajournal/ doc/1143.pdfaccessed

EFSA Journal. (2010). Scientific opinion on the influence of genetic parameters on the welfare and the resistance to stress of commercial broilers. *EFSA Journal, 8*(7), 1666. September 25, 2014. www.efsa.europa.eu/publications.htmlaccessed

Enenkel, M., See, L., Bonifacio, R., et al. (2015). Drought and food security—Improving decision-support via new technologies and innovative collaboration. *Global Food Security , 4*, 51–55. https://doi.org/10.1016/j.gfs.2014.08.005

Entomological Society of America. (2019). *Position statement on climate change*. https://www.ent soc.org/sites/default/files/files/Science-Policy/2019/ESA-Position-Statement-Climate-Change. pdf

Equator Prize. (2019). *Winners announced for local innovative climate solutions*. https://www. undp.org/content/undp/en/home/news-centre/news/2019/equator-prize-announces-20-winners-for-local-innovative-climate-.html

ESA Position Statement on Climate Change. (2019). https://www.entsoc.org/sites/default/files/files/ Science-Policy/2019/ESA-Position-Statement-Climate-Change.pdf

ESPA. (2018). *Scaling up climate-smart agriculture, lessons from ESPA research*. Working Paper 006. https://www.espa.ac.uk/files/espa/Scaling%20up%20climate-smart%20agriculture% 20final%20web.pdf

European Commission. (1999). *Council Directive 1999/74/EC laying down minimum standards for the protection of laying hens*.

European Commission. (2015). European Commission—Towards an EU Research and Innovation Policy Agenda for Nature-based Solutions & Re-Naturing Cities. Final Report of the Horizon 2020 Expert Group on Nature-Based Solutions and Re-Naturing Cities, Brussels.

FAO. (2001). *State of food insecurity*. FAO.

FAO. (2011). *Women in agriculture: Closing the gender gap*. State of Food and Agriculture Report 2010–11, Rome, FAO.

FAO. (2013a). *Climate smart agriculture sourcebook*. http://www.fao.org/3/a-i3325e.pdf

FAO. (2013b). *Agroforestry and climate change*. Emmanuel Torquebiau FAO webinar 5 February 2013. http://www.fao.org/climatechange/36110-0dff1bd456fb39dbcf4d3b211af5684e2.pdf

FAO. (2013c). *State of food insecurity in the world: The multiple dimensions of food insecurity*. FAO.

FAO. (2013d). *FAO success stories on climate smart agriculture*. http://www.fao.org/3/a-i3817e. pdf

FAO. (2015). http://www.fao.org/soils-2015/events/detail/en/c/338738/

FAO. (2017). *The future of food and agriculture—Trends and challenges*. Rome. ISBN 978-92-5-109551-5. http://www.fao.org/3/a-i6583e.pdf

FAO. (2018a). Regional conference for Asia and the pacific 2018. Climate Action for Agriculture: Strengthening the Engagement of Agriculture Sectors to Implement the Climate Change Elements of the 2030 Agenda in Asia and the Pacific. http://www.fao.org/3/MV763en/mv763en.pdf

FAO. (2018b). Submission by the food and agriculture organization of the United Nations (FAO) to the United Nations Framework Convention on Climate Change (UNFCCC) in relation to the Koronivia joint work on agriculture (4/CP.23).

FAO. (2018c). Integrating agriculture in National Adaptation Plans (NAP–Ag)—Programme Safeguarding livelihoods and promoting resilience through National Adaptation Plans, Programme highlights 2015–2018. https://www.undp.org/content/undp/en/home/librarypage/climate-and-disaster-resilience-/integrating-agriculture-in-national-adaptation-plans.html

FAO. (2019a). *Climate-smart agriculture and the Sustainable development goals: Mapping interlinkages, synergies and trade-offs and guidelines for integrated implementation*. Rome.

FAO. (2019b). *Urban agriculture*. http://www.fao.org/urban-agriculture/en/

FAO. (undated). *Floating garden agricultural practices in Bangladesh—A proposal for Globally Important Agricultural Heritage Systems (GIAHS)*. www.fao.org/3/a-bp777e.pdf

FAO Regional Conference for Asia and the Pacific. (2018). *Climate action for agriculture: Strengthening the engagement of agriculture sectors to implement the climate change elements of the 2030 agenda in Asia and the Pacific*. http://www.fao.org/3/MV763en/mv763en.pdf

FAO, IFAD, UNICEF, WFP and WHO. (2018). *The state of food security and nutrition in the world 2018*. Building Climate Resilience for Food Security and Nutrition. Rome, FAO.

FAO, IFAD, UNICEF, WHO. (2017). *The state of food security and nutrition in the world: Building resilience for peace and food insecurity*. FAO.

FAOStat. (2019). http://faostat.fao.org/. Accessed on February 14, 2019.

Foglianoa, et al. (2018). Sub-Saharan African maize-based foods: Technological perspectives to increase the food and nutrition security impacts of maize breeding programmes. *Global Food Security, 17*, 48–56. https://doi.org/10.1016/j.gfs.2018.03.007

Food Security Information Systems (FSIN). (2018). *Global report on food crises*. Rome.

Fordham, M. (2003). Gender, development and disaster: The necessity for integration. In M. Pelling (Ed.), *Natural disasters and development in a globalizing world*. (pp. 57–74). Routledge.

Frew, A., & Price Jodi, N. (2019). Mycorrhizal—Mediated plant–herbivore interactions in a high CO_2 world. *Functional Ecology, 33*, 1376–1385.

Garcia, J. (2018). *Promoting climate resilient agriculture in Nepal—Building climate change resilient communities through private sector participation.* https://www.climateinvestmentfunds. org/sites/cif_enc/files/knowledge-documents/cif_case_study_nepal_1.pdf

Gerber, P. J., Steinfeld, H., Henderson, B., Mottet, A., Opio, C., Dijkman, J., Falcucci, A., & Tempio, G. (2013). *Tackling climate change through livestock—A global assessment of emissions and mitigation opportunities.* Food and Agriculture Organization of the United Nations (FAO), Rome.

Githeko, A. K., Lindsay, S. W., Confalonieri, U. E., & Patz, J. A. (2000). Climate change and vector-borne diseases: A regional analysis. *Bulletin of the World Health Organization, 78*, 1136–1147.

Global Environmental Facility (GEF). (2018). *The Capacity Building Initiative for Transparency (CBIT).* https://www.thegef.org/sites/default/files/publications/GEF_CBIT_Nov2018_CRA. pdf&https://www.thegef.org/news/countries-meet-discuss-efforts-increase-transparency-aro und-climate-change-commitments.

Gödecke, S. Q. (2017). *The global burden of chronic and hidden hunger: Trends and determinants.* Elsevier B.V. https://doi.org/10.1016/j.gfs.2018.03.004

Godfray, H., Charles, J. R., Beddington, J. R., et al. (2010). Food security: The challenge of feeding 9 Billion people. *Science, 327*(5967), 812–818.

Godfray, H., & Garnett, T. (2014). Food security and sustainable intensification. *Philosophical Transactions of the Royal Society B, 369*, 20120273.

Gonsalves, J., Campilan, D., Smith, G., Bui, V. L., & Jimenez, F. M. (eds). (2015). Towards climate resilience in agriculture for Southeast Asia: An overview for decision-makers. International Center for Tropical Agriculture (CIAT), Hanoi, Vietnam. CGIAR Research Program on Climate Change, Agriculture and Food Security (CCAFS) (450 p).

Green Climate Fund working paper No. 1. (2019). *Adaptation: Accelerating action towards a climate resilient future* (47 p). https://www.greenclimate.fund/documents/20182/194568/ Accelerating_action_towards_a_climate_resilient_future.pdf/9215ecb7-923a-9cfb-e4ae-d8f0be 6e7b29

Gupta, H. (2015). Women and climate change: Linking ground perspectives to the global scenario. *Indian Journal of Gender Studies, 22*, 1–13. https://doi.org/10.1177/0971521515594278

Hallegatte, S., Bangalore, M., Bonzanigo, L., et al. (2016). *Shock waves: Managing the impacts of climate change on poverty.* World Bank. https://openknowledge.worldbank.org/bitstream/han dle/10986/22787/9781464806735.pdf

Havlík, P., Valin, H., Gusti, M., et al. (2015). *Climate change impacts and mitigation in the developing world: An integrated assessment of the agriculture and forestry sectors.* The World Bank.

Hein, Y., Vijitsrikamol, K., Attavanich, W., & Janekarnkij, P. (2019). Do farmers perceive the trends of local climate variability accurately? An analysis of farmers' perceptions and meteorological data in Myanmar. *Climate, 7*, 64. https://doi.org/10.3390/cli7050064

HLPE. (2016). *Sustainable agricultural development for food security and nutrition: What roles for livestock?* A report by the High Level Panel of Experts on Food Security and Nutrition of the Committee on World Food Security, Rome.

HLPE. (2019). *Agroecological and other innovative approaches for sustainable agriculture and food systems that enhance food security and nutrition.* A report by the High Level Panel of Experts on Food Security and Nutrition of the Committee on World Food Security, Rome.

Hossain, K., Quaik, S., Ismail, N., et al. (2016). Climate change-perceived impacts on agriculture, vulnerability and response strategies for improving adaptation practice in developing countries (South Asian Region). *International Journal of Agricultural Research, 11*, 1–12. https://doi.org/10.3923/ijar.2016.1.12

Howarth, R., Swaney, D., Billen, G., Garnier, J., Hong, B., Humborg, C., Johnes, P., Mörth, C-M. & Marino, R. (2012). Nitrogen fluxes from the landscape are controlled by net anthropogenic nitrogen inputs and by climate. *Frontiers in Ecology and the Environment, 10*(1), 37–43. https://doi.org/10.1890/100178. https://www.indiawaterportal.org/articles/floating-gardens-landless

Human Rights Council (HRC). (2019a). Climate change and poverty, Report of the Special Rapporteur on extreme poverty and human rights, A/HRC/41/39.

Human Rights Council (HRC). (2019b). The right to food Resolution, A/HRC/RES/40/7.

IARC. (2015). *IARC Monographs evaluate consumption of red meat and processed meat.* https://www.iarc.fr/wp-content/uploads/2018/07/pr240_E.pdf

IFAD. (2019). *Sustainable transformation for agricultural resilience in Upper Egypt (STAR).* https://www.ifad.org/en/web/latest/story/asset/41202061

IFAD October. (2018). *IFAD strategy and action plan on environment and climate change 2019–2025.* https://webapps.ifad.org/members/eb/125/docs/EB-2018-125-R-12.pdf

IFAD. (undated). *Sustainable transformation for agricultural resilience in Upper Egypt (STAR).* https://www.ifad.org/en/web/latest/story/asset/41202061

Iizumi, T., Furuya, J., Shen, Z., et al. (2017). Responses of crop yield growth to global temperature and socioeconomic changes. *Science Reports, 7*, 7800. https://doi.org/10.1038/s41598-017-08214-4

Iizumi, T., Yokozawa, M., Sakurai, G., et al. (2014). Historical changes in global yields: Major cereal and legume crops from 1982 to 2006: Historical changes in global yields. *Global Ecology and Biogeography, 23*, 346–357. https://doi.org/10.1111/geb.12120

Iizumi, T., Luo, J., Challinor, A. J., Sakurai, G., Yokozawa, M., Sakuma, H., Brown, M. E., & Yamagata, T. (2014). Impacts of El Niño Southern Oscillation on the global yields of major crops. *Nature Communications, 5*, 3712.

Integrating Agriculture in National Adaptation Plans (NAP–Ag) Programme Safeguarding livelihoods and promoting resilience through National Adaptation Plans Programme highlights 2015–2018. https://www.undp.org/content/undp/en/home/librarypage/climate-and-disaster-resilience-/integrating-agriculture-in-national-adaptation-plans.html

International Climate Initiative. (2012). Accessed on September 25, 2019. https://www.international-climate-initiative.com/en/nc/infotheque/videos/film/show_video/show/women_in_bangladesh_fight_climate_change_/

IPBES. (2019). *IPBES global assessment report.* Released from Paris, 13:00h CEDT, Monday May 6, 2019.

IPCC. (2012). Managing the risks of extreme events and disasters to advance climate change adaptation. A special report of working groups I and II of the intergovernmental panel on climate change. In C. B. Field, V. Barros, T. F. Stocker, D. Qin, D. J. Dokken, K. L. Ebi, M. D. Mastrandrea, K. J. Mach, G.-K. Plattner, S. K. Allen, M. Tignor, & P. M. Midgley (Eds.), Cambridge University Press (582 p).

IPCC. (2014). Climate change 2014: Impacts, adaptation, and vulnerability. Part A: global and sectoral aspects. Contribution of working group II to the fifth assessment report of the intergovernmental panel on climate change. Cambridge University Press.

IPCC. (2018). Summary for policymakers. In Global warming of 1.5 °C. In V. Masson-Delmotte, P. Zhai, H. O. Pörtner, D. Roberts, J. Skea, P. R. Shukla, A. Pirani, W. Moufouma-Okia, C. Péan, R. Pidcock, S. Connors, J. B. R. Matthews, Y. Chen, X. Zhou, M. I. Gomis, E. Lonnoy, T. Maycock, M. Tignor, T. Waterfield (Eds.), An IPCC special report on the impacts of global warming of 1.5 °C above pre-industrial levels and related global greenhouse gas emission pathways, in the context of strengthening the global response to the threat of climate change, sustainable development, and efforts to eradicate poverty.

IPCC. (2019). Climate change and land: An IPCC special report on climate change, desertification, land degradation, sustainable land management, food security, and greenhouse gas fluxes in terrestrial ecosystems.

Islam, M. R., Sarker, M. R. A., Sharma, N., Rahman, M. A., Collard, B. C. Y., Gregorio, G. B., & Ismail, A. M. (2016, April). Assessment of adaptability of recently released salt tolerant rice varieties in coastal regions of South Bangladesh. *Field Crops Research, 190*, 234–43. https://www.sciencedirect.com/science/article/pii/S0378429015300563. https://doi.org/10.1016/j.fcr.2015.09.012

Islam, M. T., & Nursey-Bray, M. (2017). Adaptation to climate change in agriculture in Bangladesh: The role of formal institutions. *Journal of Environmental Management, 200*, 347–358. https://doi.org/10.1016/j.jenvman.2017.05.092

IUCN. (2017). Deforestation and forest degradation issues. Brief November 2017. https://www.iucn.org/sites/dev/files/deforestation-forest_degradation_issues_brief_final.pdf

Jain, M., Naeem, S., Orlove, B., et al. (2015). Understanding the causes and consequences of differential decision-making in adaptation research: Adapting to a delayed monsoon onset in Gujarat, India. *Global Environmental Change, 31*, 98–109. https://doi.org/10.1016/j.gloenvcha.2014.12.008

Jarvis, A., Lau, C., Cook, S., Wollenberg, E., Hansen, J., Bonilla, O., & Andy, C. (2011). An integrated adaptation and mitigation framework for developing agricultural research: Synergies and trade-offs. *Experimental Agriculture, 47*(2), 185–203. https://doi.org/10.1017/S00144797 11000123

Joint Working Party on Agriculture and the Environment, Improving Energy Efficiency in the Agro-Food Chain, COM/TAD/CA/ENV/EPOC. (2016). 19/FINAL. http://www.oecd.org/officialdocuments/publicdisplaydocumentpdf/?cote=COM/TAD/CA/ENV/EPOC(2016)19/FINAL&docLanguage=En

Jones, J. W., Antle, J. M., Basso, B., Boote, K. J., Conant, R. T., Foster, I., & Wheeler, T. R. (2017). Toward a new generation of agricultural system data, models, and knowledge products: State of agricultural systems science. *Agricultural Systems, 155*, 269–288. https://doi.org/10.1016/j.agsy.2016.09.021

Kallio, M. H. (2013). Factors influencing farmers' tree planting and management activity in four case studies in Indonesia. Tropical Forestry Reports 45 Viikki Tropical Resources Institute VITR University of Helsinki, Helsinki 2013. ISBN 978-952-10-9551-1

Karim, M. R., & Thiel, A. (2017). Role of community based local institution for climate change adaptation in the *Teesta riverine* area of Bangladesh. *Climate Risk Management, 17*, 92–103. https://doi.org/10.1016/j.crm.2017.06.002

Khanal, U., Wilson, C., Lee, B., & Hoang, V.-N. (2018a). Do climate change adaptation practices improve technical efficiency of smallholder farmers? Evidence from Nepal. *Climate Change.* https://doi.org/10.1007/s10584-018-2168-4

Khanal, U., Wilson, C., Lee, B. L., & Hoang, V.-N. (2018b). Climate change adaptation strategies and food productivity in Nepal: A counterfactual analysis. *Climate Change, 148*, 575–590. https://doi.org/10.1007/s10584-018-2214-2

Khatri-Chhetri, A., Aggarwal, P. K., Joshi, P. K., & Vyas, S. (2017). Farmers' prioritization of climate-smart agriculture (CSA) technologies. *Agricultural Systems, 151*, 184–191.

Kirchmann, H., & Bergström, L. (2007). Do organic farming practices reduce nitrate leaching? *Communications in Soil Science and Plant Analysis, 32*(7–8), 997–1028. https://doi.org/10.1081/CSS-100104101

Knox, J., Hess, T., Daccache, A., & Wheeler, T. (2012). Climate change impacts on crop productivity in Africa and South Asia. *Environmental Research Letters, 7*, 34032. https://doi.org/10.1088/1748-9326/7/3/034032

Kremen, C., & Miles, A. (2012). Ecosystem services in biologically diversified versus conventional farming systems: Benefits, externalities, and trade-offs. *Ecology and Society, 17*(4), 40. https://doi.org/10.5751/ES-05035-170440

Lessons learned from three years of implementing the MDB-IDFC Common principles for climate change adaptation finance tracking. (2018). https://www.idfc.org/wp-content/uploads/2019/04/mdb_idfc_lessonslearned-full-report.pdf 28p.

Lewis, J., & Rudnick, J. (2019). The policy enabling environment for climate smart agriculture: A case study of California, *Policy & Practice Reviews. Frontiers in Sustainable Food Systems.* https://doi.org/10.3389/fsufs.2019.00031/full

Li, Y., Ye, W., Wang, M., & Yan, X. (2009). Climate change and drought: A risk assessment of crop-yield impacts. *Climate Research, 39,* 31–46. https://doi.org/10.3354/cr00797

List of on-going biodiversity projects funded by the German Federal Ministry for the Environment, Nature Conservation, Building and Nuclear Safety, February 2018. https://www.international-climate-initiative.com/fileadmin/Dokumente/2018/180213_List_of_ongoing_projects_as_of_Feb_2018.pdf

Lobell, D. B., & Field, C. B. (2007). Global scale climate–crop yield relationships and the impacts of recent warming. *Environmental Research Letters, 2,* 14002. https://doi.org/10.1088/1748-9326/2/1/014002

Lundqvist, J., de Fraiture, C., & Molden, D. (2008). Saving water: From field to fork—Curbing losses and wastage in the food chain. SIWI Policy Brief. SIWI. http://www.siwi.org/documents/Resources/Policy_Briefs/PB_From_Filed_to_Fork_2008.pdf

Mackey, B. (2009). Connecting biodiversity and climate change mitigation and adaptation; Report of the Second Ad Hoc Technical Expert Group on Biodiversity and Climate Change; CBD Technical Series No. 41; Secretariat of the Convention on Biological Diversity: Montreal, Canada; 11; September; 2009; ISBN: ISBN: 92-9225-134-1.

Marije, S., & Bell Andrew, R. (2018). Scaling up climate-smart agriculture, Lessons from ESPA research. Working Paper 006, https://www.espa.ac.uk/files/espa/Scaling%20up%20climate-smart%20agriculture%20final%20web.pdf

Mendelsohn, R. (2008). The impact of climate change on agriculture in developing countries. *Journal of Natural Resources Policy Research, 1,* 5–19. https://doi.org/10.1080/19390450802495882

Menike, L. M. C. S., & Arachchi, K. A. G. P. K. (2016). Adaptation to climate change by smallholder farmers in rural communities: Evidence from Sri Lanka. *Procedia Food Science, 6,* 288–292. https://doi.org/10.1016/j.profoo.2016.02.057

Miller, F., Osbahr, H., Boyd, E., et al. (2010). Resilience and vulnerability: Complementary or conflicting concepts? *Ecology and Society, 15*(3), 11.

Mittal, S., & Hariharan, V. K. (2018). Mobile-based climate services impact on farmers risk management ability in India. *Climate Risk Management, 22,* 42–51. https://doi.org/10.1016/j.crm.2018.08.003

Mubirua, R., et al. (2016). *Climate trends, risks and coping strategies in smallholder farming systems in Uganda.* Elsevier B.V. https://doi.org/10.1016/j.crm.2018.08.004

Multilateral Development Banks, Climate Finance Tracking Working, and the International Development Finance Club, Climate Finance Working Group. (2018). *Lessons learned from three years of implementing the MDB-IDFC Common principles for climate change adaptation finance tracking.* https://www.idfc.org/wp-content/uploads/2019/04/mdb_idfc_lessonslearned-full-report.pdf

Munang, R. T., Thiaw, I., & Rivington, M. (2011). (2011), Ecosystem management: Tomorrow's approach to enhancing food security under a changing climate. *Sustainability, 3,* 937–954. https://doi.org/10.3390/su3070937sustainabilityISSN2071-1050www.mdpi.com/journal/sustainabilityfile:///C:/Users/Faculty/Downloads/sustainability-03-00937.pdf.

NASA Earth Observatory. (2015). *Floating farms.* https://earthobservatory.nasa.gov/images/85606/floating-farms

National Academy of Sciences. (2014). *The science of science communication* II: Summary of a colloquium. Washington, DC: The National Academies Press. https://doi.org/10.17226/18478.e https://www.nap.edu/download/18478 (1–19 on 29 July 2019)

Neate, P. (2013). *Climate smart agriculture success stories.* https://cgspace.cgiar.org/bitstream/han dle/10568/34042/Climate_smart_farming_successesWEB.pdf

Nellemann, C., MacDevette, M., Manders, et al. (2009). The environmental food crisis—The environment's role in averting future food crises. A UNEP rapid response assessment. United Nations Environment Programme, GRID-Arendal. www.unep.org/pdf/foodcrisis_lores.pdf

Nicholson, K., Beloe, T., & Hodes, G. (2017). Hard choices, hard choices—Integrated approaches: A guidance note on climate change financing frameworks. UNDP. http://www.asia-pacific.undp.org/content/dam/rbap/docs/Research%20&%20Publications/democratic_governance/RBAP-DG-2017-Hard-Choices-Integrated-Approaches.pdf

Nordhaus, W. (2019, June). Climate change: The ultimate challenge for economics. *American Economic Review, 109*(6), 1991–2014.

Odhong, C., Andreas, W., Suzanne, van Dijk, Miriam, V., Samuel, N., Brian, S. & Lucy, K. (2019, 19 March). Financing large-scale mitigation by smallholder farmers: What roles for public climate finance? *Frontiers in Sustainable Food Systems.*

OECD. (2016). *Joint working party on agriculture and the environment—Improving energy efficiency in the agro-food chain.* COM/TAD/CA/ENV/EPOC(2016)19/Final. http://www.oecd.org/officialdocuments/publicdisplaydocumentpdf/?cote=COM/TAD/CA/ENV/EPOC(2016)19/FINAL&docLanguage=En

OECD. (2019). *Implementing adaptation policies: Towards sustainable development Issue Brief Prepared by the OECD as input for the 2019 G20 Process.* http://www.oecd.org/g20/summits/osaka/OECD-G20%20Paper-Adaptation-and-resilient-infrastructure.pdf

Pantanella, E., Cardarelli, M., Danieli, P. P., MacNiven, A., & Colla, G. (2011). Integrated aquaculture—Floating agriculture: Is it a valid strategy to raise livelihood? *Acta Horticulturae, 921*(921), 79–86.

Parry, M. L., Rosenzweig, C., Iglesias, A., et al. (2004). Effects of climate change on global food production under SRES emissions and socio-economic scenarios. *Global Environment Change, 14*, 53–67. https://doi.org/10.1016/j.gloenvcha.2003.10.008

Partey Samuel, T., Zougmoré Robert, B., & Ouédraogoa, M. (2018). Developing climate-smart agriculture to face climate variability in West Africa: Challenges and lessons learnt. *Journal of Cleaner Production, 187*, 285–295. https://www.sciencedirect.com/science/article/pii/S0959652618308709

Partey, S. T., Zougmore, R. B., Ouedraogo, M., & Campbell, B. M. (2018). Developing climate-smart agriculture to face climate variability in West Africa: Challenges and lessons learnt. *Journal of Cleaner Production, 187*, 285–295. https://www.sciencedirect.com/science/article/pii/S0959652618308709

Pathak, H., Aggarwal, P. K., & Singh, S. D. (2012). *Climate change impact, adaptation and mitigation in agriculture: Methodology for assessment and applications.* Indian Agricultural Research Institute, New Delhi.

Patterson, A., & Mclean, C. (2019). The precautionary principle at work: The case of neonicotinoids and the health of bees. *Science & Public Policy, 46*(3), 441–449.

Patz, J. A., Campbell-Lendrum, D., Holloway, T., & Foley, J. A. (2005). Impact of regional climate change on human health. *Nature, 438*, 310–317.

Pouyat Richard, V. (1999). Science and environmental policy—Making them compatible. *Policy Forum, Bioscience, 49*(4), 281–286.

Pretty, J., et al. (2006). Resource-conserving agriculture increases yields in developing countries. *Environmental Science and Technology, 40*(4), 1114–1119

Rahman, S., & Rahman, R. (2015). Traditional floodplain managements of the Ganges, Brahmaputra and Meghna Basin are Unique Landscape Management Practices for Climate Change Adaptation International conference on Climate Change in relation to Water and Environment (I3CWE-2015) Department of Civil Engineering DUET - Gazipur, Bangladesh 8–11 April, 2015.

Rahman, S., Nargis, S., & Rahman, M. A. (2018). Importance of landscape management in Bangladesh. In *Proceedings of the 13th International Knowledge Globalization Conference: Sustainable Development Goals – Success and Challenges.* Organized by Knowledge Globalization Institute, Boston USA.

Rao, K. P. C., Verchot, L. V., & Laarman, J. (2007). Adaptation to climate change through sustainable management and development of Agroforestry Systems World Agroforestry Center, Nairobi, Kenya SAT eJournal I ejournal.icrisat.org December 2007 I vol. 4 I Issue 1 http://oar.icrisat.org/2561/1/Adaptation_to_Climate_Change.pdf

Rare and The Behavioural Insights Team. (2019). *Behavior change for nature: A behavioral science toolkit for practitioners.* Rare. https://www.bi.team/wp-content/uploads/2019/04/2019-BIT-Rare-Behavior-Change-for-Nature-digital.pdf

Ray, D. K., Mueller, N. D., West, P. C., & Foley, J. A. (2013). Yield trends are insufficient to double global crop production by 2050. *PLoS ONE, 8,* e66428. https://doi.org/10.1371/journal.pone.0066428

Ray, D. K., West, P. C., Clark, M., et al. (2019). Climate change has likely already affected global food production. *PLOS ONE 14,* e0217148. https://doi.org/10.1371/journal.pone.0217148

Razavi, S. (2007). Liberalisation and the debates on women's access to land. *Third World Quarterly* 28(3).

Ricard, M. (2014). *Ecological principles and function of natural ecosystems.* Intensive Programme on Education for sustainable development in Protected Areas. Amfissa, Greece, July 2014. http://mio-ecsde.org/erasmus-IP-2014/trainers/day%2002-Ricard.pdf

Rosegrant, M. W., Jawoo, K., Nicola, C., et al. (2014). *Food security in a world of natural resource scarcity: The role of agricultural technologies.* International Food Policy Research Institute.

Rosegrant, M. W., Sulser, T. B., Daniel, M.-D., et al. (2017). *Quantitative foresight modeling to inform the CGIAR research portfolio.* International Food Policy Research Institute.

Rosenzweig, C., Elliott, J., Deryng, D., et al. (2014). Assessing agricultural risks of climate change in the 21st century in a global gridded crop model intercomparison. *Proceedings of the National Academy of Sciences, 111,* 3268–3273. https://doi.org/10.1073/pnas.1222463110

Roy, S. S., Ansari, M. A., Sharma, S. K., Sailol, B., Basudha Devi, Ch., Singh, I. M., Das, A., Chakraborty, D., Arunachalam, A., Prakash, N., & Ngachan, S. V. (2019). Climate resilient agriculture in Manipur: Status and strategies for sustainable development. *Current Science, 115*(7), 1342–1350.

Rutten, M., van Dijk, M., van Rooij, W., & Hilderink, H. (2014). Land use dynamics, climate change, and food security in Vietnam: A global-to-local modeling approach. *World Development, 59,* 29–46. https://doi.org/10.1016/j.worlddev.2014.01.020

Sakurai, G., Iizumi, T., Nishimori, M., & Yokozawa, M. (2015). How much has the increase in atmospheric CO_2 directly affected past soybean production? *Science Reports, 4,* 4978. https://doi.org/10.1038/srep04978

Singh, N. P., Anand, B., Singh, S., & Khan, A. (2019). Mainstreaming climate adaptation in Indian rural developmental agenda: A micro-macro convergence. *Climate Risk Management, 24,* 30–41. https://doi.org/10.1016/j.crm.2019.04.003

SmartAgri. (2016). *Status quo review of climate change and agriculture in the Western Cape Province.* Western Cape Department of Agriculture from http://www.greenagri.org.za/smartagri-2/smartagri-plan/

Soora, N. K., Aggarwal, P. K., Saxena, R., et al. (2013). An assessment of regional vulnerability of rice to climate change in India. *Climate Change, 118,* 683–699. https://doi.org/10.1007/s10584-013-0698-3

Soyatech. (2015). www.soyatech.com/soy_facts.htm

Springmann, M., et al. (2016). Analysis and valuation of the health and climate change co-benefits of dietary change. *PNAS, 113*(15), 4146–4151.

Stiglitz, J. (2019). The climate crisis is our third world war. It needs a bold response. *Guardian,* June 4. https://www.theguardian.com/commentisfree/2019/jun/04/climate-change-world-war-iii-green-new-deal.

Stinson, L. (2019). *Paris to plant mini urban forests to combat climate change.* https://www.cur bed.com/2019/6/28/18816545/paris-urban-forests-climate-change?fbclid=IwAR2fLQXgon_ wthXXjP5a8WZcLxv8A5xpbBlY7S9CFSEHt12b38akn0QZU4Q   https://www.dezeen.com/ 2019/06/26/paris-urban-forest-plant-trees-landmarks/

Stolba, A., & Wood-Gush, D. (1989). The behaviour of pigs in a semi-natural environment. *Animal Science, 48*(2), 419–425. https://doi.org/10.1017/S0003356100040411

Sunderarajan, P. (2018). *New centre to help farmers become climate resilient.* Last Updated: Tuesday 16 October 2018. https://www.downtoearth.org.in/news/climate-change/new-centre-to-help-far mers-become-climate-resilient-61899

Swaney, D. P., Hong, B., Ti, C., Howarth, R. W., & Humborg, C. (2012). Net anthropogenic nitrogen inputs to watersheds and riverine N export to coastal waters: A brief overview. *Current Opinion in Environmental Sustainability, 4*(2), 203–211. https://doi.org/10.1016/j.cosust.2012.03.004

Taylor, M. (2019). *From Madagascar to Brazil, researchers pick best spots to replant forests.* Thomson Reuters Foundation, Wednesday, 3 July 2019 18:00 GMT. http://news.trust.org/ item/20190703174042-sw5jn/?utm_source=CLUA+Quarterly+Email+Newsletter&utm_cam paign=baac8351da-Newsletter_Issue_16_Bonn_COP_11_09_2017_COPY_01&utm_medium= email&utm_term=0_b2faff125f-baac8351da-38865825

TEEB. (2010). The economics of ecosystems and biodiversity: Mainstreaming the economics of nature: A synthesis of the approach, conclusions and recommendations of TEEB; The TEEB Synthesis Report: Nagoya, Japan; 20; October; 2010.

Tesfaye, K., Kruseman, G., Cairns, J. E., et al. (2018). Potential benefits of drought and heat tolerance for adapting maize to climate change in tropical environments. *Climate Risk Management, 19*, 106–119. https://doi.org/10.1016/j.crm.2017.10.001

Tesfaye, K., Zaidi, P. H., Gbegbelegbe, S., et al. (2017). Climate change impacts and potential benefits of heat-tolerant maize in South Asia. *Theoretical and Applied Climatology, 130*, 959–970. https://doi.org/10.1007/s00704-016-1931-6

The Capacity Building Initiative for Transparency (CBIT). (2018). https://www.thegef.org/sites/ default/files/publications/GEF_CBIT_Nov2018_CRA.pdf&https://www.thegef.org/news/cou ntries-meet-discuss-efforts-increase-transparency-around-climate-change-commitments.

The Intergovernmental Science-Policy Platform on Biodiversity and Ecosystem Services (IPBES). (2019). Nature's Dangerous Decline 'Unprecedented'; Species Extinction Rates 'Accelerating'," May 6. https://www.ipbes.net/news/Media-Release-Global-Assessment

The World Bank. (2010). *World development report 2010: Development and climate change.* https:// openknowledge.worldbank.org/handle/10986/4387.

The World Bank. (2016). *Bangladesh: Building resilience to climate change.* https://www.worldb ank.org/en/results/2016/10/07/bangladesh-building-resilience-to-climate-change

The World Bank. (2018). *Climate smart agriculture.* Last updated October 05, 2018. https://www. worldbank.org/en/topic/climate-smart-agriculture

The World Bank Group. (2019a). Combined Project Information Documents/Integrated Safeguards Datasheet (PID/ISDS) Appraisal Stage | Date Prepared/Updated: 14-May-2019 | Report No: PIDISDSA25858The World Bank Odisha Integrated Irrigation Project for Climate Resilient Agriculture (P163533). http://documents.worldbank.org/curated/en/202281558084012152/ pdf/Project-Information-Document-Integrated-Safeguards-Data-Sheet-Odisha-Integrated-Irriga tion-Project-for-Climate-Resilient-Agriculture-P163533.pdf

The World Bank Group. (2019b). Summary of Brainstorming Sessions, SSKE Workshop on Climate Smart Agriculture, Nairobi, Kenya. https://cgspace.cgiar.org/bitstream/handle/10568/ 103223/Summary%20of%20potential%20collaboration%20opportunities%20between%20I ndia%20and%20Africa%20on%20CSA.pdf

The World Bank Group. (2019c). The World Bank Group's Action Plan on Climate Change Adaptation and Resilience. Managing risks for a more resilient future. http://documents.worldb ank.org/curated/en/519821547481031999/The-World-Bank-Groups-Action-Plan-on-Climate-Change-Adaptation-and-Resilience-Managing-Risks-for-a-More-Resilient-Future.pdf 24p.

The World Bank Group. (2019d). Summary of Brainstorming Sessions, SSKE Workshop on Climate Smart Agriculture, Nairobi, Kenya. https://cgspace.cgiar.org/bitstream/handle/10568/103223/Summary%20of%20potential%20collaboration%20opportunities%20between%20India%20and%20Africa%20on%20CSA.pdf

The World Bank Group's Action Plan on Climate Change Adaptation and Resilience. Managing risks for a more resilient future. http://documents.worldbank.org/curated/en/519821547481031999/The-World-Bank-Groups-Action-Plan-on-Climate-Change-Adaptation-and-Resilience-Managing-Risks-for-a-More-Resilient-Future.pdf

Thi Nhung, T., Le Vo, P., Van Nghi, V., & Quoc Bang, H. (2019). Salt intrusion adaptation measures for sustainable agricultural development under climate change effects: A case of Ca Mau Peninsula. *Vietnam. Climate Risk Management, 23*, 88–100. https://doi.org/10.1016/j.crm.2018.12.002

Tito, R., Vasconcelos, H. L., & Feeley, K. J. (2018). Global climate change increases risk of crop yield losses and food insecurity in the tropical Andes. *Global Change Biology, 24*, e592–e602. https://doi.org/10.1111/gcb.13959

Tripathi, A., & Mishra, A. K. (2017). Knowledge and passive adaptation to climate change: An example from Indian farmers. *Climate Risk Management, 16*, 195–207. https://doi.org/10.1016/j.crm.2016.11.002

UN Environment. (2018). *Emissions Gap Report 2018.* https://www.unenvironment.org/resources/emissions-gap-report-2018

UN Women. (2016). Photo essay: In Vanuatu, women tackle drought to restore livelihoods after Cyclone Pam. Accessed on 23/9/19 at https://www.unwomen.org/en/digital-library/multimedia/2016/5/photo-vanuatu-women-recover-after-cyclone-pam

UNDP (undated), UNDP and Climate Change - Zero Carbon Sustainable Development, https://www.undp.org/content/dam/undp/library/Climate%20and%20Disaster%20Resilience/UNDP_and_Climate_Change.pdf

UNEA. (2019). *Sixth global environmental outlook.* https://www.unenvironment.org/resources/global-environment-outlook-6

UNFCCC Adaptation Committee. (2019). Fifteenth meeting of the Adaptation Committee Bonn, Germany, 19 to 21 March 2019 Fostering engagement of the agri-food sector in resilience to climate change Report on the workshop. https://unfccc.int/sites/default/files/resource/ac15_8a_ps_report.pdf

United Nations. (2011). *World economic and social survey.* www.un.org/en/development/desa/policy/wess/wess_current/2011wess.pdf

United Nations. (2017). Revision of World Population Prospects.

US Energy Information Administration. (2017). *EIA projects 28% increase in world energy use by 2040.* September 14, https://www.eia.gov/todayinenergy/detail.php?id=32912

van Meijl, H., Havlik, P., Lotze-Campen, H., et al. (2018). Comparing impacts of climate change and mitigation on global agriculture by 2050. *Environmental Research Letters, 13*, 64021. https://doi.org/10.1088/1748-9326/aabdc4

Vermeulen, S. J., Aggarwal, P. K., Ainslie, A., et al. (2010). Agriculture, Food Security and Climate Change: Outlook for Knowledge, Tools and Action. CGIAR-ESSP Program on Climate Change, Agriculture and Food Security.

Waibel, H., Hoa Pahlisch, T., & Völker, M. (2018). Farmers' Perceptions of and Adaptations to Climate Change in Southeast Asia: The Case Study from Thailand and Vietnam (pp 137–160).

Wallace-Wells, D. (2019). *The uninhabitable earth—Life after warming.* Tim Duggan Books, 1st edn. ISBN-10: 0525576703.

Wang, Z., Li, J., Lai, C., et al. (2018). Drying tendency dominating the global grain production area. *Global Food Security, 16*, 138–149. https://doi.org/10.1016/j.gfs.2018.02.001

WHO. (2006). *Guidelines on Food Fortification with Micronutrients.* World Health Organization.

WHO. (2014). *Quantitative risk assessment of the effects of climate change on selected causes of death, 2030s and 2050s.* https://apps.who.int/iris/bitstream/handle/10665/134014/9789241507691_eng.pdf

WHO. (2019). *Climate change and human health.* Accessed on July 8, 2019. https://www.who.int/globalchange/ecosystems/en/

Willett, W., Rockström, J., Loken, B., et al. (2019). *Food in the Anthropocene: The EAT–Lancet Commission on Healthy Diets from Sustainable Food Systems.* Lancet 2019; published online Jan 16. https://doi.org/10.1016/S0140-6736(18)31788-4

Williams, S. B., Hochet-Kibongui, A.-M., Naue, C. E. (2005). Gender, fisheries and aquaculture: Social capital and knowledge for the transition towards sustainable use of aquatic ecosystems. ACP – EU Fisheries Research Report Number 16, Brussels.

WRI. (2019). The World Lost a Belgium-sized Area of Primary Rainforests Last Year. World Resource Institute, 25 April 2019. https://www.wri.org/blog/2019/04/world-lost-belgium-sized-area-primary-rainforests-last-year

Yen, B. T., Son, N. H., Tung, L. T., et al. (2019). Development of a participatory approach for mapping climate risks and adaptive interventions (CS-MAP) in Vietnam's Mekong River Delta. *Climate Risk Management, 24*, 59–70. https://doi.org/10.1016/j.crm.2019.04.004

Zhao, D., & Li, Y.-R. (2015). Climate change and sugarcane production: Potential impact and mitigation strategies. *International Journal of Agronomy, 2015*, 1–10. https://doi.org/10.1155/2015/547386

Dr. Mohamed Behnassi is Professor at the College of Law, Economics, and Social Sciences, Ibn Zohr University of Agadir, Morocco. He is as well a Senior Researcher of international law and politics of environment and human security focusing on some specific regions such as the MENA and the Mediterranean. He has a PhD in International Environmental Law and Governance (Hassan II University of Casablanca, 2003) and a Diploma in International Environmental Law and Diplomacy (University of Eastern Finland and UNEP, 2015). He is currently the Founding Director of the Center for Environment, Human Security and Governance (CERES)—Former North-South Center for Social Sciences (NRCS)—which is a member of MedThink 5+5 aiming at shaping relevant research and decision agendas. From 2015 to 2018, he was the Director of the Research Laboratory for Territorial Governance, Human Security and Sustainability (LAGOS) in the same university. Recently, he was appointed as Expert for the Intergovernmental Science-Policy Platform on Biodiversity and Ecosystem Services (IPBES), the National Center for Scientific and Technical Research (CNRST/Morocco), and Mediterranean Experts on Climate and Environmental Change (MEDECC). He is among the Lead Authors who elaborate the 1st Assessment Report (MAR1): *Climate and Environmental Change in the Mediterranean Basin - Current Situation and Risks for the Future* (MEDECC, 2021). Dr. Behnassi has published 15 books, including *Human and Environmental Security in the Era of Global Risks* (Springer, 2019); *Climate Change, Food Security and Natural Resource Management: Perspectives from Africa, Asia and the Pacific Islands* (Springer, 2019); *Environmental Change and Human Security in Africa and the Middle East* (Springer, 2017); *Sustainable Food Security in the Era of Local and Global Environmental Change* (Springer, 2013). In addition, Dr. Behnassi has organized many international conferences covering the above research areas and managed many research and expertise projects on behalf of various national and international organizations. Behnassi is regularly requested to provide scientific expertise nationally and internationally. Other professional activities include social compliance auditing and consultancy by monitoring human rights at work and the sustainability of the global supply chain.

Dr. Mohammed Ataur Rahman is the Director of IUBAT Institute of SDG Studies (IISS), Centre for Global Environmental Culture (CGEC) and Professor at the College of Agricultural Sciences, International University of Business Agriculture and Technology (IUBAT). He is a Plantation Crop, Forestry, and Agriculture Specialist; the Coordinator of World-Wide Opportunities on Organic Farms (WWOOF), Bangladesh; and a Member Director of Federation of WWOOF Organizations (FoWO). He worked as a Contributing Author of the *Managing the Risks of Extreme*

Events and Disasters to Advance Climate Change Adaptation, Special Report of the Intergovernmental Panel on Climate Change (IPCC) 2012 Working Group II. He is an active member of Asian Disaster Preparedness Centre (ADPC) and Disaster Management. He worked with the International Geosphere and Biosphere Program (IGBP) and Asia Pacific Network Global Research Program (APNGCR). Moreover, he is the Executive Director and Coordinator of the Regional Centre of Expertise (RCE) Greater Dhaka acknowledged by UNU-IAS. He has more than one hundred research publications, including 13 books and booklets.

Joyce D'Silva is an Ambassador Emeritus for Compassion in World Farming (CWF). Joyce has an MA from Trinity College Dublin and Honorary Doctorates from the Universities of Winchester and Keele. She was Chief Executive of Compassion in World Farming for 14 years and is currently Ambassador Emeritus for Compassion. Joyce played a key role in achieving the UK ban on sow stalls in the nineties and in getting recognition of animal sentience enshrined in the European Union Treaties. She has published widely on farm animal welfare, including co-editing *The Meat Crisis. Developing more Sustainable and Ethical Production and Consumption*, Earthscan, 2017. Joyce has received an RSPCA Award for a "very important contribution in the field of animal welfare". She has given evidence to the UK and New Zealand governments, the European Commission and the United Nations on issues ranging from genetic engineering and cloning of farm animals to meat reduction and regenerative agriculture.

Dr. Gopichandran Ramachandran is a Professor teaching various aspects of mitigation, adaptation, preventive environmental management and business communication at the NTPC School of Business in India. These are important elements of public policy, pertaining to which, Gopi has an overall work profile that spans thirty-two years. He has recently contributed to the review of the first and second order drafts of the up—coming sixth assessment report of the IPCC; with a special emphasis on chemical ecology for improved adaptation strategies. Gopi's contributions through the Compliance Assistance Programme, OzonAction of the UNEP have been quite significant and consistently so for more than two decades at the regional and global levels. This created the opportunity for him to serve as a Member of the Inter—Ministerial Empowered Steering Committee constituted by the Ministry of Environment, Forest and Climate Change, Government of India, on aspects of the Montreal Protocol. He is a well-known specialist in the areas of science and, technology management communication, with a large number of theme-specific editorials and other publications to his credit. He holds two doctoral degrees in the fields of microbial and chemical ecology respectively, with a degree in law and is an Alumnus of the International Visitors Leadership Program of the Department of State, United States of America.

Dr. Himangana Gupta received her doctorate in Environment Science from Panjab University, India in 2015. She is a JSPS-UNU Postdoctoral Fellow at the United Nations University Institute for the Advanced Study of Sustainability (UNU-IAS) and the University of Tokyo, Japan. She has worked on climate change and biodiversity policy and diplomacy, and is currently working on linkages between biodiversity, climate, and communities in socio-ecological production landscapes. Before this, she was a part of the National Communication Cell (NATCOM) of the Indian Ministry of Environment, Forest and Climate Change. She contributed to India's Second Biennial Update Report and several other publications of the Ministry. She is a Certified Expert in Climate Adaptation Finance. She is a University Gold Medallist and recipient of Academic Excellence Award. She has published three edited books with Springer and has also written research papers in reputed international and national journals on climate policy, forestry, biodiversity, and women in climate change mitigation and adaptation.

Dr. Olaf Pollmann is a civil engineering and natural scientist. He holds a Doctorate (Dr.-Ing./Ph.D.) in environmental-informatics from the Technical University of Darmstadt, Germany and a second Doctorate (Dr. rer. nat./PhD) in sustainable resource management from the North-West University, South Africa. He is a visiting scientist and extraordinary senior lecturer at the North-West University and CEO of the company SCENSO - Scientific Environmental Solutions in Germany. He is also deputy head of the section "African Climate Service Centers" in West (WASCAL) and Southern Africa (SASSCAL) at the Project Management Agency (DLR-PT) on behalf of the Federal Ministry of Education and Research (BMBF). Dr. Pollmann is regularly requested to contribute, review and evaluate research-proposals, manuscripts, and programs from national and international universities, research institutions and governments. As an environmental expert, he has written several research papers in reputed international and national journals on current state of climate, food, and environmental issues.

Chapter 2
Impacts of Climate Change on Agricultural Sector of Pakistan: Status, Consequences, and Adoption Options

Usman Mustafa, Mirza Barjees Baig, and Gary S. Straquadine

Abstract Agriculture is one of the main climate-intense economic segments that is affected both positively and negatively by climate change. Agriculture is the mainstay of Pakistan's economy. It contributes 21% to the GDP of the state. Nearly 67% of the population is linked directly or indirectly with agriculture. Agriculture is the livelihood of a large majority of agricultural communities, but it is categorized as a risky occupation. Variations in temperature or precipitation generate substantial variation in crops output and yields. It is worthwhile to mention that Pakistan contributes less than one percent of global greenhouse gas (GHGs) emissions, but it is one of the most exposed to climate change impacts. Furthermore, Pakistan has an insufficient technical and financial capacity to adjust to these adverse impacts. Due to its agrarian economy, climate change is becoming a serious issue. Pakistan has the world's largest irrigation system and the major water source for it is snow and glacial melt. Climate change is not only threatening the irrigation system due to fast melting glaciers, but also increases the risk of floods, droughts, landslides, power shortages, and avalanches. It is a continually mounting concern with unlimited importance owing to its pronounced, comprehensive socio-economic effects. Variation in temperature and rainfall patterns is a very frightening issue in the crop sector, especially the arid zone. This chapter highlights the status, consequences/risks, adoption practices, and policy recommendations to address climate change in Pakistan. The chapter also covers the role of agriculture under climate change and Sustainable Development Goals.

Keywords Climate change · Agriculture · Consequences · Adoption · Pakistan

U. Mustafa (✉)
Pakhtunkhwa Economic Policy Research Institute (PEPRI), AWK University, Mardan, Khyber Pakhtunkhwa, Pakistan

M. B. Baig
Water and Desert Research, Prince Sultan Institute for Environmental, King Saud University, Riyadh, Saudi Arabia
e-mail: mbbaig@ksu.edu.sa

G. S. Straquadine
Career and Technical Education, Utah State University, Eastern Campus, Logan, UT, USA
e-mail: gary.straquadine@usu.edu

© The Author(s), under exclusive license to Springer Nature Switzerland AG 2021   43
M. Behnassi et al. (eds.), *Emerging Challenges to Food Production and Security in Asia, Middle East, and Africa*, https://doi.org/10.1007/978-3-030-72987-5_2

1   Role of Agriculture in Pakistan's Economy

Agriculture is the mainstay of Pakistan's economy and a major source of food security. It accounts for 18.5% of Pakistan's Gross Domestic Product (GDP) and 38.5% of
national employment. Unfortunately, it remains a regressive segment of the economy;
yet there is a substantial scope of high performing agriculture for economic growth
and poverty alleviation. The agriculture sector has performed poorly, especially
during the last decade; mainly due to the stagnant production of all major crops
caused by low water availability. This stagnant output has a profound impact on the
livelihood of rural communities. Poverty in Pakistan is widespread and growing,
especially fast in rural areas. Furthermore, rural poverty is deeply focused on small
farmers and non-farmers (Arif & Farooq, 2011; Malik, 2015).

Pakistan's agriculture not only fulfils the food, feed, and fibre needs of the country,
but also supports other segments of the economy, i.e., manufacturing and services.
There are very solid forward and backward linkages with these sectors. Investments
in Pakistani agriculture has supported a six-fold growth in population since liberation
in 1947 (Mustafa & Farooq, 2020). Pakistan's agriculture has great prospects and
potential. The country has a total geographic area of 79.61 million hectares, with a
great diversity in temperature and precipitation. Out of this less than one third (27%)
is a cultivated area, 27% is the net sown area, and 22% is not even surveyed. As
revealed in Fig. 1, less than 6% of the land area is under forests and another 14% is
the culturable waste area (land suitable but not cultivated).

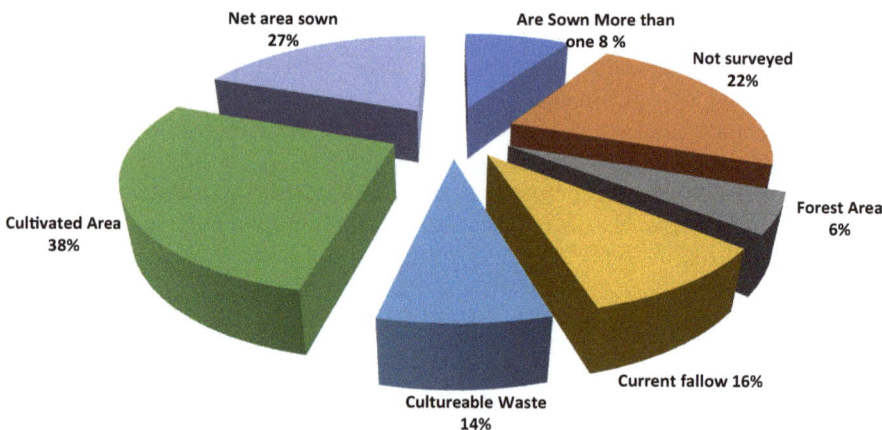

Fig. 1   Patterns of land use in Pakistan (Percentage). *Source* MONFS and R (2019)

1.1   The Diverse Climate and Agroecological Zones—A Blessing for Agriculture in Pakistan

Pakistan has been blessed with the diverse agro-ecological zones, comprising Northern Dry Mountains—High Mountain Coordinate (HMC), West Mountains—Wet Mountain Coordinate (WMC), Barani Lands—Potowar Corridor (PC), Sandy Desert—Thal Corridor (TC), Northern Irrigated Plain—Northern Irrigated Corridor (NIC), Southern Irrigated Plain—Southern Irrigated Corridor (NIC), Southern Irrigated Plan—Southern Irrigated Corridor (SIC), Sulaiman Piedmont—Sulaiman Belt Corridor (SIC), Western Dry Mountains—Dry Mountain Corridor (DMC), Dry Western Plateau—Coastal Belt Corridor (CBC), Indus delta (Fig. 2).

The Eastern areas of the Southern half mostly get precipitation (from June to September) through the Southwest summer monsoon, whereas the Northern and Western zones of the Southern half of the state receive rains largely (from December to March) through Western weather disturbances in winter. The majority (60%) of total annual precipitation falls during the summer monsoon. There is variation in climate from arid to semiarid. Three-fourths of the country receives rainfall of less than 250 milli-meters (mm) yearly. There is an exception in the Southern slopes of the Himalaya and the sub-mountain area in the Northern portion of the state, where annual rainfall varies from 760 mm to 2000 mm. The Northern area comprises many

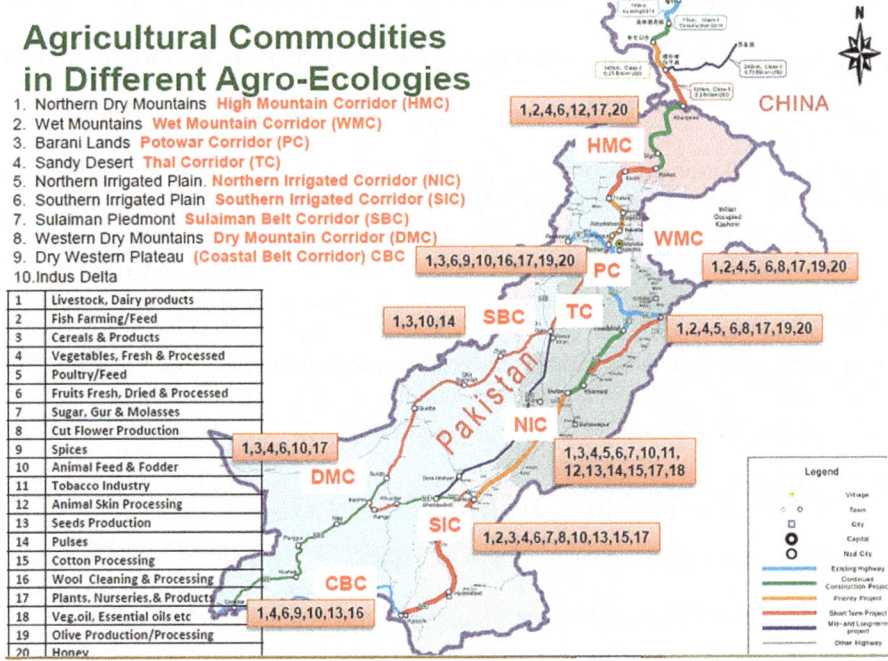

Fig. 2 Agro-ecological zones of Pakistan. *Source* Khan (2020)

of the world's highest mountains (i.e., K-2 at 8611 m [m] high) and biggest glaciers, including Siachen (70 km [km] long), and Biafo (63 km) that feed the Indus River and several of its branches (Chaudhry, 2017).

1.2 The Climate in Pakistan

Cold winters in Pakistan are characterized by extremely low temperatures like in the Northern High Mountain areas temperatures drop to as low as −50 °C and stay around 15 °C in the warmest months of May to September (McSweeney et al., 2008). The Western and Southern parts of the state denote the Indus River basin plain and Balochistan Plateau. The major (65%) of the country's total area falls in the transboundary Indus basin which shelters 520,000 km^2, including the entire provinces of Punjab, Khyber Pakhtunkhwa (KP), utmost of the Sindh territory, and the Eastern part of Balochistan (FAO, 2011). Pakistan has the world's largest contiguous irrigation system, known as the Indus Basin Irrigation System, which is 95% of the country's total irrigation system (Chaudhry, 2017). There is potential for 20 value-added agricultural commodities that can be produced in these agro-climatic zones of Pakistan (Fig. 2). More than 40 agricultural commodities are identified for promoting rural business clusters (Ahmad & Mustafa, 2018; Khan, 2020).

2 Climate Change Intensity and Impact on Pakistan's Agricultural Sector

According to National Geographic (2020), "climate change is a long-term shift in global or regional climate patterns. Often climate change refers specifically to the rise in global temperatures from the mid-twentieth century to the present. The root causes of climate change are natural events and human activities. Overtime, human population has drastically increased which requires more food, feed, and fibre. This results in an increase in per capita pressure on limited land" (Mustafa & Ahmad, 2020). The climate change, triggered by mounting greenhouse gas emissions, not only affects the atmosphere and the sea, but also the geology of the Earth.

2.1 Climate Change and Its Intensity in Pakistan

The annual average temperature in Pakistan has increased by approximately 0.5 °C in the past 50 years. A time series of area-weighted mean temperature for each year from 1960-2013 is presented in Fig. 3. The number of heatwave days per year has grown by almost fivefold in the previous 30 years (Chaudhry, 2017). The maximum temperature

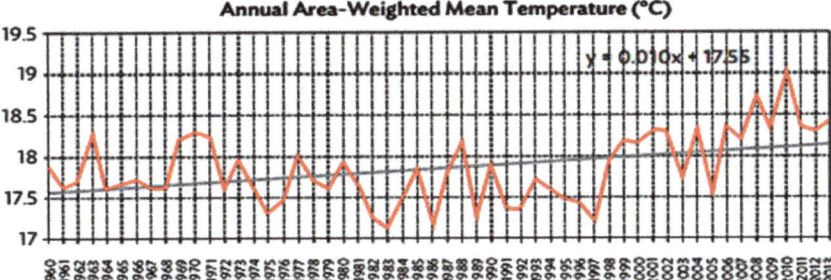

Notes: red line = area-weighted mean temperature of Pakistan, black line = linear trend (rate of change = 0.01°C), Total Change = 0.54°C).

Fig. 3 Area-weighted mean temperature from 1960 to 2013. *Source* Chaudhry (2017)

and relative humidity profile affect heatwaves. There is a major increasing tendency of both these factors, which is the crucial cause of spiky increases in heat index values in Pakistan. The total change in humidity calculated in summer from 1961 to 2007 for all Pakistan is 6.2% and change in maximum temperature is 0.25 °C (Zahid & Rasul, 2010).

Historically, annual precipitation in Pakistan has shown high inconsistency, but that variation has marginally enlarged in the past 50 years (Chaudhry et al., 2009) as depicted in Fig. 4.

As seen in Fig. 5, the sea level along the Karachi coast has risen approximately 10 cm (cm) in the last century (Rabbani et al., 2008).

By the end of this century, the annual mean temperature in Pakistan is expected to rise by 3–5 °C for a central global emissions scenario, while higher global emissions may yield a rise of 4–6 °C. Average annual rainfall is not expected to have a significant long-term trend, but is expected to exhibit large inter-annual variability. Sea level is

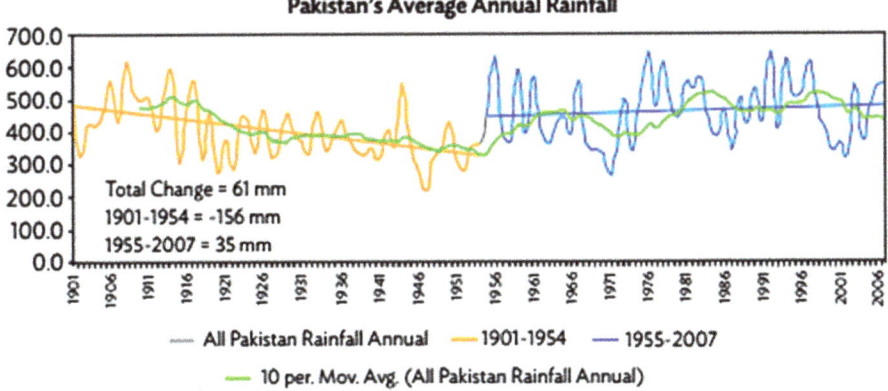

Fig. 4 Annual precipitation, 1901–2007 (Millimetres [mm])

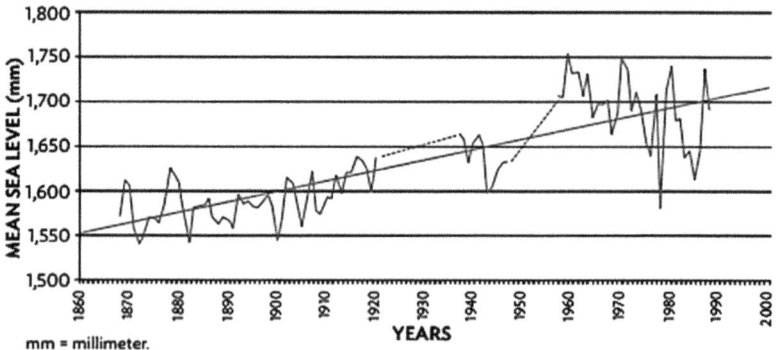

mm = millimeter.

Fig. 5 Mean sea level along the Karachi Coast, Pakistan, 1850–2000. *Source* Rabbani et al. (2008)

expected to rise by a further 60 cm by the end of the century and will most likely affect the low-lying coastal areas south of Karachi toward Keti Bander and the Indus River delta (Rabbani et al., 2008).

In Pakistan, the emissions are increasing at an annual rate of 6% or 18.5 million tonnes of carbon dioxide (CO_2) equivalent. Emissions were 147.8 million tonnes of CO_2 equivalent in 2008. It will reach 400 million tonnes of CO_2 equivalent (per year) by 2030 if the business-as-usual scenario remains intact (Government of Pakistan, 2012). Pakistan's contribution to total global greenhouse gas emissions is less than 1% (among the lowest in the world), but it is among the countries most vulnerable to climate change, and it has very limited technical and financial capacity to adjust to its adversaries (Chaudhry et al., 2015).

2.2 Climate Change Impacts on Pakistan's Agricultural Sector

Agriculture is the most climate-sensitive economic sector, and climate change has both positive and negative effects on it (NCA, 2014). There is a significant change in crop productivity and yields due to temperature or precipitation changes. Many experts have struggled to measure the effect of climate change on crop yields using international climate models (Khan & Tahir, 2018).

Climate change and agriculture are linked, and both take place on a worldwide scale. "Global warming affects agriculture in a number of ways, including through changes in average temperatures, rainfall, and climate extremes (e.g., heat waves), changes in pests and diseases, changes in atmospheric carbon dioxide and ground-level ozone concentrations, changes in the nutritional quality of some foods" (Milius, 2017: 1), and fluctuations in sea level (Hoffmann, 2013: 3).

2.2.1 Crop Sector

The consequence of climate change on crop yield varies by agro-climatic zone due to differences in their climate environments. Temperature and rainfall are the main determinants affecting the yield of major crops across different agro-climatic zones throughout Pakistan. Hussain and Bangah (2017) reported that wheat productivity has been impacted more in the Northern Irrigated Plan zone by an average temperature and in the Northern Dry Mountain region by rainfall changes than in other zones. Rice yield has been more affected in dry mountain regions by average temperature and in the Indus Delta by rainfall changes than in other zones. Sugarcane productivity has been affected more by average temperature and rainfall changes in the Indus Delta as of Zone IV. In the Northern dry mountains zone, maize productivity has been impacted more than in other zones.

Similarly, findings indicate that if the increase in temperature is less than 30 °C, the wheat yield will be decreased by 5–7% (Aggarwal & Sivakumar, 2010); decrease by 7% in Swat district and 14% in Chitral district (Hussain & Mudasser, 2007) and decrease by 6–9% in the sub-humid, semiarid and arid zones (Sultana & Ali, 2006). If the temperature increase is greater than 30 °C, the wheat yield will decrease in arid, semiarid and sub-humid zones, but increase in the humid zone (Aggarwal & Sivakumar, 2010); it would decrease by 21% in Swat district and 23% in Chitral district (Hussain & Mudasser, 2007). The consequences of a shift in the rainfall pattern on rice, wheat, and maize yield are negative and non-significant; but it is negative and significant for wheat (Ali et al., 2017).

A rise in temperature of 0.50–20 °C will decrease agricultural productivity by 8% to 10% by 2040 (Chaudhry, 2017). Iqbal et al. (2009) run different simulations using crop growth simulation models to project the production of major crops in four agro-climatic zones of Northern and Southern Pakistan. The results of the wheat-growing season and production in different climate regions of Pakistan by 2080 are presented in Table 1 (Iqbal et al., 2009). As the temperature rises, the length of the growing season falls. The biggest reduction in the growing season is 14 days for every one-degree rise in temperature in the Northern Mountainous region compared to Southern Pakistan.

2.2.2 Livestock Sector

Livestock is a vital sector in Pakistan's economy and delivers about 11.2% of Pakistan's GDP. The livestock sector is a source of stable livelihood for rural and small-agri-business holders. Moreover, it can play a major role in poverty alleviation in rural areas of Pakistan. It contributed about 60.5% of the output from the agricultural sector in 2018–19. The sector also contributes about 40% of the agricultural sector's GHG emissions (Mir & Ijaz, 2015).

Livestock emissions by a source show that feed production and processing add almost 45% of the whole sector's emissions (3.2 Gigatonnes of carbon dioxide equivalents). Enteric fermentation creates about 2.8 Gigatonnes (39%) and manure storage

Table 1 Change in length of wheat growing season and production in different climate regions of Pakistan by 2080

Temperature (Increase over baseline)	Length of growing season (Days)			
	Northern Pakistan		Southern Pakistan	
	Mountainous (Humid)	Submountainous (Subhumid)	Plains (Semiarid)	Plains (Arid)
Baseline	246	161	146	137
1	232	155	140	132
2	221	149	135	127
3	211	144	130	123
4	202	138	125	118
5	194	133	121	113
Climate Regions	Impact of climate change on wheat production by 2080			
	% Share in national production	Current yield (kilogram/hectare)	% Change in yield in 2080	Plains (Arid)
			A2 Scenario	B2 Scenario
Northern mountainous	2	2658	+50.0	+40.0
Northern submountainous	9	3933	−11.0	−11.0
Southern semiarid plains	42	4306	−8.0	−8.0
Southern arid plains	47	4490	−5.0	−6.0
Pakistan	100	4326	−5.7	−6.4

Source Iqbal et al. (2009)

accounts for 0.71 Gigatonnes (10% of the total). The rest (6% or 0.42 Gigatonnes of carbon dioxide equivalents) is attributable to the processing and transportation of animal products (Gerber et al., 2013, Fig. 6).

"The impact of climate change on livestock production systems, especially in developing countries, is not known, and although there may be some benefits arising from climate change, however, most livestock producers will face serious problems. Climate change may manifest itself as rapid changes in climate in the short term (a couple of years) or more subtle changes over the decades. The ability of livestock to adapt to a climatic change is dependent on a number of factors. Acute challenges are very different from chronic long-term challenges, and in addition animal responses to acute or chronic stress are also very different. The extents to which animals are able to adapt are primarily limited by physiological and genetic constraints. Animal adaptation then becomes an important issue when trying to understand animal responses" (Gaughan & Cawdell-Smith, 2015: 50).

Similarly, Sejian et al. (2015) and Giampiero et al. (2019) reported that the direct impacts of climate change on livestock are associated with loss in body weight,

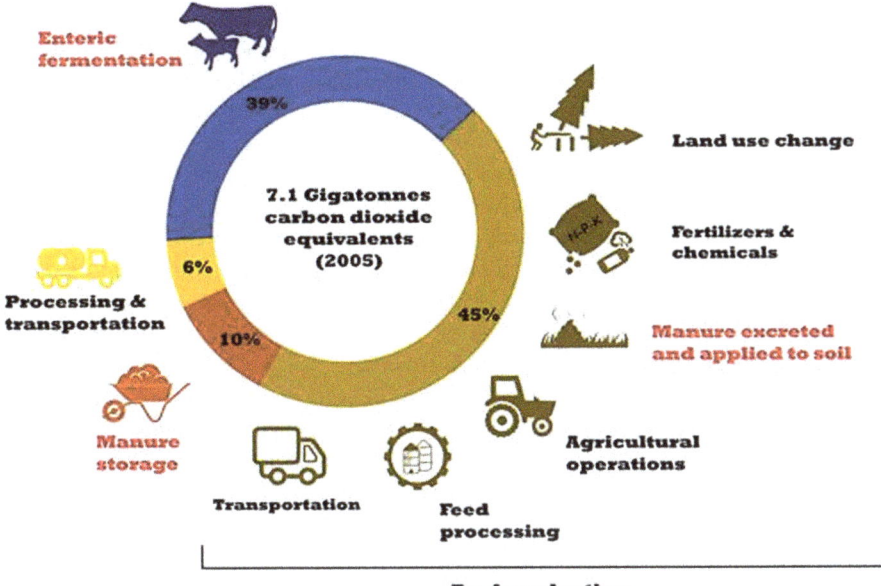

Fig. 6 Livestock emissions by source—Direct livestock emissions are shown in red. *Source* Adapted from Gerber et al. (2013)

disease occurrences, reproduction problems, metabolic activity, and reduced milk production. The indirect impacts are reductions in their feed resources, pasture availability, and water.

2.2.3 Forest Sector

Forestry is an important sector for Pakistan; it provides fuel, wood, lumber, paper, latex, medicine, food, wildlife, and ecotourism. According to total forest cover assessed by the consultant in 2012 is only 5.4% of total land area (REDCO + Pakistan, 2018). This has intensive impacts on biodiversity, environment and agriculture of the country (UNDP, 2018). Climate change disturbs forests by altering the frequency, duration, intensity, and timing of fires, drought, new species introductions, insect and pathogen outbreaks, hurricanes, windstorms, ice storms, and landslides (Virginia et al., 2000). According to Dale et al. (2009), "Climate change can affect forests by altering the frequency, intensity, duration, and timing of fire, drought, introduced species, insect and pathogen outbreaks, hurricanes, windstorms, ice storms, or landslides".

Global, regional, and local changes in temperature and precipitation can impact the incidence, duration, extent, timing, frequency, and intensity of turbulences in deforestation (Baker, 1995; Turner, 1998).

There is a scarcity of literature on the consequences of climate change for forests in Pakistan. Siddiqui and Ayaz (1999) predicted (by using the BIOME3 model) reduced forest cover for many plant species and relocation of some forest types to a new forest biome. They further found an increase in the net primary productivity of all biomass. "The study assessed nine dominant plant types or biomes for the climate change impact. Out of nine biomes selected, three biomes (alpine tundra, grassland or arid woodlands, and deserts) showed a reduction in their area, and five biomes (cold conifer or mixed woodland, cold steppe or mixed forests, temperate conifer or mixed forest, warm conifer or mixed forest, and steppe or arid shrublands) showed an increase in their area. Net primary productivity exhibited an increase in all biomes and scenarios" (Siddiqui & Ayaz, 1999: 8). Table 2 shows the change in forest types in Northern Pakistan from the 1961–1990 average.

In Pakistan, the major negative influence of climate change on forest and range-lands is longer drought durations. This has serious impacts on plants. In arid and semiarid regions, it results in desertification. As forage availability is badly affected, it also affects ruminant's health and body weight; and, ultimately, the livelihood of

Table 2 Change in forest types in northern Pakistan from 1961–1990 average

Biome type	2020			2050			2080		
	−P	0P	+P	−P	0P	+P	−P	0P	+P
Alpine tundra	−16.7	−16.7	−16.7	−31.5	−31.5	−31.5	−38.9	−38.9	−38.9
Cold conifer/Mixed woodland	22.2	22.2	11.1	5.6	44.4	22.2	44.4	33.3	11.1
Cold conifer/Mixed forest	10.3	10.3	13.8	13.8	20.7	24.1	6.9	10.3	13.8
Temperature conifer/Mixed forest	5.6	5.6	5.6	2.8	5.6	5.6	16.7	19.4	19.4
Warm conifer/Mixed forest	22.0	22.0	34.1	43.9	56.1	63.4	51.2	68.3	85.4
Xerophytic wood/Scrub	0.0	0.0	0.0	−1.8	−3.6	−1.8	0.0	0.0	0.0
Grassland/Arid shru bland	−42.9	−28.6	−28.6	−57.1	−57.1	−57.1	−57.1	−57.1	−28.6
Steppe/Arid shrubland	5.2	8.1	10.7	9.1	13.0	17.6	10.1	16.9	20.5
Desert	−7.6	−7.6	−16.3	−12.9	−19.3	−26.1	−14.4	−25.4	−33.0

Note The assumptions of the study are: Precipitation change (P) of 0, +3, −3% with 0.9 °C increase in temperature, and carbon dioxide concentrations of 425 parts per million (ppm) in 2020; 0, +3, −3 P, 500 ppm of carbon dioxide, and 1.80 °C increase temperature for 2050; and 0, +9, −9 for P, 575 ppm of carbon dioxide, 2.7 ° C increase in temperature for 2080 with 1990 as the base year
Source Adopted from Siddiqui et al. (1999)

poor people living in these areas (Lyons et al., 2018). Dry weather also causes light-ning strikes in forest and triggers rangeland fires, which causes serious biodiversity losses.

Government of Pakistan realizing the important of forest in Pakistan, Pakistan Tehreek-e-Insaf (PTI) government in their manifesto (PTI Manifesto 2018), the PTI says that it "will expand and restore the fractured forests of Pakistan through a '10 Billion Tree Tsunami' spread over 5 years under principles of true forest valuation, community stewardship as well as public–private partnerships". The PTI govern-ment has started the five-year project to plant 10 billion trees across Pakistan from 2018 to 2023 (Gul, 2018). Prime Minister Imran Khan kicked off the drive on 2 September 2018 with approximately 1.5 million trees planted on the first day (Tribue, 2018). The drive was based on the thriving'Billion Tree Tsunami' movement of the former PTI government, also led by Imran Khan, in the province of Khyber Pakhtunkhwa in 2014 (Hutt, 2018).

3 Climate Related Disasters in Pakistan

Climate change is the biggest common threat faced by Pakistan and India (Ashan, 2018). Over the past two decades, Pakistan has witnessed many climate-related catas-trophes. The intensity and frequency of extreme climate events have increased (Fraaqi et al., 2005). Climate change has profound impacts on glacier melting, flood, drought, land value, etc. in Pakistan.

3.1 Glacier Melting

Pakistan has more glaciers than anyplace else on Earth except for the Polar Regions with almost 7253 glaciers located in its high mountain ranges. Climate change (temperature rise seems to be the primary cause of glacier melting) is "eating away Himalayan glaciers at a dramatic rate" (Khan, 2019). Glaciers are enormously impor-tant for the nation as the rivers that feed on glaciers make up about 75% of the stored water supply of the country. Therefore, the importance and dependency of the country is on its glaciers (Siddiqui, 2019). Pakistani agriculture's major source of irrigation is from canals, which are linked with rivers and their major source of water is from glaciers. "Over the last seven years, several places, including Gilgit Baltistan and Chitral Valley, have experienced numerous major floods that many studies have given credits to the climate change. Countless lives have been laid to waste because of the floodwaters and has also taken away a living place from thousands of people. The undercutting of the once-flamboyant tourist industry is also a major negative impact caused by the overflow of water caused by the melting of glaciers" (Siddiqui, 2019: 1).

Overflow of flash floods caused by the melting of glaciers poses a threat to the tourist industry that was once very significant economically (Siddiqui, 2019). In the Northern area of Pakistan, Chitral Valley experienced glacier catastrophes. The resulting super flood destroyed their land and property and displaced people. They had to move from their ancestral homes to find an alternative place to live. Aisha Khan, the Executive Director of Civil Society Coalition for Climate Change (CSCCC), who also heads the Mountain and Glacier Protection Organization (MGPO), explains why there is an increase in the formation of glacial lakes reported by Abubakar (2020: 3): "Global warming and changes in mountain ecosystem will increase the formation of glacial lakes and exert pressure on human and natural systems. Most of these glacial lakes in the Hindu Kush Himalayan region of Pakistan were formed in the last five decades and the number of glacial lake outburst flood events is likely to increase, as new lakes are being formed and old lakes become more unstable".

3.2   Floods Due to Climate Change

One of the major climate-induced extreme events in Pakistan is floods. In recent years, flooding has become a regular feature with devastating effects. Pakistan received 170.5% of its normal rainfall in the month of July 2010 while Khyber Pakhtunkhwa (KP) province received 279.1% of its normal rainfall in July 2010 (Paknet, 2010). Pakistan faced 'mega' floods in 2019, which affected more than 20 million people, and severe flooding in 2011 and 2012 wreaked massive destruction in Sindh and Balochistan provinces. Unpredictable monsoons in 2014 provoked floods in several districts of Punjab province and a large number of districts of KP province were seriously affected as flash floods killed dozens and triggered substantial long-term challenges (Government of Pakistan, 1916). According to the Federal Flood Commission, Annual Report 1917 (Government of Pakistan, 2017), "Pakistan is one of the most disaster-prone countries in the world. Flood constitutes one of the world's most serious environmental hazards associated with global climate change, as witnessed by Pakistan during past 6–7 years. Under future climate change scenarios, the country is likely to experience increased variability of river flows due to increased variability of precipitation and melting of glaciers. Demand for irrigation water may increase due to higher rates of evaporation. Urban storm drainage networks may also be stressed by torrential rainfall and flash floods during monsoon season. Sea-level rise and storm surges may adversely affect coastal infrastructure and livelihoods in the coming years" (Government of Pakistan, 2017: 5).

3.3   Frequent and Prolonged Droughts

Drought is a prolonged period of dryness, one that causes extensive damage to crops or prevents their successful growth. The drought conditions in Sindh and Balochistan provinces developed into one of the worst disasters in Pakistan. Sindh in the south and Upland Balochistan are the utmost severely affected by drought. These affected parts of Sindh and Balochistan have been prone to water deficiencies. Rainfall recorded in these areas over the last years has reached a new low, with occasionally little or no rainfall (IFRC, 2019).

A sever heat wave hilted Pakistan in June 2015, resulted in high number of causalities particularly in Karachi. The majority of the country was in the grab of heat wave in 17 to 24 June. As on 20th June, soaring temperatures were witnessed in the Southern parts of Pakistan. The temperature ranges commencing 49 °C in Larkana and Sibi to 45 °C in Karachi (Chaudhry, 2017).

3.4 Land Slides

Torrential rain during the monsoon season has triggered extensive landslides and provoked severe devastation across Northern Pakistan, KP province, and Asad Jammu and Kashmir (AJK). The Northern mountain ranges of Pakistan are very extensive and at high risk of landslides as a result of rainfall and poor land management practices. Deforestation, poor road construction and management practices, and limited accessibility in mountainous areas are important causes of landslides besides natural ones. These parts of the country are also prone to land degradation caused by soil erosion when the watershed is not properly conserved. The intensity of extreme climate events—i.e., heavy precipitation over short periods—raises the risk of frequent and quick landslides and avalanches (WFP, 2018).

3.5 Agricultural Land Value

There is a complex relationship between land use and climate. Climate is mainly affected by the land cover—as moulded by land-use practices. It affects the global absorption of GHGs and land use variation is a key driver of climate change. An altering climate can lead to changes in land use and land cover: for instance, growers might move from their regular crops to crops that will have higher economic profit under fluctuating climatic circumstances. Higher temperatures affect mountain snowpack and vegetation cover as well as water needed for irrigation. The understanding of the interactions between climate and land-use change is improving, but the continued scientific investigation is needed.

The present level of precipitation is inadequate for agricultural production which results in downward pressure on land value. In case of humidity, it is evident that humidity level has a positive impact on agricultural crop pattern adding an amount of around PKR 302,798.7 per hectare (Hussain & Mustafa, 2017).

4   Latitudinal Redistribution of Precipitation in Pakistan

Precipitation variability is the degree to which rainfall quantities vary across an area or over time; it is a significant characteristic of an area's climate. Precipitation variability can have grave impacts on the livelihood of people living in these areas. Hanif et al. (2013) examined latitudinal precipitation features and movements in Pakistan to find areas vulnerable to climate change. They found that precipitation is shifting toward high latitudes and no significant trends over lower latitudes of the country. These areas required more attention with regard to water management and mitigation of flood disasters in the future.

Over time, monthly rainfall is shifting from dry months to wet months and even from wet months to wetter months (March and July). On the other side, dry months are becoming drier and drier months are becoming the driest as depicted in Fig. 7.

Due to significant increases in seasonal and annual precipitation in Pakistan, and the westward shift (80–100 km) of monsoonal rainfall, rainfall over the catchment areas of eastern rivers has decreased (moved away). There is a higher probability of heavy rainfall events, which lead to flash floods over the Western rivers, and lessened in eastern rivers of Pakistan. Northwest Pakistan (Central parts of KP and North-western parts of Punjab) are extremely vulnerable to flash floods (Hanif, 2014).

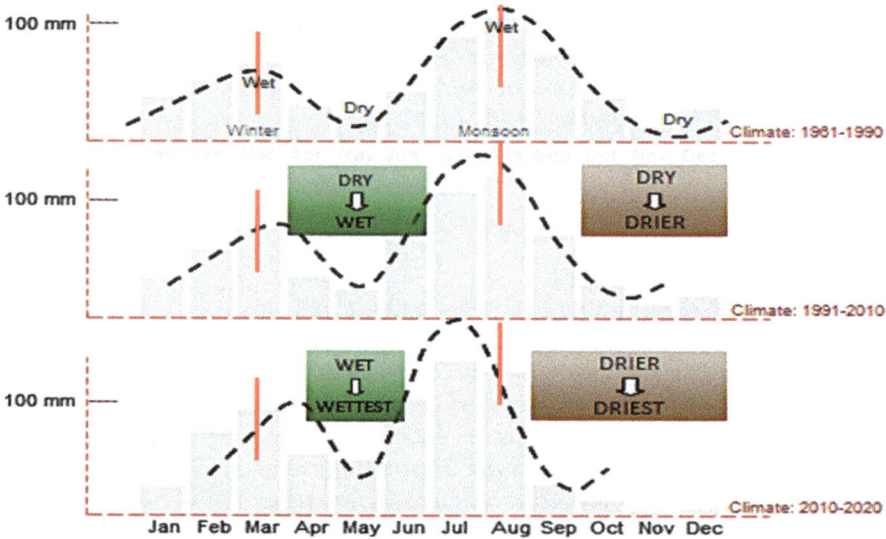

Fig. 7 Overtime monthly rainfall shift in Pakistan. *Source* Hanif (2014)

5 Shift in Cropping Pattern Due to Rainfall

There is inadequate variation in cropping patterns amongst all farm classifications in Pakistan. The large variability in crop yields across Agro-climatic Zones in Pakistanis depicted in Fig. 8 (HIES, 2011; Malik, 2015). Agricultural growth is facing serious challenges, and this will affect employment and rural poverty reduction (Malik, 2015), which are further aggravated by climate change (Mustafa & Farooq, 2020):

- Low and stagnant crop yields and large yield gaps relative to potential;
- Low productivity of irrigation water and unreliability of water services (inefficient allocation and use of irrigation water);
- Under-performance of rural inputs and output markets;
- Serious under-investment in agricultural R&D and dissemination/extension, especially public investment, and it is declining;
- Unequal land distribution and resultant skewed distribution of power and policy biases;
- Government intervention in markets;
- Neglect of agriculture in all policy and resource allocation decisions unless the decision leads to elite capture;
- A serious disconnection between the central government and the provinces in decision making and implementation—one size fits all policies –overly-focused on wheat and fixated on four crops only; and
- A regulatory environment that discourages investment and reduces market efficiency.

The above-mentioned challenges result in underemployment in agriculture and force farmers to face a vicious cycle of poverty.

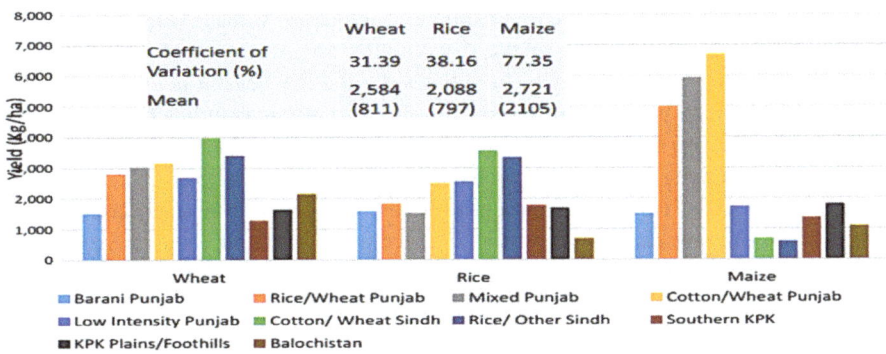

Fig. 8 Large variability in crop yields across agro-climatic zones in Pakistan. *Source* Malik (2015)

6   Role of Agriculture Under Climate Change and Sustainable Development Goals (SDGs)

There is tremendous pressure from increases in population and demand for food, feed, and fibre. Per unit pressure on limited land has increased, resulting in mounting social and environmental costs. Sustainable production of food, feed, and fibre has great potential to revitalize the rural landscape and provide inclusive growth to states. It is cost-effective and efficient because sustainable and higher agriculture production does not require better technology and huge investment (Mustafa & Farooq, 2020). The country can improve agriculture and reduce social and environmental costs by adopting simple measures, technologies, and methods (FAO, 2018).

Food and agriculture play a major role in achieving many SDGs and targets through "5Ps" i.e., people, planet, prosperity, peace, and partnership (UNDP, 2015). Agriculture is the common thread that holds many SDGs together and is directly linked with 10 of the 17 SDGs (Mustafa & Farooq, 2020). Therefore, agriculture must be a central part of national development agendas (Annan, 2017). Spending on improvements in agriculture not only addresses hunger, malnutrition, and food security but also other challenges such as poverty alleviation, water, energy efficiency, climate change, and responsible production and consumption (Rametsteiner, 2016).

7   Concluding Remarks: Various Adaptability and Compatibility Strategies in Relation to Climate Change

Climate change has profound impacts on Pakistan's agriculture. Increases in temperature melt glaciers which are the main source of irrigation water. Overtime, monthly rainfall is shifting in Pakistan so the dry months are becoming wet and the wet months are becoming wettest. On the other side, the dry season is becoming drier and the drier season is becoming even drier. So, the wet season has become wetter while the dry season has become drier. Consequently, there is an increase in flood and drought intensity in the country. Therefore, there is a serious need to formulate agricultural and rural development policies, strategies, programs, and projects to deal with climate change scenarios. The knowledge-based, innovative agriculture and rural development remain vital to cope with these challenges. Agricultural productivity is sensitive to these climate change impacts, which are also affecting the livelihood of poor smallholder families linked with farming.

Climate change also poses a serious challenge to Pakistan's water availability and food security. The government is trying to help farmers by providing agriculture inputs at affordable prices and ensuring better prices for their produce. To ensure food security, it is imperative to enhance domestic agricultural production through increased productivity (increasing per acre yield). Although Pakistan has rich production potential in agriculture, livestock, and fisheries, water is an important ingredient.

Therefore, for sustainable economic growth and prosperity, the development of these sectors on a long-term basis is of fundamental importance for the country's growth and prosperity. This calls for efficient utilization of production resources by adopting modern technologies and the establishment of a realistic, efficient marketing system. The following are some of policies and strategies to address agricultural issues, especially with regard to climate change: These structural alignments must recognize the need for Pakistan to identify key weaknesses, challenges, and opportunities of its agricultural innovation system to formulate a reform agenda to cope with climate change challenges.

- Increase investment in as well as linkages between demand-driven agriculture education, research, and effective extension services to cope with agricultural issues including climate change.
- Develop climate-resilient dryland crops using climate-smart technologies that can support farmers to cope with climate change. These methodologies should be formulated to encourage smallholder farmers in semi-arid tropics to use climate-smart technical interventions to increase agricultural production.
- Invest in agricultural sector not only to address hunger and malnutrition, but also other challenges including poverty, water and energy use, climate change, and unsustainable production and consumption.
- SDGs are well prepared and interrelated: reaching SDGs targets will not be possible without a resilient, robust, and sustainable agricultural sector development. Agriculture has a significant role in 10 out of 17 SDGs. Moreover, it is the only sector that has a common thread among all 17 SDGs together.
- Formulate a holistic policy approach for agriculture as part of an integrated overall growth-promoting policy framework. Agriculture should be seen as a whole system and not just as a set of five major crops. Similarly, food security should not revolve around sustaining wheat self-sufficiency as the only key target.
- Allocate a reasonable budget for agricultural and rural development to break the inertia in the productivity of crop and livestock sectors.
- Strengthen and adopt participatory research and development for rural support programs in the countries. Public–private partnerships should be encouraged and strengthened to improve their interaction in the innovative system and, thereby, support farm technology adoption.
- Disseminate new and improved innovative technologies and agricultural R&D through modern Information Technology (IT). This technology will facilitate the exchange of information in faster and easier ways. This knowledge dissemination will transform the nation and reduce the gap between Pakistan and the rest of the world.

Transforming Pakistani agriculture to make it more effective, efficient, and productive is not a difficult task. Further, it is cost-effective as there are tested, proven technologies and methods that are readily available. Therefore, by simply adopting them, agricultural production can increase and reduce the social and environmental costs.

Acknowledgements The authors are extremely thankful to Dr. Michael R. Reed, Emeritus Professor, and former Director—International Programs for Agriculture at the University of Kentucky, USA for reviewing the initial drafts, making helpful comments and offering valuable suggestions. His sincere cooperation is highly appreciated.

References

Abubakar, S. M. (2020). Can Pakistan's mountain communities protect themselves against melting glaciers. Climate change is triggering glacial melt in Pakistan and resulting in glacial lake outburst floods. Dawn Up dated May 07, 2020 01:24 pm. Viewed in Mayat https://www.dawn.com/news/1532235/can-pakistans-mountain-communities-protect-themselves-against-melting-glaciers

Ahmad, R., & Mustafa, U. (2018). Impact of CPEC projects on agriculture sector of Pakistan: Infrastructure and agricultural output Linkages. In *Proceeding of 32th Annual General Meeting (AGM) of Pakistan Society of Development Economic (PSDE)*, November 13–15, 2016, Merit Hotel, Islamabad. Viewed in March 2020 at https://www.pide.org.pk/psde/pdf/AGM32/papers/Impact%20of%20CPEC%20Projects%20on%20Agriculture%20Sector%20on%20Pakistan.pdf

Aggarwal, P. K., & Sivakumar, M. V. (2010). Global climate change and food security in South Asia: An adaptation and mitigation framework. *Climate change and food security in South Asia.* (pp. 253–275). Springer.

Ali, A., Rahut, D. B., Mottaleb, K. A., & Erenstein, O. (2017). Impacts of changing weather patterns on smallholder well-being: Evidence from the Himalayan region of northern Pakistan. *International Journal of Climate Change Strategies and Management, 9*, 225–240.

Ali, A., Somana, R., & Shahid, I. (2014). Deforestation and its impacts on climate change: An overview of Pakistan. *Global Change, 21*, 51–60. Viewed on July 3, 2020 at file:///C:/Users/Win%2010/Downloads/Deforestation_And_Its_Impacts_On_Climate_Change_An.pdf

Annan, K. (2017). The importance of agriculture for meeting the sustainable development goals. Remarks made by Mr. Annan at the Forum for the Future of Agriculture in Brussels, Belgium, 28 March 2017. Kofi Annan Foundation, P.O.B. 157, 1211 Geneva 20, Switzerland. Viewed in January 2020 at https://www.kofiannanfoundation.org/combatting-hunger/agriculture-sdgs/

Arif, G. M., & Farooq, S. (2011). Poverty, inequality and unemployment in Pakistan. Background Paper for the IDB Group MCPS Document for Pakistan. Pakistan Institute of Development Economics Islamabad, Pakistan. Islamic Development Bank, Jeddah, KSA.

Ahsan, A. (2018). Pakistan and India face common threats. Climate change is the biggest one. Dawn, Sep 18, 2019 09:40 am. Viewed on 3 July 2020 at https://www.dawn.com/news/1505534/pakistan-and-india-face-common-threats-climate-change-is-the-biggest-one

Baker, W. (1995). Long-term response of disturbance landscapes to human intervention and global change. *Landscape Ecology, 10*, 143–159

Chaudhry, Q. Z., Mehmood, A., Rasool, G., & Afzaal, M. (2009). Climate change indicators of Pakistan. Technical Report. No. 22. Islamabad: Pakistan Meteorological Department. Viewed in April 2020 at http://www.pmd.gov.pk/CC%20Indicators.pdf

Chaudhry, Q. U. Z. (2017). Climate change profile of Pakistan. Asian Development Bank 6 ADB Avenue, Mandaluyong City, 1550 Metro Manila, Philippines. Viewed in April 2020 at https://www.adb.org/sites/default/files/publication/357876/climate-change-profile-pakistan.pdf

Chaudhry, Q. U. Z., Ghulam, R., Ahmad, K., Munir, A. M., & Shahbaz, M. (2015). Karachi heat wave event June 2015. Islamabad: Technical Report for the Ministry of Climate Change Pakistan. Viewed at March 2020 at http://www.ndma.gov.pk/files/heatwave.pdf

Dale Virginia, H., Joyce, L. A., S. Mcnulty, Neilson, R. P., Ayres, M. P., Flannigan, M. D., Hanson, P. J., Irland, L. C., Lugo, A. E., Peterson, C. J., Simberloff, D., Swanson, F. J., Stocks, B. J., & Michael Wotton, B. . (2009, September). Climate change and forest disturbances. *BioScience*.

https://doi.org/10.1641/0006-3568(2001)051[0723:CCAFD]2.0.CO;2. Viewed in July 3, 2020 at file:///C:/Users/Win%2010/Downloads/Climate_Change_and_Forest_Disturbances.pdf

FAO. (2011). AQUASTAT Transboundary River Basins—Indus River Basin. Food and Agriculture Organization of the United Nations (FAO). Rome, Italy. Viewed at March 2020 at http://www.fao.org/3/CA2136EN/ca2136en.pdf

FAO. (2018). Transforming food and agriculture to achieve the SDGs: 20 interconnected actions to guide decision-makers. Food and Agriculture Organization of the United Nations Rome.

Fraaqi, A. B., Khan, A. H., & Mir, H. (2005). Climate change perspective in Pakistan. *Journal of Metrology, 2*(3), 11–21

Gaughan, J., & Cawdell-Smith, A. J. (2015). Impact of climate change on livestock production and reproduction. In V., Sejian, J. Gaughan, L. Baumgard, & C. Prasad (Eds.), *Climate change impact on livestock: Adaptation and mitigation.* Springer.

Giampiero, G., Pietro, G., Andrea, V., & Williams, A. G. (2019, January). Livestock and climate change: Impact of livestock on climate and mitigation strategies. *Animal Frontiers, 9*(1), 69–76. https://doi.org/10.1093/af/vfy034

Gerber, P. J., Steinfeld, H., Henderson, B., Mottet, A., Opio, C., Dijkman, J., Falcucci, A., & Tempio, G. (2013). Tackling climate change through livestock: A global assessment of emissions and mitigation opportunities. Rome: FAO. Available from http://www.fao.org/3/a-i3437e.pdf

Government of Pakistan. (2012). National climate change policy. Ministry of Climate Change, Government of Pakistan. Viewed in April 2020 at http://www.gcisc.org.pk/National_Climate_C hange_Policy_2012.pdf

Government of Pakistan. (2016). Annual flood report 2016. Federal Flood Commission, Ministry of Water Resources Government of Pakistan. Viewed in April 2020 at http://mowr.gov.pk/wp-content/uploads/2018/06/Annual-Flood-Report-of-FFC-2016.pdf

Government of Pakistan. (2017). Annual flood report 2017. Federal Flood Commission, Ministry of Water Resources Government of Pakistan. Viewed in April 2020 at http://mowr.gov.pk/wp-content/uploads/2018/06/Annual-Flood-Report-of-FFC-2017.pdf

Government of Pakistan. (2019). Pakistan economic survey 2018–2019. Finance Division, Government of Pakistan.

Gul, A. (2018). Pakistan's incoming Government to plant '10 billion trees. August 06, 2018 06:20 PM. Voice of America. Viewed on 3 July 2020 at https://www.voanews.com/south-central-asia/pakistans-incoming-government-plant-10-billion-trees

Hanif, H., Azmat, H. K., & Shahzada, A. (2013). Latitudinal redistribution of precipitation in Pakistan. *Journal of Hydrology, 492*, 266–272. Viewed in April 2020 at https://www.sciencedirect.com/science/article/pii/S0022169413002643

Hanif, M. (2014). Redistribution of precipitation (Seasonal shift) in Pakistan and super flood in Pakistan-2010. National Weather Forecasting Centre, Pakistan Meteorological Department, Government of Pakistan.

HIES. (2011). Households Income and Expenditure Survey 2010–11. Pakistan Bureau of Statistics, Government of Pakistan, Islamabad.

Hoffmann, U. (2013). Section B: Agriculture—A key driver and a major victim of global warming. In Lead Article (Chap. 1, pp. 3, 5) Hoffmann.

Hussain, A., & Bangah, R. (2017). Impact of climate change on crop productivity across selected agro-climatic zones in Pakistan. *Pakistan Development Review, 56*(2) Summer 2017, 163–167.

Hussain, S. S., & Mudasser, M. (2007). Prospects for wheat production under changing climate in mountain areas of Pakistan—An econometric analysis. *Agricultural Systems, 94*, 494–501

Hussain, A., & Mustafa, U. (2017). Impact of variable climatic patterns on the values of agricultural lands in Punjab Pakistan. *Journal of Agricultural Research, 55*(4):, 679–692. http://apply.jar.punjab.gov.pk/upload/1515405650_129_10._JAR_698_Final_%28Rezwan_Javed%29.pdf

Hutt, R. (2018). *Pakistan has planted over a billion trees.* World Economic Forum. Retrieved on 3 July 2020 at https://www.weforum.org/agenda/2018/07/pakistan-s-billion-tree-tsunami-is-ast onishing/

Iqbal, M. M., Arif, G. M., & Khan, A. M. (2009). Climate change aspersion on food security of Pakistan. *Science Vision, 15*(1), 15–23

IFRC. (2019). Pakistan: Drought Information bulletin. International Federation of Red Cross and Red Crescent Societies (IFRCARCS). IFRC, 28 Jan 2019. Viewed in April 2020 at https://relief web.int/report/pakistan/pakistan-drought-information-bulletin

Khan, R. S. (2019). Pakistan's dangerous melting glaciers and why we should be concerned. Tribune June 27, 2019. Viewed in April 2020 at https://blogs.tribune.com.pk/story/84753/pakistans-dan gerous-melting-glaciers-and-why-we-should-be-concerned/

Khan, M. A. (2020). Presentation on "Agricultural in Pakistan: Challenges and Opportunities" at Pakistan Institute of Development Economics (PIDE), Pakistan Agricultural Research Council (PARC), Ministry of Food National Security and Research, Islamabad. February 20th, 2020. Viewed on February, 2020 at https://pide.org.pk/index.php?option=com_content&view=article& id=682

Khan, M. A., & Tahir, A. (2018). Economic effects of climate change on agriculture productivity by 2035: A case study of Pakistan, 2018 Conference, July 28-August 2, 2018, Vancouver, British Columbia 275969, International Association of Agricultural Economists.

Lyons Robert, K., Rick, M., & Forbes, T. D. A. (2018). *Understanding forage intake in range animals.* Texas A & M, Agricultural extension, USA. E-393, 1–99. Viewed on July 3, 2020 at http://agrilifeextension.tamu.edu/wp-content/uploads/2018/12/E-393-understanding-for age-intake-in-range-animals.pdf

Malik, S. J. (2015). Putting the constraints to agriculture growth within the poverty reduction agenda for Pakistan. PSSP Third Annual Conference—Islamabad. April 14, 2015. Pakistan Strategy Support Program. IFPRI, USAID, Government of Pakistan.

McSweeney, C. M., New, M., & Lizcano, G. (2008). UNDP climate change country profile. New York. http://ncsp.undp.org/sites/default/files/Pakistan.oxford.report.pdf

Milius, S. (2017). Worries grow that climate change will quietly steal nutrients from major food crops. Science News. 13 December 2017. Retrieved 21 January 2020 at https://en.wikipedia.org/ wiki/Climate_change_and_agriculture

Mir, K. A., & Ijaz, M. (2015). Greenhouse gas emission inventory of Pakistan for the year 2011–2012. GCISC PR-19, Islamabad. Global Change Impact Analysis Centre (GCISC).

MONFS and R. (2019). Agriculture statistics of Pakistan, 2017–18. Ministry of National Food Security and Research. Islamabad. Government of Pakistan.

Mustafa, U., & Farooq, U. (2020). Pakistan national innovation system for agriculture and rural development: challenges and opportunities. In L. Singh, & A. Singh (Eds.), *Agriculture innovation system in Asia: towards inclusive rural development.* Routledge India member of the Taylor and Francis Group, 912 Tolstoy House, 15–17 Tolstoy Marg, 2 and 4 Park Square, Milton Park, Abingdon, Oxfordshire OX14 4RN. https://www.routledge.com/Agriculture-Inn ovation-Systems-in-Asia-Towards-Inclusive-Rural-Development/SinghGill/p/book/978036714 6665?fbclid=IwAR3IWVjsJIYnRwLWxOlplZ8aDQIhMw6AnWAzjmKpYazJQtS1R1Rp9mQ tris

National Geographic. (2020). Climate change. National Geographic. Resource Library. Ency-clopaedic Entry. Viewed on February, 2020 at https://www.nationalgeographic.org/encyclope dia/climate-change/

NCA. (2014). Climate disruptions to agriculture have been increasing and are projected to become more severe over this century. Explore agriculture. National Academy of Sciences, 2014 National Climate Assessment. U.S. Global Change Research Program, 1800 G, Street, NW, Suite 9100, Washington, D.C. 20006, USA. Viewed in April 2020 at https://nca2014.globalchange.gov/hig hlights/report-findings/agriculture

Paknet. (2010). Pakmet.com.pk: Monthly Statement for the Month of July, 2010. Pakmet.com.pk. Archived from the original on 20 August 2010. Retrieved April 2020 at https://web.archive.org/ web/20100820185911/ and http://www.pakmet.com.pk/FFD/index_files/rainfalljuly10.htm

Rabbani, M. M., Inam, A., Tabrez, A. R., Sayed, N. A., & Tabrez, S. M. (2008). The Impact of sea level rise on Pakistan's coastal zones. In A climate change scenario. 2nd International Maritime Conference at Bahria University, Karachi. Pakistan.

Rametsteiner, E. (2016). Transforming agriculture for sustainable development. IMPAKTER. Philanthropy, SDG Series, United Nations, November 8, 2016. Viewed February 2020 at https://impakter.com/transforming-agriculture-sustainable-development/

REDCO + Pakistan. (2018). REDCO + Pakistan, Forest reference emission levels 2019. Submission by Islamic Republic of Pakistan Forest Reference Emission Levels. In the Context of Decision 1/CP.16 para 70 UNFCCC. Viewed on 3 July 2020 at https://redd.unfccc.int/files/frel_paki stan_nro_06january_finalsubmitted.pdf

Sejian, V., Veerasamy, S., Raghavendra, B., Soren, N. M., Malik, P. K., Ravindra, J. P., Cadaba, S., Prasad, & Rattan, L. (2015). Introduction to concepts of climate change impact on livestock and its adaptation and mitigation. In V. Sejian, J. Gaughan, L. Baumgard, & C. Prasad (Eds.), *Climate change impact on livestock: Adaptation and mitigation*. Springer.

Siddiqui, K. M., Iqbal, M., & Ayaz, M. (1999). Forest ecosystem climate change impact assessment and adaptation strategies for Pakistan. *Climate Research, 12*, 195–203

Siddiqui, L. (2019). The melting glaciers of Pakistan. *Daily Times*. April 17, 2019. Viewed in April athttps://dailytimes.com.pk/377810/melting-glaciers-of-pakistan/

Sultana, H., & Ali, N. (2006). Vulnerability of wheat production in different climatic zones of Pakistan under climate change scenarios using CSM-CERES-Wheat Model. In *2nd International Young Scientists' Global Change Conference* (pp. 7–9). Beijing, China.

Tribue. (2018). Plant for Pakistan' campaign kicks off across the country. *The Express Tribune*. The Express Tribune. 1 September 2018. Retrieved on 3 July 2020 at https://tribune.com.pk/story/1793014/1-plant-pakistan-drive-kicks-off-tomorrow

Turner, M. G. (1998). Factors influencing succession: Lessons from large, infrequent natural disturbances. *Ecosystems, 1*, 511–523

UNDP. (2015). Transforming our world: the 2030 Agenda for Sustainable Development. United Nations – Sustainable Development knowledge platform. Retrieved February 1, 2020 at https://sustainabledevelopment.un.org/post2015/transformingourworld

UNDP. (2018). Sustainable forest management to secure multiple benefits in Pakistan's high conservation Areas. Project Implementation Review (PIR) 2018. United Nations Development Program (UNDP), Pakistan. Viewed in July 3, 2020 at https://info.undp.org/docs/pdc/Documents/PAK/Donor%20Report-Susatainable%20Forest%20Management%202018.pdf

Virginia, H. D., Linda, A. J., Steve, M., & Nielsen, R. P. (2000). The interplay between climate change, forests, and disturbances. *Science of the Total Environment, 262*(3), 201–204.

WFP. (2018). Climate Risks and Food Security Analysis: A Special Report for Pakistan Islamabad, December 2018. World Food Program (WFP), Government of Pakistan, Sustainable Development Policy Institute (SDPI). Viewed on 3 July 2020 at https://reliefweb.int/sites/reliefweb.int/files/res ources/Climate_Risks_and_Food_Security_Analysis_December_2018.pdf

Zahid, M., & Rasul, G. (2010). Rise in summer heat index over Pakistan. *Pakistan Journal of Meteorology, 6*(12).

Professor Dr. Usman Mustafa is presently working as Senior Consultant with Pakhtunkhwa Economic Policy Research Institute (PEPRI), AWK University, Mardan, Pakistan. Before this, he served Pakistan Institute of Development Economics (PIDE) in different positions, i.e., Professor, Joint Director, Chief (Project), and HoD Management Studies. He has more than 36 years of meritorious research, teaching, and development services records in different international/national organizations programs/projects including International Rice Research Institute (IRRI), UNDP, FAO, World Bank, IUCN, Pakistan Agricultural Research council (PARC), etc. He got his Ph.D. and MS in Economics from IRRI/University of Philippines at Los Banos (UPLB) during 1991 and 1987, respectively. He also has an MBA (Mkt.) from Pakistan. Beside these, he completed his Post-Doctorate from AVRDC—The World Vegetable Center, Taiwan. Dr. Mustafa has expertise in economics, management, marketing, supply change/value addition, policy analysis, training, M&E, HRM, environment, planning, co-ordination and collaboration, team building, development, and formal and informal diagnostic survey. He is also a visiting faculty member in number of universities and teaching and training institutes. He is a certified trainer from the World Bank Group. He has also been an active member of different international/national social and academic organizations. He has more than 75 research publication on his credit. He serves on the Editorial Boards of many international journals and as a member of many international professional organizations.

Dr. Mirza Barjees Baig is a Professor at the Prince Sultan Institute for Environmental, Water and Desert Research, King Saud University, Saudi Arabia. He earned his MS degree in International Agricultural Extension in 1992 from the Utah State University, Logan, Utah, USA and was placed on the 'Roll of Honor'. He completed his Ph.D. in Extension for Natural Resource Management from the University of Idaho, USA and was honored with the '1995 outstanding graduate student award'. Dr. Baig has published extensively on the issues associated with natural resources in the national and international journals. He has also made oral presentations about agriculture and natural resources and role of extension education at various international conferences. Food waste, water management, degradation of natural resources, deteriorating environment and their relationship with society/community are his areas of interest. He has attempted to develop strategies to conserve natural resources, promote environment and develop sustainable communities. Dr. Baig started his scientific career in 1983 as a researcher at the Pakistan Agricultural Research Council, Islamabad, Pakistan. He served at the University of Guelph, Ontario, Canada as the Special Graduate Faculty from 2000–2005. He served as a Foreign Professor at the Allama Iqbal Open University (AIOU), Pakistan through the Higher Education Commission from 2005–2009. He served as a Professor of Agricultural Extension and Rural Society at the King Saud University, Saudi Arabia from 2009–2020. He serves as well on the Editorial Boards of many international journals and the member of many international professional organizations.

Prof. Dr. Gary S. Straquadine serves as the Interim Chancellor, Vice Chancellor for Academic Programs, and Vice Provost at the Utah State University—Eastern, USA. He did his Ph.D. from Ohio State University, USA. He also leads the applied sciences division of the USU Eastern campus. He is responsible for faculty development and evaluation, program enhancement, and accreditation. He is also teaching undergraduate and graduate courses and supervise the research projects of his graduate students. He has a passion for the economic development of the communities through education, and has successfully developed significant relations with agricultural leadership in the private and public sectors. He has also served as the Chair, Agricultural Comm, Educ, and Leadership, at Ohio State University, USA. He served on many positions as the Department Head, Associate Dean, Dean and Executive Director, USU-Tooele Regional Campus, and the Vice-Provost (Academic). His professional interests include extension education, sustainable agriculture, food security, statistics in education, community and international development, the motivation of youth, and outreach educational programs. He has also helped several under developing countries to improve their agriculture and educational programs. He is seen as an administrator, educator, extension expert, community developer, and an international development professional.

Chapter 3
Impacts of Climate Change on Agriculture and Food Security in Tunisia: Challenges, Existing Policies, and Way Forward

Mohamed Ouessar, Abderrahman Sghaier, Aymen Frija, Mongi Sghaier, and Mirza Barjees Baig

Abstract Tunisia is a small North-African country with a dry Mediterranean climate. It is among the poorest countries in terms of water availability (450 m³/capita/year). With 516,000 farms, the agricultural sector occupies 65% of the country's land, contributes with 12.6% to the GDP, and employs 15% of the labor force. The agricultural production represents 9% of the total export earnings. Besides cereals (wheat and barley), Tunisian farmers grow olives, dates, and fresh fruits for both export and domestic consumption. Meat and vegetables are also important consumption commodities. The main agricultural exports are olive oil, dates and citrus. The EU is the principal trading partner for Tunisia. Agricultural products, mainly cereals, represent 8% of total imports and agricultural trade balance is negative (about 40% deficit). Most of climate change projection models foresee a significant decrease (10–35%) in rainfall and an increase (1.9–2.9 °C) in temperature during this century. It is expected that climate change will affect negatively many sectors (i.e., tourism, health, etc.) but the heaviest impacts will affect the agricultural sector as it depends largely on natural resources (soil, water, etc.) and meteorological conditions. The main impacts of climate change on Tunisian

M. Ouessar (✉) · A. Sghaier
Laboratory of Eremology and Combating Desertification, Institute of Arid Regions (IRA), University of Gabes, Medenine, Tunisia
e-mail: Ouessar.Mohamed@ira.rnrt.tn

A. Frija
North Africa Regional Program, International Center for Agricultural Research in the Dry Areas (ICARDA), Tunis, Tunisia
e-mail: a.frija@cgiar.org

M. Sghaier
Laboratory of Economics and Rural Societies, Institute of Arid Regions (IRA), University of Gabes, Medenine, Tunisia
e-mail: m.sghaier@ira.rnrt.tn

M. B. Baig
Prince Sultan Institute for Environmental, Water and Desert Research, King Saud University, Kingdom of Saudi Arabia, Riyadh, Saudi Arabia
e-mail: mbbaig@ksu.edu.sa

agriculture would be a disruption of the cropping cycle for the main agricultural products. The country's food security is severely threatened by such impacts, particularly in both terms of quantity and quality of agricultural products. About 83% of the severely food-insecure households belong to the poorest (50%) and poor (33%) sections of the Tunisian population. The objective of Tunisia's food security strategy is to make improvements in the living standards of the population while taking full advantage of the contributions of the agricultural sector to economic and rural development, poverty reduction, and employment generation. A focus on improving household nutrition and food security among disadvantaged groups is also made. To cope with these challenging issues, Tunisia has been engaged for the last three decades in elaborating various climate mitigation and adaptation strategies. A special focus has been devoted to the agriculture sector as it is the main pillar of the food security through implementing institutional and technical measures. Nevertheless, the integration with other sectors as well as mainstreaming the climate imperative in economic development plans of the country are highly needed.

Keywords Agriculture · Climate change · Food security · Tunisia · Vulnerability

1 Introduction

The mean global air temperatures have been increasing gradually with the advent of the industrial and agricultural revolutions in the mid-nineteenth century. However, it followed particularly an exponential rate during the last five decades causing continuous accumulation of greenhouse gases in the atmosphere and huge impacts on agriculture and land (FAO, 2008; IPCC, 2019).

Worldwide, rapid weather changes are already ongoing with multiple expected consequences, but the degree of impact depends on which part of the world is concerned. According to the IP climate change (2007, 2019), the Middle East and North African (MENA) regions are the most vulnerable to climate change. Climatic changes aggravate the water constraints in the MENA region which already suffers from weather variability—the region experiences increasingly frequent droughts and a looming water supply shortage. It is predicted that environmental stresses will increase due to climate change (Drine, 2011; Mutin, 2009). In fact, the impacts of human-induced climate change are often considered a future prospect, yet for the MENA region indications of a changing climate are clearly evident. Most of the projection outcomes associated with many global climate models are already occurring in the region, adding complexity to the ongoing problems of water scarcity, land degradation, desertification, salinization, and sea-level rise (Sowers & Weinthal, 2010).

According to IPCC (2019), the stability of food supply in the world will decrease as a result of food chains disruptions caused by the magnitude and frequency of extreme weather events. Climate change will affect all four dimensions of food security: food availability, food accessibility, food utilization, and food systems stability. It will have an impact on human health, livelihood assets, food production, and distribution channels as well as changing purchasing power and market flows (FAO, 2008). Its impacts will be both short term, resulting from more frequent and more intense extreme weather events, and long term, caused by changing temperatures and precipitation patterns. The most vulnerable and food-insecure people are likely to be the first affected. Agriculture-based livelihood systems that are already vulnerable to food insecurity face an immediate risk of increased crop failure, new patterns of pests and diseases, lack of appropriate seeds and planting material, and loss of livestock. People living on the coasts, floodplains, mountains, drylands, and the Arctic are most at risk. As an indirect effect, low-income people everywhere, but particularly in suburban areas, will be at risk of food insecurity owing to loss of assets and lack of adequate insurance coverage. This may also lead to shifting vulnerabilities in both developing and developed countries. Food systems will also be affected through possible internal and international migration, resource-induced conflicts and civil unrest triggered by climate change and its impacts. Agriculture, forestry, and fisheries will not only be affected by climate change, but also contribute to it through emitting greenhouse gases. However, they can contribute to climate change mitigation through the reduction of emissions by changing agricultural practices. At the same time, it is important to strengthen the resilience of rural population and help them cope with additional threats to their food security. Particularly in the agriculture sector, climate change adaptation can go hand-in-hand with mitigation.

In Tunisia, similarly to other North-African countries, the population dynamics and their respective livelihood activities are significantly linked with seasonal and annual climate fluctuations (Agoumi, 2003). Water, agriculture, coastal, and tourism were identified as the most vulnerable sectors in Tunisia to climate change (MARH-GTZ, 2007). Two-thirds of the country are arid to semiarid, making agriculture and water supply vulnerable to the predicted climate changes. The multiyear and recurrent episodes of drought generally affect most of the country and can lead to serious problems caused by water scarcity (Iglesias et al., 2011). The spatial and temporal changes in climatic variables relating to climate change have resulted in much debate worldwide and several studies are ongoing worldwide (Mansour & Hachicha, 2014).

Fully aware of these challenges, Tunisia ratified the United Nations Framework Convention on Climate Change (UNFCCC) in 1993, the Kyoto Protocol in 2002, and the Paris Agreement on October 17th 2016 (MALE, GEF-UNDP, 2017). Many initiatives have been undertaken by the Tunisian government, particularly with the support of international development agencies (GIZ, UNDP, World Bank, etc.). These initiatives dealt with both mitigation and adaptation to climate change of many sectors (tourism, agriculture, health, coastline, energy, biodiversity etc.) (MRH-GTZ, 2007) as well as cross-cutting issues (capacity building, positioning of Tunisia in international negotiations, institutional, legal and regulatory framework, etc.).

Climate change adaptation and mitigation measures need to be integrated into the overall development approaches and agenda. In this context, Tunisia will continue its efforts to reduce the climate change impacts on food security through mobilizing financial resources to face the investments required to achieve the mitigation target and adaptation goals. Tunisian government estimated these requirements over the 2017–2030 period to 31.5 billion of dinars (13 billions of dollars) and 848 millions of dinars (353 millions of dollars), respectively (MALE, GEF-UNDP, 2017).

2 Tunisia's Context Analysis

2.1 Biophysical Features

Tunisia is located at the northernmost tip of Africa, a strategic location that throughout history has made it a crossroads between Europe and the Middle East. Jebel Chaambi at 1544 m of elevation is the highest point in the country, it is located in the western central region, near the town of Kasserine (Khila et al., 2018). Tunisia's lowest point is Chott El Gharsah at 17 m under sea level. It is a sedimentary basin located in the central part of Tunisia, near its border with Algeria.

Tunisia has a varied topography. In the rainy north, the landscape is mostly mountainous, and soils are relatively fertile. Historically, this area was serving as the 'breadbasket' of the Roman Empire, as it claims one of the major Tunisia's rivers, Medjerda. The central part of the country features a semi-arid highland with poor soil quality and low rainfall. The southern part of Tunisia is semiarid in the east and becomes arid and desert as you move south-west closer to the Sahara Desert, except for occasional oases on the fringe of the Sahara. Tunisia has 1148 km of Mediterranean coastline. It has fertile coastal plains along the eastern coast. This area is famous for its olive production. Across the country, three main morpho-climatological zones can be distinguished:

- *The desert zone*: It is a collection of vast depressions whitened by salt efflorescences. Some of these depressions can form endorheic salt lake such as: Chott El Gharsa, Chott El Jerid, and Chott El Fejaj (Kamel et al., 2014). To the east of these depressions, large expanses of hard rocky plateaus (called locally hamadas) are found, which rise flowing a gentle slope to meet with a hard rocky basin (called locally Serir). These basins are bordered by a collection of small mountain ranges: Dahar (713 m) and Djebel Tebaga (469 m). To the south, these depressions merge with the Grand Erg Oriental, which is a large field of sand dunes.
- *The coastal zone*: This zone can be divided into two main sub-regions: the Sahel and the Jeffara. It extends from the gulf of Hammamet in the north to the Zarzis peninsula in the south. The two dominant landforms in this region are dunes and beaches. The latter can form tombolos, lagoons, and basins (sebkhet Kelbia, sebkhet Sidi El Hani, Sebkhet En Noual, Sebkhet El Melah). Along this coastal zone, some islands emerge, of which the Kerkennah and Djerba are the prominent ones (Mtimet, 2001).
- *Mountainous zone*: In Tunisia, mountainous landforms can be found in both the northwest and the south. The northern mountainous formations are an extension of two mountain ranges: the Tell Atlas and the Saharan Atlas. Both ranges

stretch from Morocco through Algeria. The Tell Atlas runs parallel to the Northern African coast and forms a natural barrier between the Mediterranean and the Sahara (Hezzi, 2014). It is rocky and hard to access these mountains, with a forest cover made mainly from pine-trees and cork oak. This massif is mainly composed of three mountain chains: Khroumire (north of Wadi Medjerda) culminating at around 1000 m; Mogods (located between the northern coast and the Medjerda River) at 500 m; and Nefza (lies between the two mountainous chains Khroumire and Mogods) at 600 m. Toward the south, we can see the fertile alluvial plain of the Medjerda, followed by an area of irregular hills (the Téboursouk Mountains chain) (El-Farhati et al., 2020). The Saharan Atlas lies at the northern edge of the Sahara and descends to the level of Cap Bon to the north, forming the northeastern extension, and the Gulf of Gabes to the south, forming the southeastern extension. Where, the latter mountainous chain is known colloquially as 'Djebel Tebaga'. The northeastern extension of the Saharan Atlas in Tunisia is designated by 'The Dorsal'. This mountain range crosses Tunisia diagonally, delimiting the Tell to the north and the steppe zone to the south. The Tell is characterized by rolling hills and plains which, alongside many depressions, facilitate its access and its exploitation (Hezzi, 2014). The Steppe essentially consists of a plateau located at an altitude of about 500 m. Djebel Tebaga extends from Kebeli in the west toward the town of El-Hamma in the east. Its highest point rises to 469 m a.s.l. In southernmost part of Tunisia lies the 'Jebel Dahar' mountain chain. It runs perpendicular to the previously mentioned ranges; therefore, it is oriented along a north–south axis, where it creates a barrier between the Sahara to the west and the relatively fertile Jeffara coastal plain to the east (Hezzi, 2014).

2.2 Socio-economic Characteristics

The agricultural sector in Tunisia plays important social, economic, and environmental roles. It directly contributes to food security, to exports volume, and to balance the trade deficits. It has also a positive impact on the socio-economic conditions through employment creation and reduction of rural exodus. According to the 2014 national census, Tunisia population reached 11.6 million. The total GDP has increased from 29,142.3 to 51,515.3 million US$ (constant 2010 US$) during the period 2000–2019 (Fig. 1). The GDP annual growth is unstable but positive; it is situated between 1 and 7% except in the year 2011 where it declined below −1.9%. The 2011 year corresponds to the revolution in Tunisia when the economic system has shown a huge shock. The economic growth has averaged 2.23% since 2012, which reveals a certain recovery and resilience of the economy thanks in particular to the contribution of agriculture, pillar of the economic system in Tunisia (Fig. 2). The GDP per capita is estimated in 2019–3447 current US$. The trade deficit is estimated

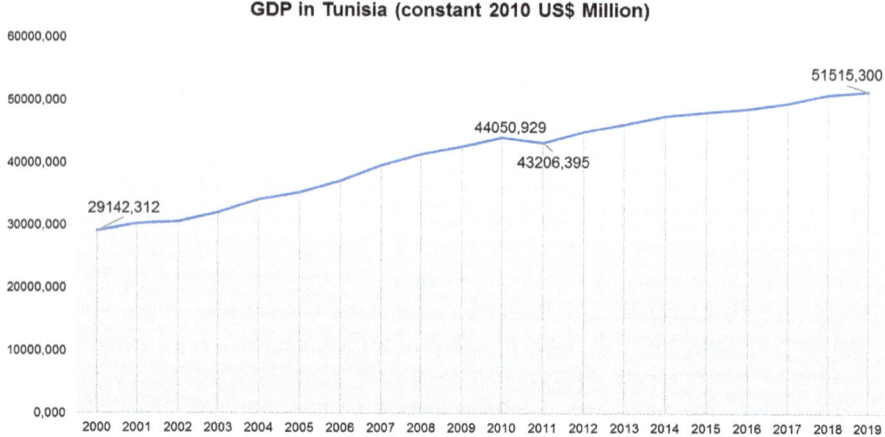

Fig. 1 GDP in Tunisia (constant 2010 US$ Million). *Source* The World Bank (2020)

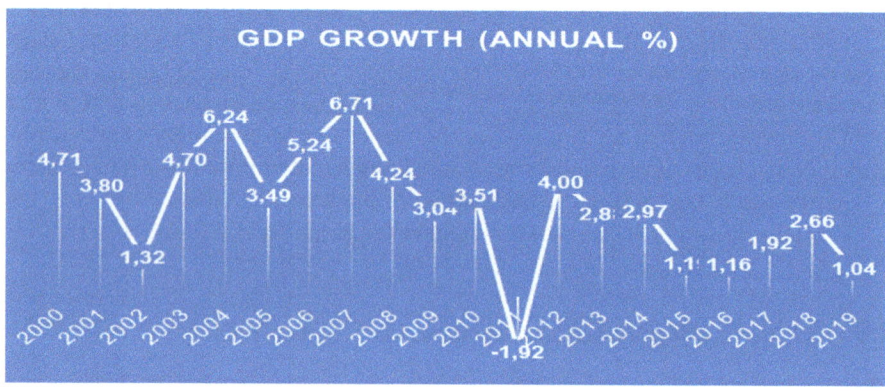

Fig. 2 GDP growth in Tunisia (annual %). *Source* The World Bank (2020)

at 11.3% of GDP (TND[1] 12.86 billion) during the first eight months of 2019.[2] The unemployment rate decreased from 19% in 2011 to 15.3% in 2015, which is considered relatively high. Youth (15–29-year-olds) suffer the most from unemployment with a rate of 30%. The national poverty rate is around 15.2% according to the lower middle-income poverty rate fixed at $3.2. The Gini index is estimated at 30.9 and life expectancy at birth is 75.9 years (World Bank, 2019).

Agriculture sector in Tunisia is very important for the national economy and the society due to its large contribution to national food security. Agriculture occupies about 10 million ha representing 62% of the total area of Tunisia and concerns about

---

[1] Tunisian national dinars.

[2] https://www.trademap.net/itc1/en/country_figure.htm?typetrade=E&selctry=788&product=TOTAL&reporter=Y.

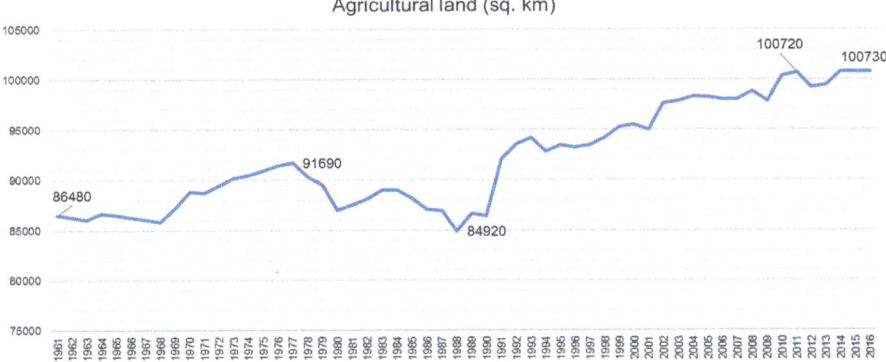

Fig. 3 Agricultural land in Tunisia. *Source* FAO (2019)

516,000 farms and 60,000 fishermen. The total agricultural land has increased from 86,480 km^2 in 1961 to 100,730 km^2 in 2016 (Fig. 3), which represents 63% of the national territory. The arable and cultivated land occupy in 2016 4.9 million ha and 4 million ha, respectively. This development is due to the land rehabilitation and water and soil conservation programs implemented since the early sixties. Agriculture employs more than 15% of the country's labor force and provides permanent income for 470,000 farmers, contributing to the stability of the rural population (MALE, GEF-UNDP, 2017). It contributes with 9% of the gross domestic product (GDP) and represents 7.5% of the economic investments. It represents 9% of exports and 8% of imports (Guesmi, 2018). Unfortunately, the agricultural trade balance is negative due to a low trade coverage situated around 61% (Alvarez-Coque, 2012 cited in Radhouane, 2018a; ITES, 2017).

Permanent cropland in Tunisia (% of land area) has increased from 7.4% in 1961 to 15% in 2016 (Fig. 4). The conditions of agricultural production are relatively

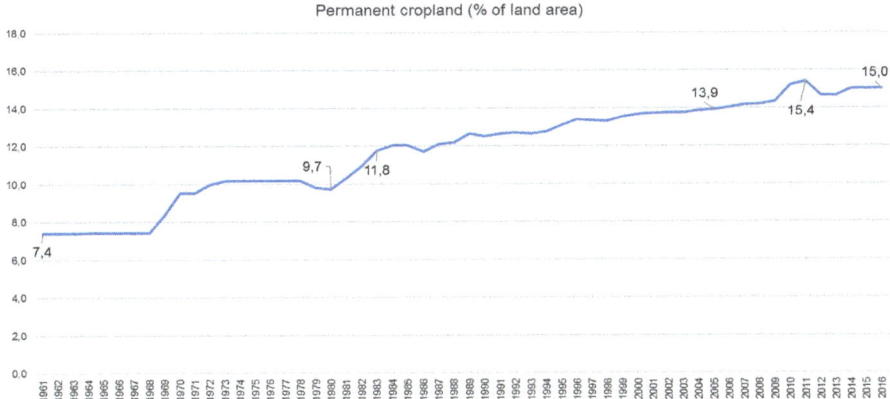

Fig. 4 Permanent cropland in Tunisia (% of land area). *Source* The World Bank (2020)

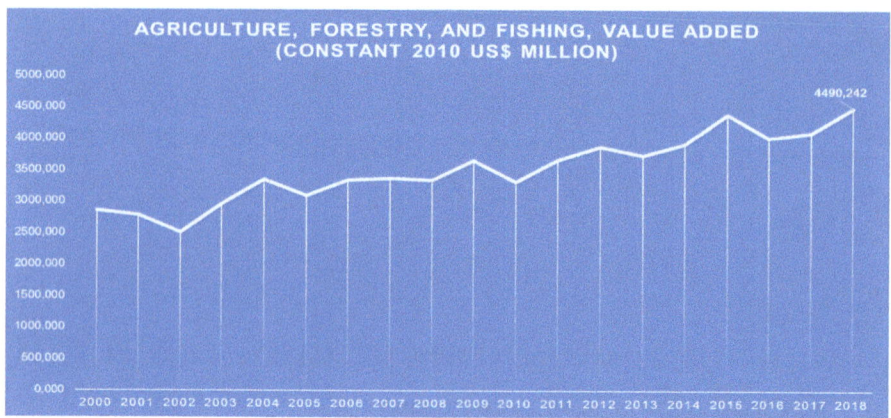

Fig. 5 Agriculture, forestry, and fishing, value added (constant 2010 US$ Million). *Source* The World Bank (2020)

difficult and do not offer great possibilities for intensification in dry land over large areas (Chebbi et al., 2019).

The agriculture, forestry, and fisheries value added reached 4,490,242 US$ Million, at constant 2010 prices (Fig. 5), which is equivalent to 10.4% of the GDP. Average agricultural contribution to total GDP is about 8.9% during 2000–2010 and 9.4% afterward (Fig. 6).

The main crops in Tunisia are cereals, olives, dates, fresh fruits, and seafood. Livestock production, especially small ruminants (sheep and goats) and dairy cattle, is also important. Vegetables—such as tomatoes and potatoes—are also produced

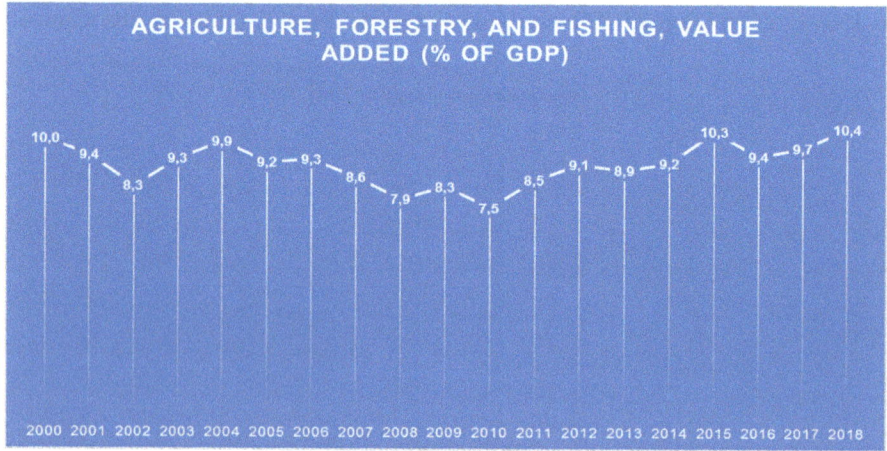

Fig. 6 Agriculture, forestry, and fishing, value added in Tunisia (as % of GDP). *Source* The World Bank (2020)

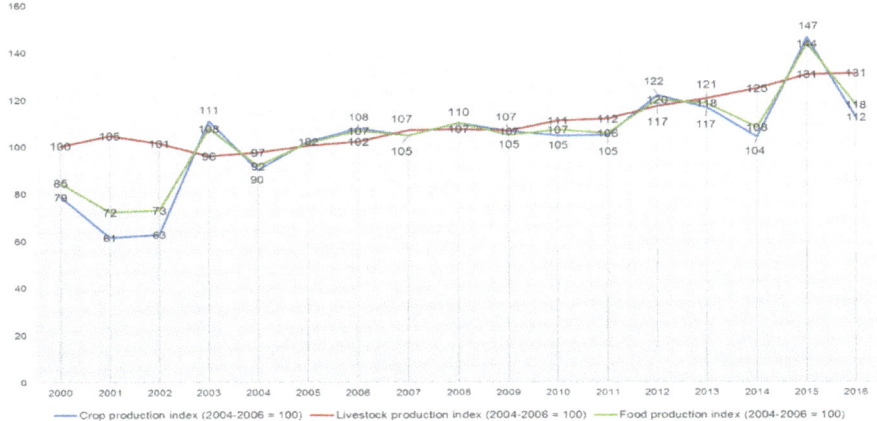

Fig. 7 Crop, livestock and food production indexes in Tunisia (2004–2006 = 100). *Source* The World Bank (2020)

to supply local market but also for export to neighboring regions such as European countries and Libya (Chibani, 2018).

The World Bank figures show that crop, livestock, and food production indexes in Tunisia (2004–2006 = 100) have increased from 79, 100, and 86 in 2000 to 112, 131, and 118 in 2016, respectively. Small ruminants (sheep and goat), cattle, and poultry dominate livestock, which occupies more than 400,000 breeders (MALE, GEF-UNDP, 2017). Livestock production index is characterized by a stable increase due mainly to the intensification and modernization programs based on imported grains, unlike the crop production and food indexes, which are more fluctuating due to climate variability (Fig. 7).

With regards to import/export of main of agricultural products in Tunisia, INS figures show that both import and export are increasing from 2000 to 2018 with a general trend of accelerating negative gap of the trade balance, which reached its maximum in 2017, i.e. TND −3843.7 million. The export of main agricultural commodities in Tunisia reaches 3355.4 Million TND (current TND value) in 2018, and is dominated by olive oil and dates to mainly European countries with TND 2128.0 million (63.3%) and TND 744.1 million (22.2%), respectively (Fig. 8).

In Tunisia, the agricultural and fishing sectors have a relatively high priority in terms of public and private investments. Total investment value was about TND 1250 million in 2016. Hydraulic and irrigation infrastructure accounts for the largest share of these investments. However, foreign direct investment (FDI) is still very low and volatile after the 2011 revolution due to the political and social instabilities. They are estimated to about 26 million TND in 2017 (1.9% of total FDI) (Chebbi et al., 2019).

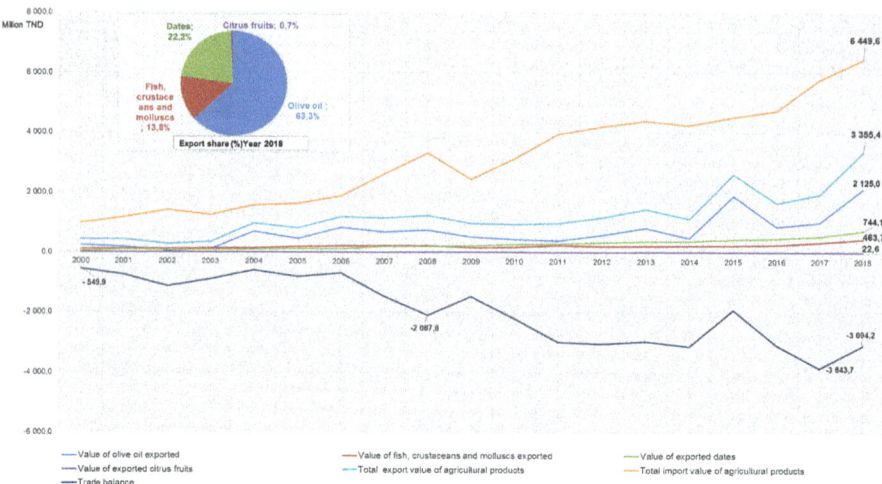

Fig. 8 Import/export of main of agricultural commodities in Tunisia (value, Current Million TND). *Source* Adapted by the authors based on data from INS (2020)

2.3 Food Security

The World Food Summit held in 1996 by the Food and Agriculture Organization of the United Nations (FAO) states that *"Food security exists when all people, at all times, have physical and economic access to sufficient, safe and nutritious food that meets their dietary needs and food preferences for an active and healthy life"* (FAO, 2006; Pinstrup-Andersen, 2009). Figure 9 describes the framework of food and nutrition security and the influence of food prices on the different components. In addition, Sustainable food and agriculture (SFA) contributes to all four pillars of food security and the dimensions of sustainability (environmental, social and economic)

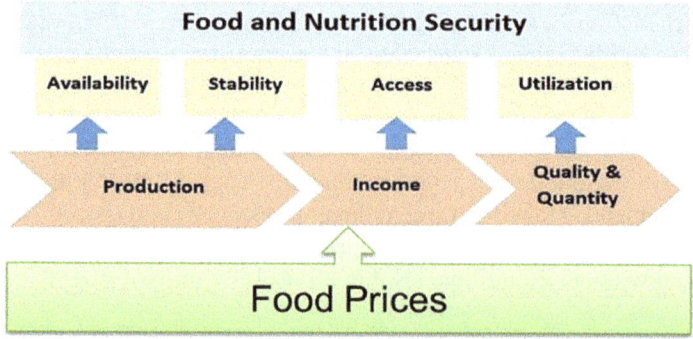

Fig. 9 Pillars of food and nutrition security. *Source* Bengoumi et al. (2018)

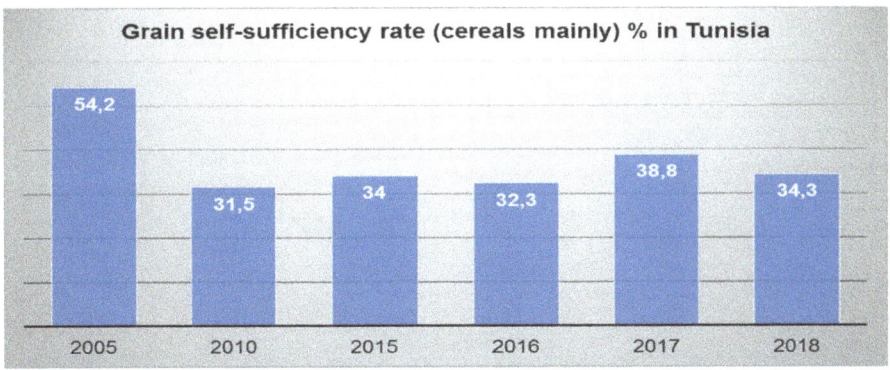

Fig. 10 Grain (cereals mainly) self-sufficiency rate % in Tunisia. *Source* Adapted by the authors based on the data from ONAGRI (2019a)

(FAO, 2020a). SFA contributes also to achieve the Sustainable Development Goals (SDG 2[3]) Zero Hunger (FAO, 2020b).

The concept of food and nutrition security adopted in Tunisia is based on the FAO definition with some adaptations that consider two additional components: (i) the provision of food should primarily be based on national production to the extent this is possible; and (ii) seeking to achieve a sustainable food trade balance.

According to the study of the National Observatory of Agriculture (ONAGRI) (2019a), Tunisia was ranked 51st out of 113 countries in terms of the Food Security Index during the year 2018. But the food security situation remains relatively fragile considering the economic instability and limited natural resources, especially water.

Tunisia's food security strategy (TFSS) focuses on improving household nutrition and food security among poor and disadvantaged groups. The target objective of TFSS is to achieve self-sufficiency in vegetables, cereals, red meat, poultry, and milk products. This objective seems highly ambitious given the current situation of domestic food production which only covers 20% of durum wheat needs, 75% of soft wheat needs, 100% of vegetable oil needs (without olive oil), and 100% of rice and seed needs. With regards to animal feeding, the national production provides an average of 0.5 million tons of barley, against the import of 462,000 tons in 2019.

With regard to the abundance of food products, the situation is improving given that the coverage rate of food needs through domestic production is upgraded for many basic products, such as milk, meat, vegetables, and grains while the situation remains relatively critical for grains where the dependency ratio on the supply was approximately 61% overall (ONAGRI, 2019a). The Grain self-sufficiency rate (cereals mainly) in Tunisia is highly volatile and correlated to climate variability. The 2018-value was about 34.4% with an average of 34.85% for the period 2015–2018 (Fig. 10).

---

[3] SDG 2: End hunger, achieve food security and improved nutrition and promote sustainable agriculture.

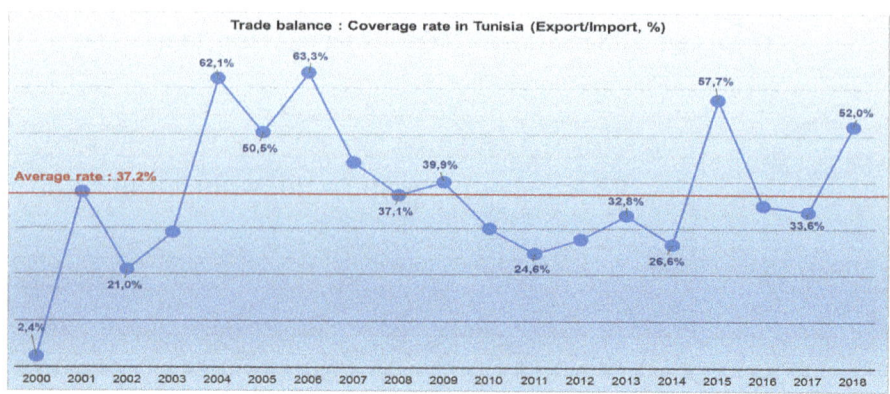

Fig. 11 Trade balance coverage rate in Tunisia (Export/Import of agricultural products, %). *Source* Adapted by the authors based on the data from ONAGRI (2019a)

As mentioned above, one of the major instruments to enhance self-sufficiency and food security in Tunisia is through sustainable food trade balance. This objective seems ambitious given the current situation characterized by a deep trade deficit during the last two decades. Indeed, the coverage rate of import by the export of food commodities is averaging 37.2% with remarkable fluctuations. The lowest coverage rate was recorded in 2000 and was due to an important drought event; while the highest rate was recorded in 2006 with 63.3% (Fig. 11).

According to ONAGRI (2019b), the food trade balance for December 2019 was a deficit of 1398.2 MD against a deficit of 476.1 MD during December 2018; thus, it is recording a coverage rate of 75.3% in 2019. However, this deficit represents only 7.2% of the overall food trade balance during the year 2019. This situation is due to the decrease of the export by 13%, while importations increased with 5.4%, in particular those of cereals.

An analysis of the food security situation in recent years shows many weaknesses and risks. During the period 2010–2018, the deterioration of the situation and the aggravation of the deficit in relation to the food trade balance are continuously recorded.

Overall, major issues facing Tunisian agriculture in terms of food security challenges are twofold:

- Seasonal and annual climate variability which affects the stability of domestic agricultural supply; and
- Food price fluctuations and volatility: Fluctuations in foreign markets and international prices of imported agricultural products have a significant impact on the food trade balance over the years. Extreme price volatility implies insecurity and financial risks for all the commercial operators involved. It puts food supplies (and FS) at risk during times of low supply and high demand.

On the other side, the main factors leading to food price volatility in Tunisia are:

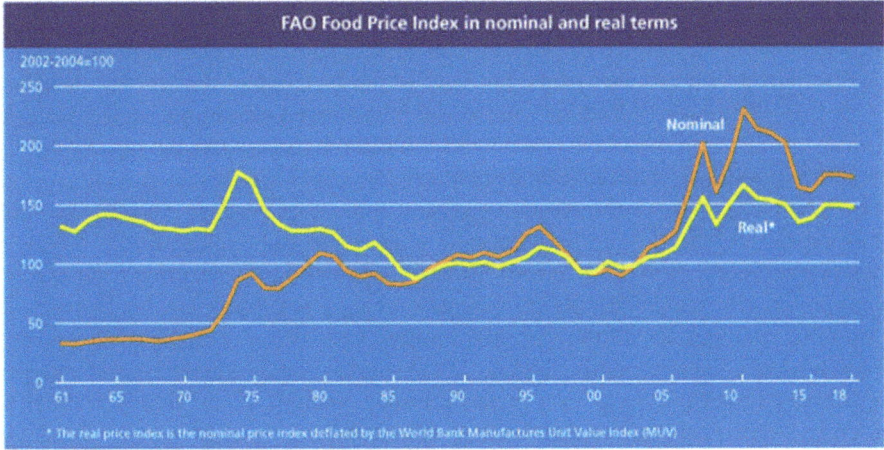

Fig. 12 Evolution of the FAO Food Price Index between 1960 and 2018. *Source* Guesmi (2018)

- Fluctuation in national production (due to climate variability and change, speculations, storage, etc.);
- Increase/decrease in demand;
- Macroeconomic dynamics and financial markets (currency exchange rates, inflation, etc.);
- Stockpiling policies (agricultural markets and trade); and
- Market distortions and imperfections.

Price fluctuations are a common feature of agricultural product markets but when these become large and unexpected—volatile—they can have a negative impact on food security (Fig. 12).

3 Climate Change and Its Impacts on Food Security in Tunisia

3.1 *Climate and Climate Change*

3.1.1 Introduction

There is no internationally agreed definition of the term 'climate change' (see Annex II for internationally agreed terminology on climate and climate change). This term can refer to: (i) long-term changes in average weather conditions (World Meteorological Organization (WMO) usage); (ii) all changes in the climate system, including the drivers of change, the changes themselves and their effects (Global Climate Observing System (GCOS) usage); or (iii) only human-induced changes in the climate system (UNFCCC usage). There is also no agreement on how to define the term 'climate variability'. Climate has been in a constant state of change throughout the Earth's 4.5-billion-year history, but most of these changes occur on astronomical or geological timescales, and are too slow to be observed on a human scale. Natural climate variation on these scales is sometimes referred to as 'climate variability', as distinct from human-induced climate change. The UNFCCC has adopted this usage (e.g., UNFCCC, 1992). For meteorologists and climatologists, however, climate variability refers only to the year-to-year variations of atmospheric conditions around a mean state (WMO, 1992). To assess climate change and food security, FAO prefers using a comprehensive definition of climate change that encompasses changes in long-term averages for all the essential climate variables. For many of these variables, however, the observational record is too short to clarify whether recent changes represent true shifts in long-term means (climate change), or are simply anomalies around a stable mean (climate variability) (FAO, 2008).

Climate warming is now 'unequivocal', according to the Intergovernmental Panel on Climate Change (IPCC). Despite a 'hiatus' in the early 2000s, global temperatures have risen again since 2012, with 2016 and 2017 being the Earth's hottest and the second hottest years on record. Similar trends were recorded for the entire MENA region and Tunisia, specifically. In fact, Tunisia's climate, including both rainfall and temperatures, varies considerably from north to south. It is strongly influenced by the Mediterranean Sea to the north and east and by the Sahara Desert to the south and southwest. Similar to the western Mediterranean, most of the rainfall occurs between October and May. This is because extra tropical weather systems from Europe and the Atlantic Ocean bring colder air and cloudiness. This also reduces the rainfall gradient from north to south. Temperatures in the arid and semiarid southern and southwestern parts of the country are generally high, whereas rainfall is significantly higher from November to April in areas further to the north. Tunisia's major bioclimatic zones are depicted in Fig. 13 Tunisia's Bioclimatic Zones. In the northern and central parts of the country, where agricultural activity predominates, the climate is characterized by hot dry summers and cool moist winters. These seasonal variations determine the growing period. Rainfall is irregular and varies considerably from the relatively humid coastal area to the desert like conditions in the south. The past decade has seen a variety of noteworthy climate events in Tunisia, including numerous droughts, heat waves, and the occasional heavy rainfall and flooding (Verner et al., 2018).

3.1.2 Climate Trends in Tunisia

The National Meteorological Center (INM) conducted studies on the evolution of climate in Tunisia during the period 1951–2010 (Belghrissi, 2018; Ben Rached et al., 2015). Regarding temperature, these studies found that:

- The annual and seasonal temperatures (min, avg, max) have increased significantly averaging around +2.1 °C. (Figs. 14 and 15).

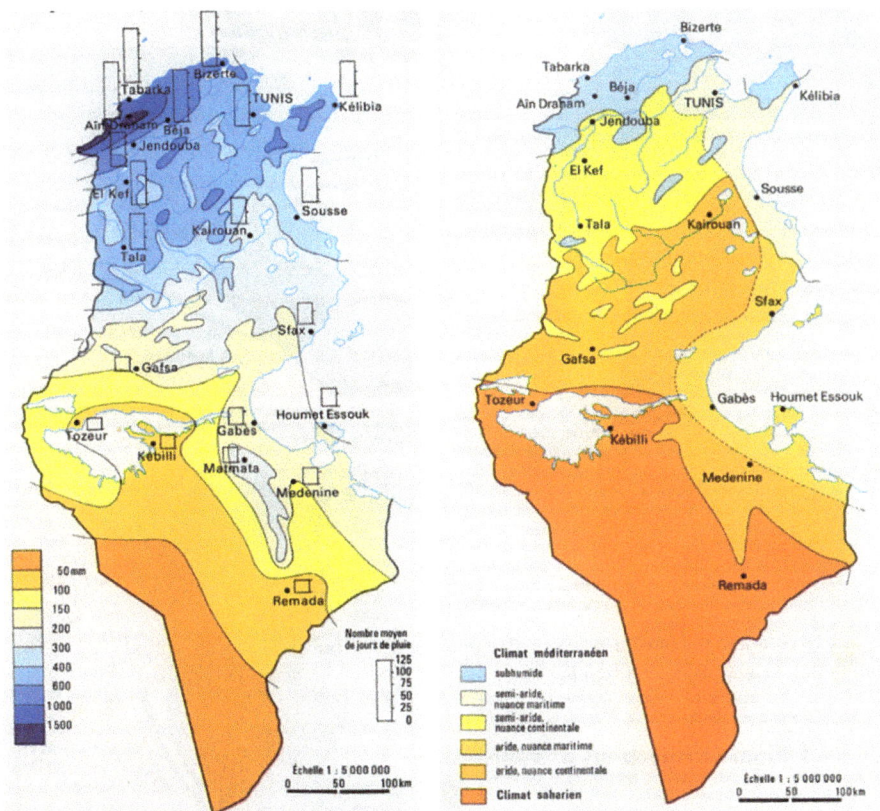

Fig. 13 Average annual precipitation and rainy days (left) and bioclimatic zones (right) of Tunisia. *Source* MEAT (1998)

- There was also an increase in the number of warm days and nights in parallel with a decrease in the number of cold days and nights (Figs. 16 and 17).

Regarding precipitation, the same authors found that:

- Except for a small part in the eastern side, there was a general tendency of decrease in the annual precipitation. The same trend was observed in the seasons but there was an increase in precipitation along the eastern part of the country (Figs. 17 and 18).
- Drought has become more long, frequent and intense as illustrated by the evolution of drought classes and SPI (Standardized Precipitation Index) (Fig. 19).

The combined effect of the increase in temperature and the reduction of precipitation resulted in the intensification of aridity which shows for example a clear expansion from south to north between 1971 and 2010 (Fig. 20).

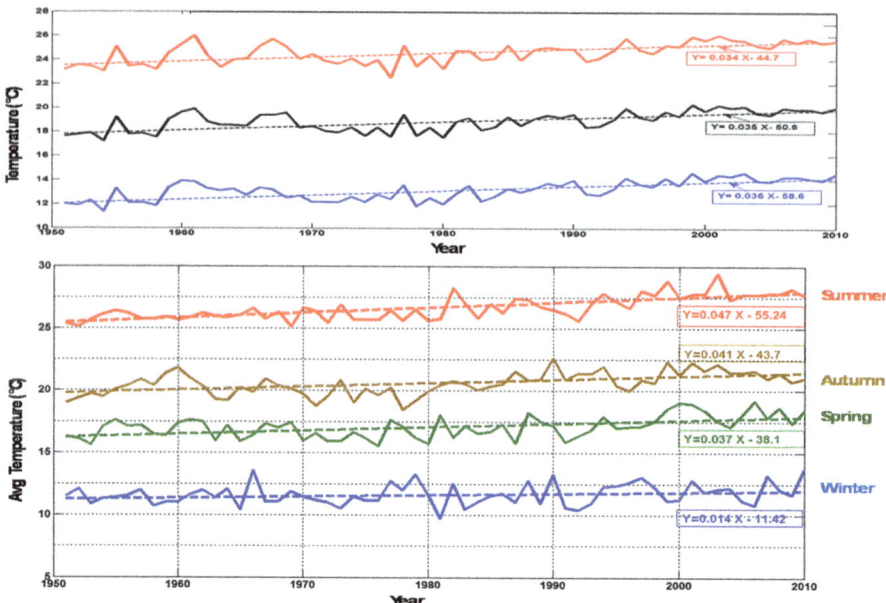

Fig. 14 Evolution of annual and seasonal temperatures in Tunisia during the period (1951–2010). *Source* Belghrissi (2018) and Ben Rached et al. (2015)

3.1.3 Climate Projections in Tunisia

In order to perform future climate projections for Tunisia, multiple model validation approaches using historical climate data (1961–1990) and 9 global circulation models have been retained (Belghrissi, 2018; Ben Rached et al. 2015). Based on the median scenario, the model projections for the horizon 2050 show that (Fig. 21):

- The average temperature will increase by 1.4–2.1 °C across the country compared to the reference period. However, this increase is more significant in the western and southern parts of the country; and
- A drop in precipitation of 2–16% over the entire territory compared to the reference period. Nevertheless, the western and southern parts will be more affected than the eastern coasts.

Based on the median scenario, the model projections for the horizon 2100 show that (Fig. 22):

- The average temperature will increase from 1.9 to 2.9 °C across the country compared to the reference period. However, this increase is more significant in the western and southern parts of the country; and
- A drop in precipitation of 10–35% over the entire territory compared to the reference period. Nevertheless, the western and southern parts will be more affected than the eastern coasts.

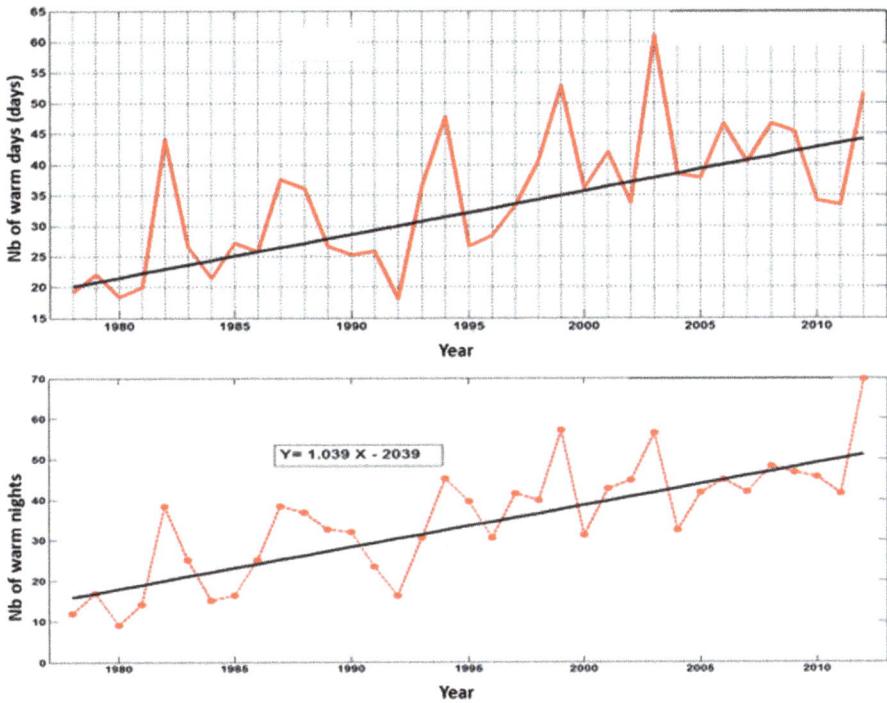

Fig. 15 Evolution of the annual number of warm days and nights in Tunisia during the period 1951–2010. *Source* Belghrissi (2018) and Ben Rached et al. (2015)

These projections are in concordance with the results of many studies. For example, Nasr et al. (2008) adopted the HadCM3 Global Climate Model (GCM) with the A2 greenhouse effect scenario to analyze the future dry projections in Tunisia by comparison to the period 1961–1990. They suggest a fall of 5–10% in precipitation amounts for the 2020–2050 period. Paeth et al. (2009) estimate that by 2050 surface temperatures in North Africa will increase by approximately 1.5–2 °C and precipitation will decrease by 10–30% across many of the desert areas of the region, with larger precipitation decreases along the coasts of Morocco, Algeria, and Tunisia. Radhouane (2013) mentioned that the model projections indicate a clear increase in temperature over the next decades that is expected to continue throughout the twenty-first century, probably at a rate higher than the estimated global average. Recently, Verner et al. (2018) reported that IPCC change projections indicate a high likelihood of continued warming in Tunisia, where a 0.50 °C per decade increase since the 1970s greatly exceeds the global average of 0.15 °C. Under a business-as-usual scenario, 3–7 °C increases are projected by the end of the twenty-first century, with the largest increases (4–7 °C) occurring during summer months (June, July, and August). Under the same scenario, precipitation is projected to decline 10–40% annually, including 10–30% in the wet season, from October to April, and 10–40% in the dry season, from May to September.

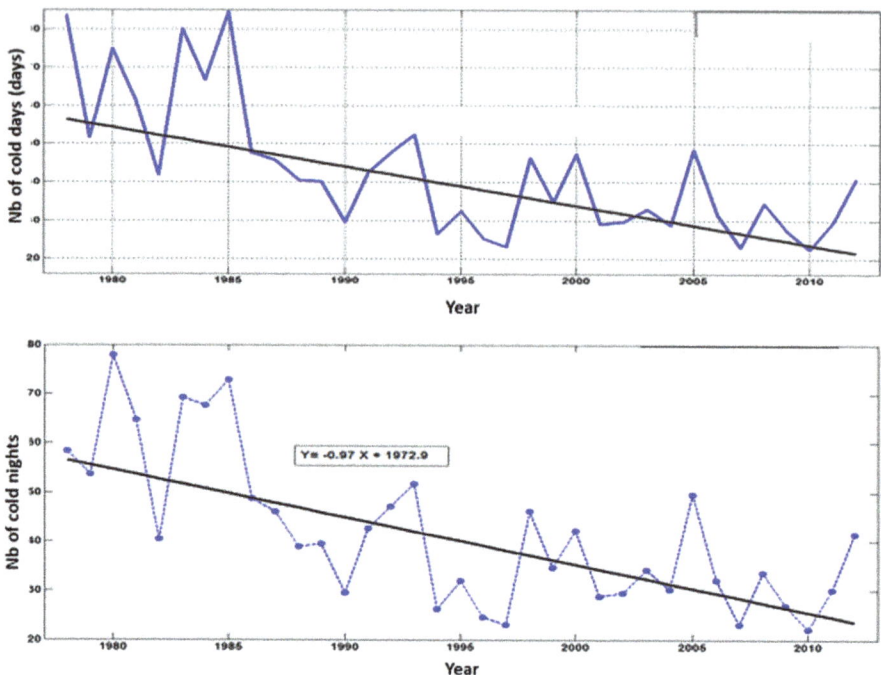

Fig. 16 Evolution of annual cold days and nights in Tunisia during the period (1951–2010). *Source* Belghrissi (2018) and Ben Rached et al. (2015)

3.2 Impacts of Climate Change

3.2.1 Biophysical Impacts

The negative impacts of the climate chnage will affect all sectors (Radhouane 2018a) but the analysis will be limited to the water resources and the main agricultural products.

Water Resources

The projected impacts on the various water resources of the country are illustrated in Fig. 23.

According to the MARH-GTZ (2007), the decrease in precipitation is expected to induce a 21% reduction in the conventional water resources. However, the surface aquifers will be the most affected (59% reduction) because of the lower recharge rates. The storage capacities of the dams are expected to be reduced with 11% due to the reduction of the inflow and increased siltation. On the other hand, the deep aquifers will be subject to overexploitation. Although the non-conventional water resources (treated waste water and desalination) are expected to increase by 218%, total water resources of the country will still be reduced by 11%. In addition, groundwater

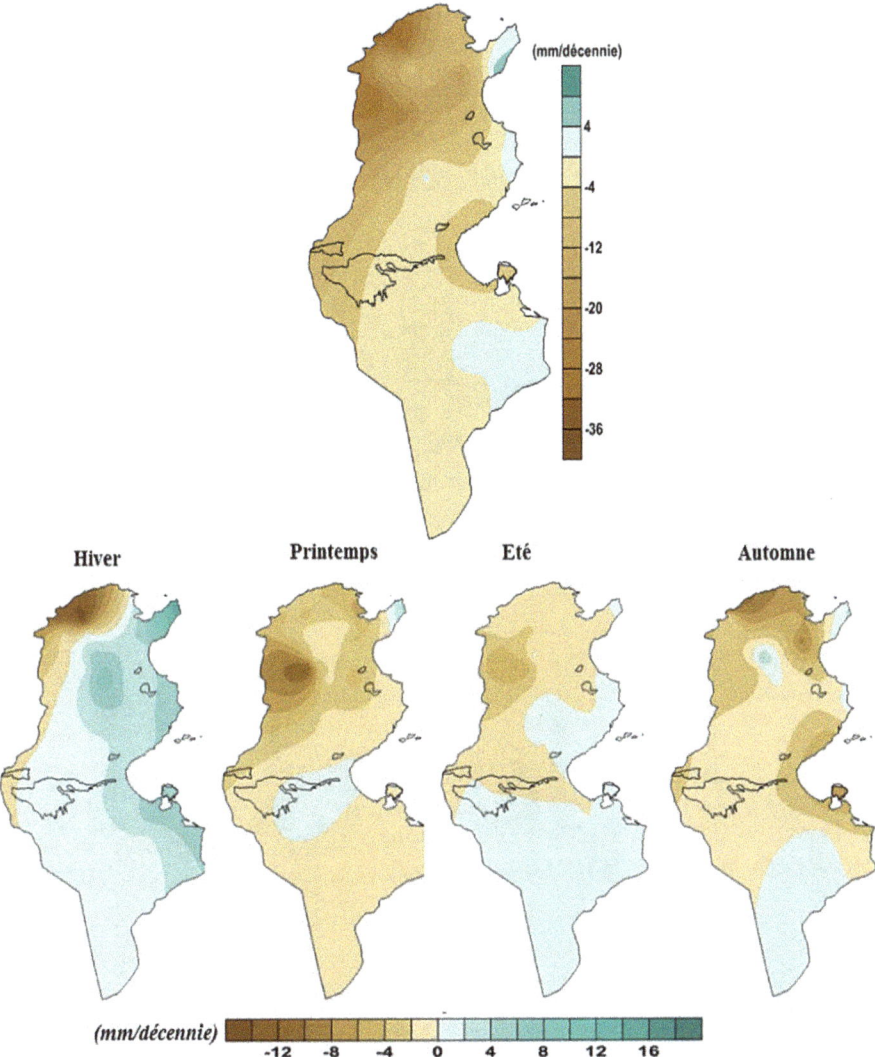

Fig. 17 Evolution of the annual and seasonal precipitation in Tunisia during the period 1951–2010. *Source* Belghrissi (2018) and Ben Rached et al. (2015)

salinity is expected to increase due to the combined effects of precipitation decrease and increase in temperature and exploitation, but also due to the rise in sea level (marine intrusion).

Cereals

More than 95% of cereals cropping in Tunisia is conducted under rainfed conditions. Therefore, they will be subject to adverse climate change impacts as demonstrated

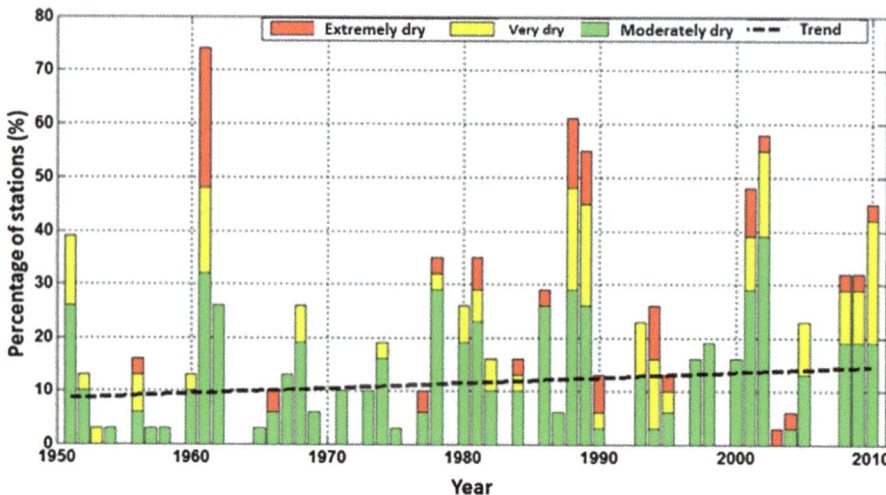

Fig. 18 Evolution of the annual drought classes in Tunisia during the period (1951–2010). *Source* Belghrissi (2018) and Ben Rached et al. (2015)

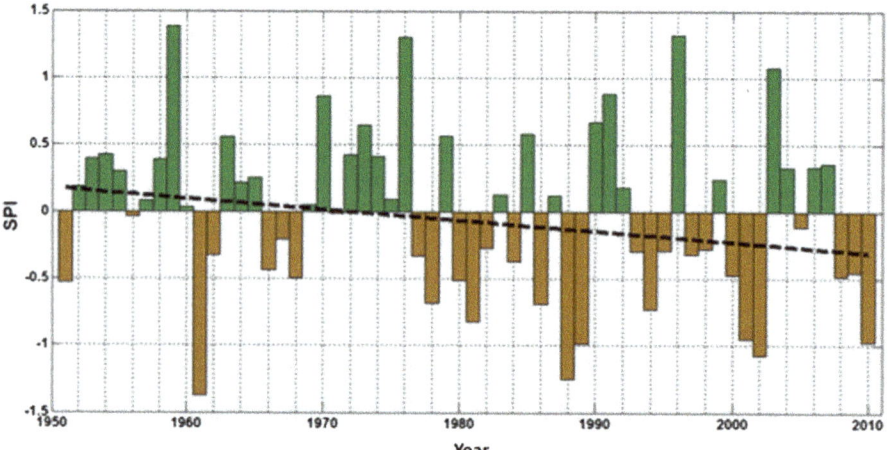

Fig. 19 Evolution of the annual SPI in Tunisia during the period (1951–2010). *Source* Belghrissi (2018) and Ben Rached et al. (2015)

by research for various cereal cropping regions in the country. Grami and Ben Rejeb (2015) used an econometric model to show that there will be a reduction in yield of about 5% under climate change in the main wheat production region in the sub-humid northern western part (Beja). In the semi-arid central part (Kairouan plain), Mougou et al. (2011) found that climate change will result in an increase of 15.2% in the wheat water requirements using the DSSAT model. A decrease of only 10%

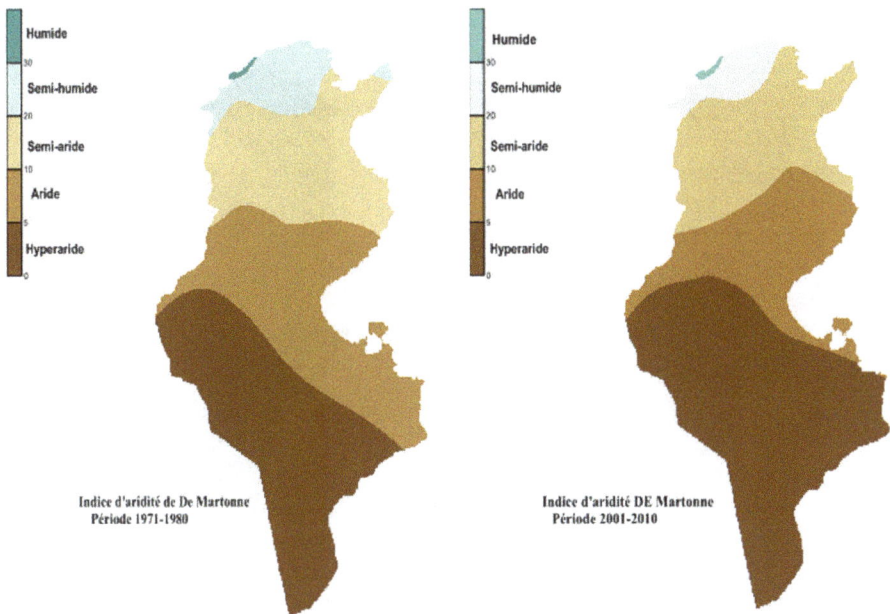

Fig. 20 Expansion of aridity in Tunisia (left: 1971–1980; right: 2001–2010). *Source* Belghrissi (2018) and Ben Rached et al. (2015)

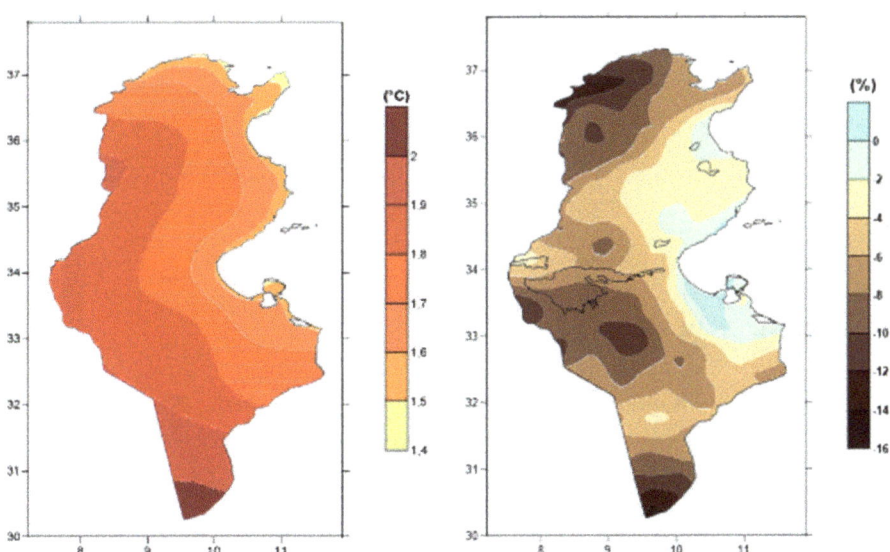

Fig. 21 Projected changes of temperature (in °C) (left) and precipitation (in %) (right) at 2050 horizon. *Source* Belghrissi (2018) and Ben Rached et al. (2015)

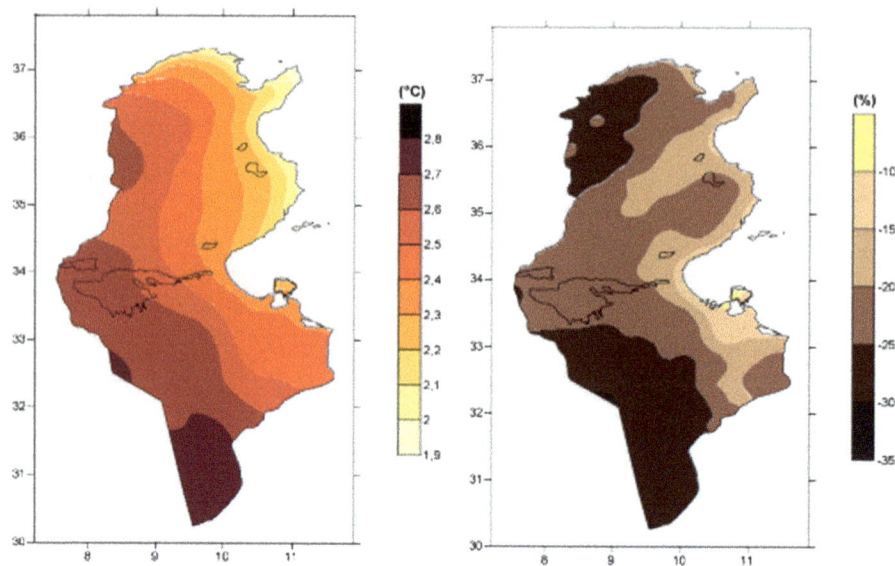

Fig. 22 Projected changes of temperature (in °C) (left) and precipitation (in %) (right) at 2100. *Source* Belghrissi (2018) and Ben Rached et al. (2015)

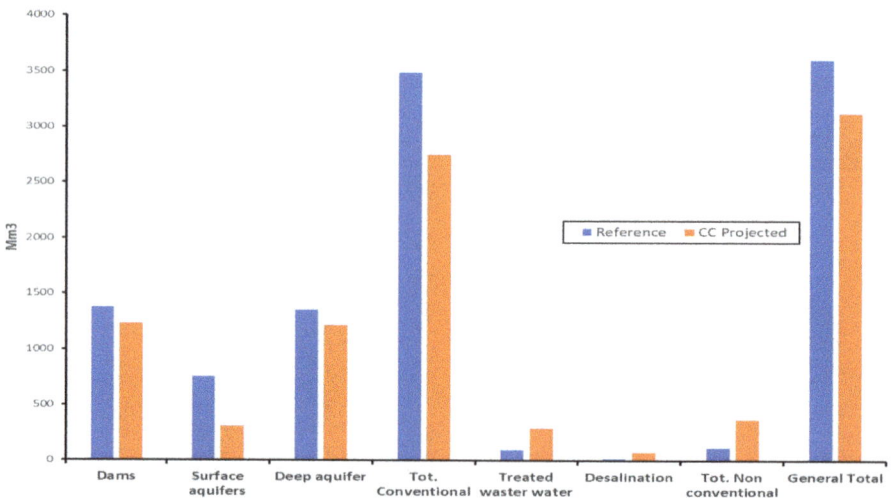

Fig. 23 Projected impacts of climate change on water resources in Tunisia. *Source* Compiled by the authors from MARH-GTZ (2007)

in precipitation and an increase of 1.3 °C in temperature would result in a 50% fall in yield.

Olive Cultivations

Tunisia is the second largest olive oil producer, after the European Union, with an average annual production of 165,000 tons. More than 30% of its arable lands are devoted to the olive crop, comprising about 70 million trees. The olive-growing sector occupies a strategic place in the Tunisian economy. Unfortunately, despite the strong resilience of the rainfed olive orchard systems, climate change is expected to influence negatively the possibilities of expanding the plantation of this tree under rainfed conditions (Zlaoui et al., 2019).

Ben Zaied and Zaouabi (2016) developed a panel cointegration model to analyze the impacts of climate change on olive production in Tunisia. They used a dataset covering 1980–2012 in 24 Tunisian regions. The long-run analysis reveals that rising temperatures and inappropriate working practices would reduce the olive output in semi-arid areas. Therefore, they recommended to carry out an appropriate training for workers in order to develop their skills and a public policy subsidizing the innovation of used capital stock, at least in the southern regions. They proposed also to encourage the development of drought-tolerant olive trees, especially in the south of Tunisia where global warming has caused a severe drought.

Sghaier and Ouessar (2013) studied the potential impacts of climate change on the spatial distribution of suitable olive growing orchard areas in the arid governorate of Medenine (southern Tunisia) using a GIS water balance-based approach. They concluded that the non-suitable and suitable areas for growing olive trees will evolve (from reference to climate change impact projection situations) from 8 to 25% and from 92 to 75%, respectively. Consequently, they recommended that planners and decision makers discourage and warn farmers that growing olive trees would be increasingly risky in large areas of the governorate where the annual rainfall amount is expected to fall considerably in the coming decades.

Citrus

Citrus is one of the major fruit crops covering significant agricultural areas globally (Liu et al., 2012). The citrus industry contributes considerably to GDPs of several countries, and particularly Tunisia. While the citrus industry significantly contributes to local, national, and global economies, its production consumes considerable amounts of fresh water resources, which makes the industry compete for available fresh water resources with other major water users, e.g., production of other crops, domestic and industrial uses, and ecosystems services. The citrus industry would face a decline in available (both quantity and quality) irrigation water as well as an increase in CO_2 as a result of climate change. In a study on potential climate change impacts on citrus water requirement across major producing areas in the world using the Irrigation Management System (IManSys) model (Fares & Fares, 2012), Fares et al. (2017) showed that though irrigation requirements (IRR) will be increased as a result of rainfall decrease and temperature in Nabeul governorate, the CO_2 concentration increase played a mask role and, therefore, the IRR will be reduced by about 11%.

3.2.2 Socio-economic Impact (Impact on Food Security, Etc.)

The IPCC has made evidences that climate change has an important socio-economic impact on the vulnerability and adaptive capacities of the human society. The international community is fully aware that the harmful effects of climate change are not only limited to ecosystems but also cover the higher-level sustainable development and food security issues (Maerh et al., 2003).

Climate change represents a serious threat to food security, sustainable development, and poverty levels in developing countries, particularly in Africa and South Asia. The estimated annual cost of climate change by 2030 for the African continent is around 3% of the continent's GDP, about US$ 40 billion (AFDB, 2012; Buchner et al., 2011). Indeed, climate change has important direct consequences on food security and hunger due to its impact on food production and availability in terms of quantity and quality, market price volatility, household's income, utilization, and stability of food systems. Patterson et al (2019: 27) argue that "food production is likely to fall in response to higher temperatures, water scarcity, greater CO_2 concentrations in the atmosphere, and extreme events such as heat waves, droughts, and floods. Already, yields of major food crops such as maize and wheat are declining owing to extreme events, epidemics of plant diseases, and declining water resources".

A study carried out by the Arid Land Institute (IRA) in the governorate of Medenine, Tunisia has revealed that climate change has direct and indirect impacts on the agricultural production system and livelihood conditions (Ouled Belgacem et al., 2011). The direct impacts are in different levels: (i) agronomic (includes high change in yield and productivity, reduction of appropriate area, limitation of cultivated area, loss of olive area, decrease of production, decrease of quality and loss on competitiveness); (ii) soil (land degradation); and (iii) ecosystem (vegetation cover). Indeed, severe droughts have led to important negative effects such as: huge decrease of yield and productivity of olive oil; a decline of suitable areas for crops mainly olive trees; quality degradation of main fruits; and new pests and diseases infections. Some indirect impacts of climate change on food security in the most vulnerable areas (arid regions) were cited by Ouled Belgacem et al. (2011), mainly (i) the substantial decrease of farmer's income mainly and for the most vulnerable households (marginal population, smallholders), (ii) higher rural exodus and migration of youth and vulnerable households as consequences of degradation of wellbeing conditions (less income, poverty, warmer climate).

Climate change has a crucial impact on health, nutrition, and hunger. The Global Hunger Index (GHI)[4] declines in Tunisia from 10.7 in 2000 to 7.9 in 2018, showing

---

[4] The GHI scores are based on a formula that captures three dimensions of hunger: insufficient caloric intake, child undernutrition, and child mortality. The GHI ranks countries on a 100-point scale. Values less than 10.0 reflect low hunger; values from 10.0 to 19.9 reflect moderate hunger;

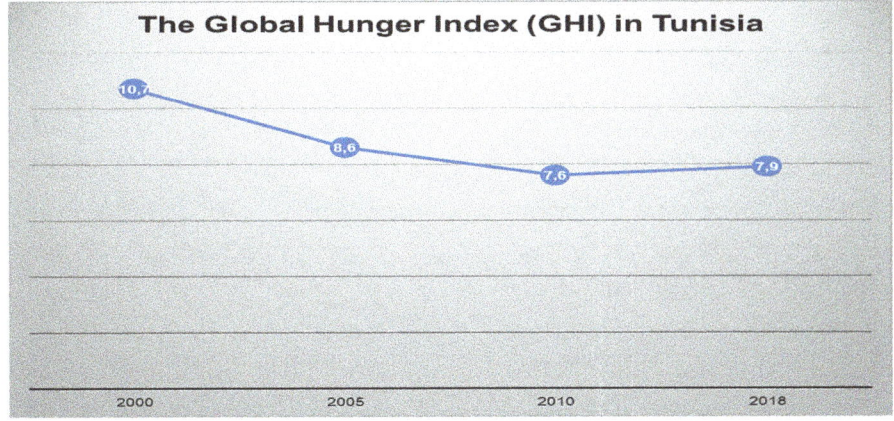

Fig. 24 The Global Hunger Index (GHI) in Tunisia during the period 2000–2018. *Source* Adapted from Patterson et al. (2019)

a low 'level of hunger' (Fig. 24). Tunisia was then ranked 28th out of 119 countries in 2018 (Patterson et al., 2019).

The average protein supply in Tunisia shows a decline from 2000s to 2013 reaching 90 g/capita/day in 2013 (Fig. 25).

The prevalence of undernourishment has been on a downward trend since 2003, but resumed on upward trend in 2015, reaching around 4.9% (0.6 million people) in 2017, similar situation to year 2000 (Fig. 26).

The undernourishment situation is closely linked to the capacity of the most vulnerable population (small-scale farmers, urban poor, etc.) to find a way to feed

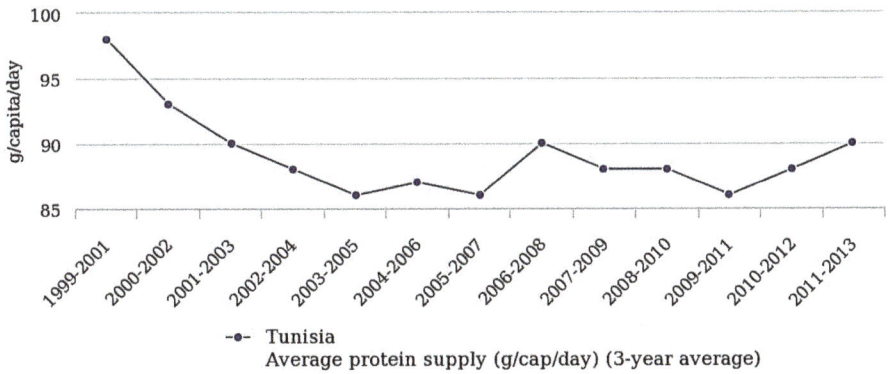

Fig. 25 The average protein supply in Tunisia. *Source* FAOSTAT (June 08, 2019)

---

values from 20.0 to 34.9 indicate serious hunger; values from 35.0 to 49.9 are alarming; and values of 50.0 or more are extremely alarming.

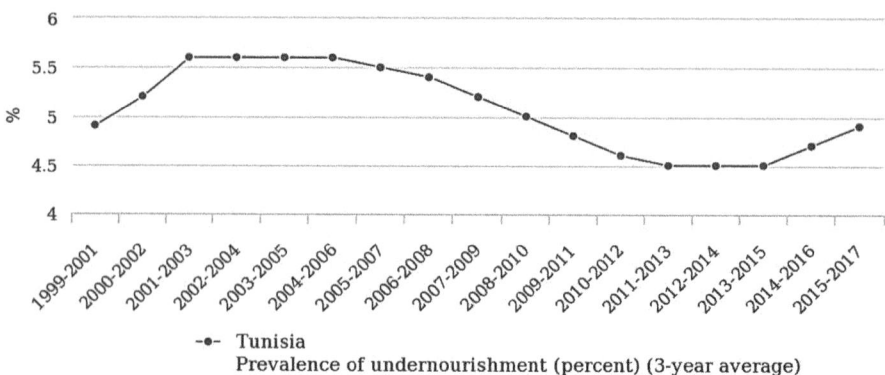

-•- Tunisia
Prevalence of undernourishment (percent) (3-year average)

Fig. 26 The prevalence of undernourishment (%) in Tunisia. *Source* FAOSTAT (June 08, 2019)

themselves and to face the substantial decrease in yield productivity, and food availability under climate change.

Food dependency of vulnerable people in Tunisia is increasing due to the impacts of climate change on low yields and the greater exposure of the country to the increasing world market volatility (Alvarez-Coque, 2012). In addition, a national study conducted by the Ministry of Agriculture (MARH-GTZ, 2007) indicated that climate change in Tunisia is expected to have serious impacts on socio-economic and environmental conditions. According to the simulations of the climate and economic forcing scenarios[5] adopted by the Tunisian climate change adaptation strategy, the assessment of the impacts of climate change on the agricultural sector and agrosystems yielded the following results:

- The assessment of the impacts of climate changes on the agricultural sector in this study was carried out using a sectoral model of Tunisian agriculture. Agricultural GDP would decrease by 5.1% in 2030 as a result of areas and yield decreases. This decrease would be 9.6% in 2030 according to a second scenario.
- Projections of the climate change impacts on farmers' incomes show a general decrease of up to 7.2%. The types of farms that would be most affected are small farms in general. However, those located in the South and Dorsal regions seem to resist better. The most vulnerable production systems to climate change, in terms of their income reduction (ranging from 10 to 20%), are the small farms in the north-west region, the episodic cereal farming systems and sheep farming on rangelands in the Center, the extensive crop systems of the Center, the intensive crop systems of the Center, and the mixed systems of the Center (intensive cattle).

---

[5] Two main types of forcing have been adopted: an economic forcing reflected in the opening of the Tunisian economy to international markets (two opening scenarios: slow and fast); and climate forcing linked to the pressures exerted by climate change on the agricultural sector (two climate scenarios: drought and good rainfall).

The simulation of the effects of climate forcing (Drought scenario) gave the following results:

- Both in the case of the two economic forcing scenarios (slow and rapid opening), the areas of cereals will show a decrease of 16% (from 1,229,000 ha to around 1,027,000) in 2030. Rainfed cereal production will decrease by 44% in 2030. It will drop from 1,971,720 to 1,095,610 tons. Olive production would drop from 856,710 tons to 411,220 tons in 2030, a decrease of 52%. The production of meat would ultimately fall by 66%, depending on the animal species.
- In the case of scenario 1 (slow opening), agricultural GDP will experience a reduction from TND 3294–2547 million in 2030.
- In the case of scenario 2, agricultural GDP would decrease by around 22.5% in 2030 (from TND 3161 to 2567 million).
- In the case of a rapid opening-up scenario, the effect of drought would be accompanied by a drop in cereal production in rainfed conditions by 44% in 2030. Olive oil production in dry conditions would fall by 52% in 2030. Animal production would decrease by 36–49%for ovine and caprine species, respectively.

4 Mitigation and Adaptation Strategies

4.1 Water Resources and Cropping

As stressed earlier, water availability is already low and expected to be worsened by climate change while demands are exponentially increasing as a result of population growth and improvement of living standards. The Tunisian government have been particularly engaged during the last three decades in setting up and implementing appropriate strategies to mitigate and adapt to climate change. The main considered adaptation actions can be listed as follows (MARH-GIZ, 2007):

- Taking into account the ecosystem components in order to maintain the water cycle (groundwater recharge, harness rainfall, soil conservation, infiltration) and good water quality;
- Continuing review of water pricing from a climate change perspective while preserving ecosystem services;
- Improving control of water demand and strengthening the national water saving program (especially in the irrigated sector);
- Increasing the water storage capacity (dams, etc.) and mobilization of all available resources (waste water, desalination, water harvesting, etc.);
- Reducing energy consumption during water production and expanding use of renewable energies;
- Upgrading the water resources monitoring networks (rain gages, piezometers, floods, etc.);
- Developing related research programs; and

- Addressing legal and institutional aspects.

Regarding cropping systems, the below few strategic orientations have been considered (MARH-GTZ, 2007; Radhouane, 2018c):

- Breeding and selection of more water stress and salt tolerant varieties;
- Preserving biodiversity; and
- Encouraging agro-ecology based cropping practices.

4.2 Agricultural Sector and Food Security

The 'Stern Review' report highlighted the opportunity (favorable benefit/cost ratio) of action to mitigate climate change effects, compared to non-action, specifying that the overall economic losses in the very long term could be very high (5–20 points of GDP against 2–9 points according to the IPCC) (Maerh et al., 2003). In order to reduce the risk of food insecurity in Tunisia, the national strategy for adapting to climate change (NSCC) in Tunisia should implement actions to cope with the negative impacts of climate change. The NSCC has three main strategic objectives (Troudi, 2013):

- Short-term social and economic development (in particular for social and spatial rebalancing) integrating a certain number of safeguards in view of medium-term ecological development (resource management, etc.);
- A reduction target of around 60% in carbon intensity by 2030 compared to 2009 and a proactive policy by 2050 allowing to achieve stabilization of carbon emissions; and
- A proactive and preventive adaptation policy supported by international aid as part of the financing and technology transfer mechanisms set up by the global climate governance (green funds, etc.).

Addressing the above challenges should also make a significant contribution to achieving the objectives of the national food security strategy. The FAO encourages countries to incorporate the objectives of food security into national poverty reduction strategies targeting impacts at the national, sub-national, household and individual levels. They suggest a particular emphasis on reducing hunger and extreme poverty (FAO, 2006). Bengoumi et al. (2018) adds some actions to mitigate food insecurity:

- Effective and sustainable management of natural resources given their scarcity, fragility, and increasing pressure on them in the coming years;
- Control of price inflation; and
- Rationalization of consumption (reducing quantity and increasing quality).

Some policy-related recommendations are cited by Patterson et al. (2019):

- Prioritize resilience and adaptation among the most vulnerable groups and regions;
- Better prepare for and respond to disasters;

- Transform food systems and address inequalities: governments and donors must significantly increase investments in rural development, social protection, health services, and education;
- Take action to mitigate climate change without compromising food and nutrition security: countries must harmonize climate policy with food and trade policies to consolidate mitigation and CO_2 removal measures; and
- Commit to fair financing: governments must increase their financial support to the most vulnerable people and regions.

In Tunisia, the prioritized actions which should be taken to modernize agriculture and achieve food security are as follows:

- Continue to develop and support the strategic sectors which contribute the most to achieving the objectives of self-sufficiency and food security such as: cereal production, olive oil, dates, citrus fruits, dairy farming, red meat, poultry, and fishing;
- Implement adaptive actions to climate change within the main agricultural and cropping systems (adapted varieties and species, resistance to climate stress);
- Promote the integration of agriculture into the agri-food industry in a value chain approach targeted on food security;
- Develop agricultural production infrastructure in a territorial and rural development approach aimed at improving the social well-being and livelihood of rural populations;
- Continue to promote the export of agricultural products to competitive and fair markets in order to reduce the dependency on dominant traditional markets (the European Union for example) and to achieve a better balance of food and commercial balance;
- Reduce dependency on foreign markets for traditional imported products (vegetable oil, sugar, etc.) through the promotion of national productions; and
- Promote the empowerment of peasant agriculture and small agricultural holdings, which hare more vulnerable to climate change and market access.

5 Conclusions and Future Directions

Tunisia is a typical Mediterranean country which will be faced by various adverse impacts of climate change. Agriculture remains vital for the economy of the country and its development has significant implications for domestic food security and poverty reduction. In this regard, developing adaptation mechanisms to deal with climate change is considered a high priority action of any government in power. Various strategies have been already developed but they need to be fully enforced on the ground.

Tunisia is facing two main challenges which interact as a nexus of food security-climate change adaptation in a post-revolution transition context, characterized by profound changes at the institutional, political, and socio-economic levels. Indeed, faced with the negative impacts of climate change resulting in increasingly scarce natural resources and strong social pressures driven by a high rate of unemployment and increase in poverty, the country must promote priority actions of its national food security strategy. The following directions may be relevant to boost such as process in the future (Guesmi, 2018):

- Develop national production and adopt economic viability and differential advantages in the framework of an approach aimed at developing natural resources and maintaining their sustainability;
- Develop and rationalize the exploitation of natural resources to meet the increasing needs of water by continuing the implementation of new facilities to securing future needs of water resources in terms of quantity and quality;
- Develop the contribution of the irrigation sector to total production by intensification and expanding the area of irrigated areas, while respecting the limits of available water resources and meeting the imperatives of sustainability;
- Continue to increase the forest and pastoral system, and to push for the implementation of water and soil conservation programs, the protection of water installations, and the strengthening of desertification control programs;
- Continue to implement sectoral strategies for the various production sectors in order to further develop basic products of cereals, oils, meats, dairy products, vegetables, and sea products with a view to ensuring sustainable food security for the country;
- Provide advanced and efficient agricultural support services based on a research system capable of developing and testing technologies consistent with the reality and specifics of the sector;
- Create the conditions for achieving a better income for the farmer that is less vulnerable to fluctuations, relying on the continuation of an encouraging price policy at the farm level for basic products such as grains, olive oil, and dairy, and further strengthening the economic policy adjustment;
- develop and diversify production and promote additional conditions to increase livelihoods and farmer income and enhance food security; and
- Organize farmers in the framework of professional structures to make them more effective and able to defend their interests and to improve their framing and surrounding conditions.

Other future directions recommended by international organizations such as the FAO, could also be investigated: (i) managing agricultural transformations that enables sustainable livelihoods and provides social security; (ii) encouraging climate-smart agriculture, resistant to drought, to saline soils, and to crop pests and diseases, that can help small-scale farmers and enhance their capacity to adapt; (iii) managing livelihood transitions through supporting poverty, hunger, and food security aims; (iv) enforcing agricultural policies mainly in the areas of natural resource and land tenure policies in order to reduce the overexploitation of natural resources and land degradation (Lewis et al. 2018); and (v) developing local production systems which have sufficient capacity to adapt to climate change and different stressors (Alvarez-Coque, 2012).

References

AfDB. (2012). Les solutions pour le changement climatique, la réponse de la Banque africaine de développement aux impacts en Afrique. 48 pages.

Agoumi, A. (2003). Vulnérabilité des pays du Maghreb face aux changements climatiques. Besoin réel et urgent d'une stratégie d'adaptation et de moyens pour sa mise en oeuvre. IISD/Climate Change Knowledge Network.

Alvarez-Coque, J. M. G. (2012). Agriculture in North Africa: A chance for development. Policy Brief. The German Marshall Fund of the United States. Mediterranean Policy Program-Series on the Region and the Economic Crisis, Washington, Accessed at: https://www.gmfus.org/publicati ons/agriculture-north-africa-chance-development

Belghrissi, H. (2018). Etude des tendances et des projections climatiques en Tunisie. INM, Tunis.

Ben Rached, S., Belghrissi, H., Zammel, A., & Mairech, H. (2015). Régionalisation des changements climatiques en Tunisie. In: Amélioration de la gestion des ressources en eaux et adaptation aux changements climatiques en Tunisie. Rapport de synthèse, CRTEAN, LDAS Project, Tunis, pp. 81–107.

Ben, Z. Y., & Zouabi, O. (2016). Impacts of climate change on Tunisian olive oil output. *Climatic Change, 139*(3–4), 535–549

Bengoumi, M., Wannes, I., & Amrani, M. (2018). Food security and food price fluctuations in the global markets. FAO Office for North Africa. International Experts forum on Food Security in GDA Members, IRA Médenine, Tunisia, June 27–28, 2018.

Buchner, B., Falconer, A., Hervé-Mignucci, M., Trabacchi, Ch., & Brinkman, M. (2011). The Landscape of Climate Finance, Climate Policy Initiative (CPI) Report, 27 octobre 2011., in Banque africaine de développement, 2012. Les solutions pour le changement climatique, la réponse de la Banque africaine de développement aux impacts en Afrique. 48 pages.

Chebbi, H. E., Pellissier, J.-P., Khechimi, W., & Rolland, J.-P. (2019). Rapport de synthèse sur l'agriculture en Tunisie. [Rapport de recherche] CIHEAM-IAMM. pp. 99. hal-02137636.

Chibani, A. (2018). Climate change mitigation in Tunisia: Challenges and progress. Climate Change, Middle East, Sustainable Development.

Drine, I. (2011). Climate variability and agricultural productivity in MENA region. UNU-WIDER Working paper N.2011/96.

El-Farhati, H., Jaziri, B., Hizem, M. W., & Nouira, S. (2020). Distribution, bioclimatic niche and sympatry of two Erinaceidae in Tunisia. *African Journal of Ecology, 58*(2), 193–210.https://doi. org/10.1111/aje.12671

FAO. (2006). Food security. Policy Brief, June 2006 Issue 2. 4 pages. https://www.fao.org/filead min/templates/faoitaly/documents/pdf/pdf_Food_Security_Cocept_Note.pdf

FAO. (2008). *Climate change and food security: A framework document.* (p. 110). FAO.

FAO. (2019). FAOSTAT, https://www.fao.org/faostat/fr/#home

FAO. (2020a). https://www.fao.org/sustainability/en/

FAO. (2020b). https://www.fao.org/climate-smart-agriculture-sourcebook/concept/module-a1-int roducing-csa/chapter-a1-1/en/

Fares, A., Bayabil, H. K., Zekri, M., Mattos-Jr, D., & Awal, R. (2017). Potential climate change impacts on citrus water requirement across major producing areas in the world. *Journal of Water and Climate Change*, 576–591.

Fares, A., & Fares, S. (2012). Irrigation management system, IManSys, a user-friendly computer based water management software package. In: *The Irrigation Show and Education Conference*, Orlando, FL.

Grami D., & Ben Rejeb, J. (2015). L'impact des changements climatiques sur le rendement de la céréaliculture dans la Région du Nord-Ouest de la Tunisie (Béja). *NEW MEDIT, 4*, 36–41.

Guesmi, A. (2018). Food security in Tunisia: Challenges and priorities. International Experts forum on Food Security in GDA Members, Médenine, Institut des Régions Arides, Tunisia June 27–28, 2018 (communication in Arabic).

Hezzi, I. (2014). Caractérisation géophysique de la plateforme de Sahel, Tunisie nord-orientale et ses conséquences géodynamiques.

Iglesias, A., Mougou, R., Marta Moneo, M., & Quiroga, S. (2011). Towards adaptation of agriculture to climate change in the Mediterranean. *Regional Environmental Change, 11*(Suppl 1), S159–S166.

INS. (2020). https://www.ins.tn/fr/themes/commerce-ext%C3%A9rieur

IPCC. (2007). Impacts, adaptation, and vulnerability. M. L. Parry, O. F. Canziani, J. P. Palutikof, P. J. van der Linden, & C. D. Hanson (Eds.), *Contribution of the working group II to the fourth assessment report of the IPCC.* Cambridge University Press.

IPCC. (2019). *Climate change and land: An IPCC special report on climate change, desertification, land degradation, sustainable land management, food security, and greenhouse gas fluxes in terrestrial ecosystems.* In P. R. Shukla, J. Skea, E. Calvo Buendia, V. Masson-Delmotte, H.-O. Pörtner, D. C. Roberts, P. Zhai, R. Slade, S. Connors, R. van Diemen, M. Ferrat, E. Haughey, S. Luz, S. Neogi, M. Pathak, J. Petzold, J. Portugal Pereira, P. Vyas, E. Huntley, K. Kissick, M. Belkacemi, J. Malley (Eds.) (In press).

ITES. (2017). Revue Stratégique de la sécurité alimentaire et nutritionnelle en Tunisie. Institut Tunisien des Etudes Stratégiques – Programme Alimentaire Mondial. 245 pages.

Kamel, S., Moumni, L., & Jedoui, Y. (2014). Environmental isotopes as the indicators of the ground-water recharge in Tunisian Chotts region. *Carbonates and Evaporites, 29*(2), 221–238. https://doi.org/10.1007/s13146-013-0183-0

Khila, M., Zaabar, W., & Achouri, M. S. (2018). Diversity of terrestrial isopod species in the Chambi National Park (Kasserine, Tunisia). *African Journal of Ecology, 56*(3), 582–590. https://doi.org/10.1111/aje.12502

Lewis, P., Monem, M. A., & Impiglia, A. (2018). *Impacts of climate change on farming systems and livelihoods in the Near East and North Africa—With a special focus on small-scale family farming.* (p. 92). FAO.

Liu, Y., Heying, E., & Tanumihardjo, S. A. (2012). History, global distribution, and nutritional importance of citrus fruits. *Compr. Rev. Food Sci. Food Saf., 11*, 530–545. https://doi.org/10.1111/j.1541-4337.2012.00201.x

MAERH, MIE, ANER, CIEDE (2003). Guide d'information sur les changements climatiques. Ministère de l'Agriculture, de l'Environnement et des Ressources Hydrauliques et Ministère de l'Industrie et de l'Energie, Agence Nationale des Energies Renouvelables, Centre d'Information sur l'Energie Durable et l'Environnement. 55 pp.

MALE (Ministère des Affaires Locales et de l'Environnement), GEF-UNDP. (2017). Tunisia's Third National Communication as part of the United Nations Framework Convention on Climate Change (UNFCCC). 36 pages.

Mansour, M., & Hachicha, M. (2014). The vulnerability of Tunisian agriculture to climate change. In P. Ahmad (Ed.), *Emerging technologies and management of crop stress tolerance*, Vol. 2, pp. 449–470.

MARH-GTZ (2007). Stratégie nationale d'adaptation de l'agriculture tunisienne et des écosystèmes aux changements climatiques, COPA consultants and ExA Consult Tunisie.

MEAT (Ministère de l'Environnement et de l'Aménagement du Territoire). (1998). Programme d'action national de lutte national de lutte contre la désertification. MEAT, Tunis, 102 pp.

Mougou, R., Mansour, M., Iglesias, A., Chebbi, R. Z., & Battaglini, A. (2011). Climate change and agricultural vulnerability: a case study of rain-fed wheat in Kairouan, Central Tunisia. *Regional Environmental Change, 11*(Suppl 1): S137–S142.

Mtimet, A. (2001). Soils of Tunisia. *Options De Mèditerranéennes, Séries B, 34*, 243–268

Mutin, G. (2009). Le Monde arabe face au d´efi de l'eau. Enjeux et Conflits. GREMMO.

Nasr, Z., Almohammed, H., Gafrej, L. R., Maag, C., & King, L. (2008). Drought modelling under climate change in Tunisia during the 2020 and 2050 periods. *Option Méditerranéennes, Séries A, 80*, 365–369

ONAGRI. (2019a). Etude prospective sur la sécurité alimentaire à l'horizon 2030. Commission nationale chargée de l'élaboration d'une stratégie et étude prospective sur la sécurité alimentaire et développement des exportations en Tunisie à l'horizon 2030. Tunis, 21 pages.

ONAGRI. (2019b). La balance commerciale alimentaire en Tunisie, Décembre 2019.

Ouled Belgacem, A., Sghaier, M., Ouessar, M., Taamallah, H., & Khatteli, H. (2011). Patterns of vulnerability in the agriculture and water sector in the southern region of Tunisia. Case of olive production sector in the governorate of Médenine. Methodological approach. Publié par Deutsche Gesellschaft für Internationale. Zusammenarbeit (GIZ) GmbH Siège de la société: Bonn et Eschborn, en partenariat avec le Ministère de l'Equipement, de l'Aménagement du Territoire et du Développement Durable (Tunisie). 51 pages.

Paeth, H., Born, K., Girmes, R., Podzun, R., & Jacob, D. (2009). Regional climate change in tropical and Northern Africa due to greenhouse forcing and landuse changes. *Journal of Climate, 22*(1), 114–132

Patterson, F., Wiemers, M., Chéilleachair, R. N., Foley, C., von Grebmer, K., Bernstein, J., Fritschel, H., Gitter, S., Ekstrom, K., & Mukerji, R. (2019). Global Hunger Index, the challenge of hunger and climate change, 2019 Synopsis. Editors: Welthungerhilfe, Concern Worldwide. Bonn, Germany, 8 pages. https://reliefweb.int/sites/reliefweb.int/files/resources/Synopsis%202 019%20Global%20Hunger%20Index.pdf

Pinstrup-Andersen. (2009). Food security: Definition and measurement. *Food Security, 1*, 5–7. https://doi.org/10.1007/s12571-008-0002-y

Radhouane, L. (2013). Climate change impacts on North African countries and on some Tunisian economic sectors. *Journal of Agriculture and Environment for International Development (JAEID), 107*(1), 101–113

Radhouane, L. (2018a). Why don't adapt Tunisian agriculture to climate change? 1. Climate change and agriculture in Tunisia. *International Journal of Science, Environment and Technology, 7*(5), 1495–1508.

Radhouane, L. (2018b). Why don't adapt Tunisian agriculture to climate change? 2. How climate affects agriculture. *International Journal of Science, Environment and Technology, 7* (5), 1542–1562.

Radhouane, L. (2018c). Why don't adapt Tunisian agriculture to climate change? 3. Mitigation and adaptation. *International Journal of Science, Environment and Technology, 7*(5), 1615–1630.

Sghaier, M., & Ouessar, M. (2013). L'oliveraie tunisienne face au changement climatique : Méthode d'analyse et étude de cas pour le gouvernorat de Médenine. GIZ, Tunis, 40 pp.

Sowers, J., & Weinthal, E. (2010). Climate change adaptation in the Middle East and North Africa: Challenges and opportunities. The Dubai Initiative, Dubai, 32 pp.

Troudi, V. (2013). Stratégie Nationale sur le Changement Climatique Synthèse. GIZ, Tunis. 8 pages. https://www.environnement.gov.tn/PICC/wp-content/uploads/Strat%C3%A9gie-Nation ale-%E2%80%93-Synth%C3%A8se.pdf

UNFCCC. (1992). United nations framework convention on climate change. UN, New York, pp. 24

Verner, D., Tréguer ,D., Redwood, J., Christensen, J., Mcdonnell, R., Elbert, C., & Konishi, Y. (2018). Climate variability, drought, and drought management in Tunisia's agricultural sector. World Bank Group, 114 pp. https://doi.org/10.1596/30604

WMO. (1992). International meteorological vocabulary, 2nd edn. Publication No. 182. Available at: https://meteoterm.wmo.int/meteoterm/ns?a=T_P1.start&u=direct=yes&relog=yes#expanded

World Bank. (2019). Economic development Tunisia. (WDI World Development Indicators, 2018) Macro Poverty Outlook, and official data.

World Bank. (2020). World Development Indicators. https://databank.worldbank.org/databases.

Zlaoui, M., Dhraief, M., Jebali, O., & Benyoussef, S. (2019). Assessment of the Tunisian olive oil value chain in the international markets: Constraints and Opportunities. *FARA Research Result, 4*(2), PP36.

Mohamed Ouessar has a Doctorate in Applied Biological Sciences (Land management) from the University of Ghent in Belgium. He is currently a Senior Researcher at the Laboratory of Eremology and Combating Desertification of the Arid Zones Research Institute (*Institut des Régions Arides-IRA*), Médenine, Tunisia. He has actively contributed, as coordinator or team member, to the realization of numerous (more than 25) joint research projects funded and/or conducted by national and international agencies. His research programs focus mainly on: water harvesting, water resources mobilization and management, geospatial-based hydrological modeling, decision-support systems, impact assessment, watershed management, land degradation and rehabilitation, climate change impacts and adaptation, combating desertification, and drylands management. He has been solicited to provide consultancy/expertise for studies and/or projects conducted by national (ministries of agriculture, environment, and planning) and international organizations. He published several scientific papers in national and international journals (over 60) in addition to his contribution to the edition of books (over 10) and book chapters (over 30). He is also reviewer for several international journals (over 15).

Abderrahman Sghaier has a Ph.D. in Biological Science from the University of Gabes, Tunisia. Currently, he is a Postdoctoral Researcher working on water balance modelling of olive trees in arid regions. He has been involved in several research projects as a team member and has published original articles.

Aymen Frija has a Ph.D. in Agricultural Economics from the Ghent University, Belgium and an MSc in Agricultural Economics from SupAgro Montpellier (ENSAM), France. He is a researcher affiliated with the International Center for Agricultural Research in the Dry Areas (ICARDA), specializing in economic modelling with a focus on natural resources policies and governance. His current research interests include agricultural water management, institutional performance analysis, and conservation agriculture economics.

Mongi Sghaier has a Ph.D. and is currently Professor in agricultural and natural resources and environment economics. He is a Senior Researcher in the Arid Regions Institute (IRA) of Medenine, Laboratory of Economy and Rural Societies, Tunisia. He coordinates and takes part in several programs and international projects in the areas of natural resource management and local development. He organized many national and international scientific meetings and coordinated and took part in workshops and training programs specialized in the economy of natural resources, local development, stakeholders' engagements, and combating desertification. He published more than 70 papers in national and international journals and publish five scientific books.

Mirza Barjees Baig is a Professor at the Prince Sultan Institute for Environmental, Water and Desert Research, King Saud University, Saudi Arabia. He earned his MS degree in International Agricultural Extension in 1992 from the Utah State University, Logan, Utah, USA and was placed on the 'Roll of Honor'. He completed his Ph.D. in Extension for Natural Resource Management from the University of Idaho, USA and was honored with the '1995 outstanding graduate student award'. Dr. Baig has published extensively on the issues associated with natural resources in the national and international journals. He has also made oral presentations about agriculture and natural resources and role of extension education at various international conferences. Food waste, water management, degradation of natural resources, deteriorating environment and their relationship with society/community are his areas of interest. He has attempted to develop strategies to conserve natural resources, promote environment and develop sustainable communities. Dr. Baig started his scientific career in 1983 as a researcher at the Pakistan Agricultural Research Council, Islamabad, Pakistan. He served at the University of Guelph, Ontario, Canada as the Special Graduate Faculty from 2000–2005. He served as a Foreign Professor at the Allama Iqbal Open University (AIOU), Pakistan through the Higher Education Commission from 2005–2009. He served as a Professor of Agricultural Extension and Rural Society at the King Saud University, Saudi Arabia from 2009–2020. He serves as well on the Editorial Boards of many international journals and the member of many international professional organizations.

Chapter 4
Food Security in the MENA Region: Does Agriculture Performance Matter?

Assil El Mahmah and Amine Amar

Abstract This chapter focuses on how the agriculture sector could affect—or not—food security in the Middle East and North Africa (MENA). Empirical evidence on the macroeconomic determinants of food security, based on the FAO's four pillars, namely availability, accessibility, utilization, and stability, has been provided. In addition, a panel data model for each pillar over the period 1990–2017 has been constructed, by adopting the Generalized Method of Moments (GMM) approach to estimate the relationship between all variables. It was found that the determinants of food security in the MENA vary according to the adopted pillars. Also, food security in this region depends not only on agricultural factors, but essentially on various macroeconomic factors, such as the trade openness and international food prices, which may affect directly or indirectly the country's food security position, regardless of its dependency on agriculture or oil sector.

Keywords Food security · Agriculture sector · Macroeconomic factors · MENA region · GMM model

1 Introduction

According to the recent definition of the Food and Agriculture Organization (FAO), 'food security' is a situation that exists when all people, at all times, have physical, social and economic access to sufficient, safe and nutritious food that meets their dietary needs and food preferences for an active and healthy life. This concept is based on four dimensions, namely availability, accessibility, utilization, and stability.

Each food crisis confirmed that climate change is the greatest threat to food security in the twenty-first century, particularly in agriculture-based countries (FAO, 2003,

---

The views expressed in this chapter are those of the authors and should not be interpreted as those of the Moroccan Agency for Sustainable Energy and the National Bank of Kuwait.

A. E. Mahmah
National Bank of Kuwait, Kuwait, Kuwait

A. Amar (✉)
Moroccan Agency for Sustainable Energy (MASEN), Rabat, Morocco

2006a, b, 2008, 2013). Climate change affects food security by directly reducing food production through changes in agro-ecological conditions and, indirectly, by affecting growth and distribution of incomes.

Many studies—as described below—showed that food security depends not only on climate and agricultural factors, but on various macroeconomic factors as well, which may affect directly or indirectly the country's food security position. Some countries, with the worst climate conditions and weak agricultural sector, succeeded to sustain food security in the long-run, thanks to their economic development and trade openness strategies. Conversely, many developing countries could not reach food security goals, despite the importance of the agricultural sector in their economies.

There has been a large volume of studies (i.e., Carter et al., 2013; Maitra, 2018) on the determinants of food security in low-income countries. But, according to our literature review, very few studies (Efron et al., 2018; Keulertz, 2019) have been focused on the food security situation in the Middle East and North Africa (MENA) countries. Thus, the role of macroeconomic factors in improving food security in this region is poorly understood.

The MENA region has its particularities. On one hand, many countries, especially those around the Mediterranean Sea, are highly dependent on agriculture, which plays an important role in their economies. On the other hand, about half of the countries are net oil exporters,[1] with a weak agricultural sector. The net impact of the economic aggregates on achieving food security objectives is still ambiguous.

The main purpose of this chapter is to determine the key drivers of the food security in MENA countries, by constructing a panel data model and using annual series over the period 1990–2017. The selected variables are sourced from different government entities, under the constraint of data availability and study objective. This adopted approach takes into consideration the interconnectedness of different MENA countries and their specific characteristics, in order to obtain plausible and reliable results.

The remainder of this chapter is organized as follows: Sect. 2 provides a review of related literature; Sect. 3 gives a brief background on food security in MENA countries; Sect. 4 describes the data used and discusses the adopted methodology; Sect. 5 presents the empirical results; and finally, Sect. 6 concludes with some policy recommendations.

2   Literature Review

Although food security is largely considered as very important for a country's economic and social stability, there is no clear definition or universal indicator about

---

[1] In this chapter, the agriculture-dependent countries are Morocco, Tunisia, Egypt, Jordan, and Lebanon, while the net oil exporters are Algeria, Bahrain, Kuwait, Oman, Qatar, Saudi Arabia, and United Arab Emirates.

how it should best be assessed (Seed, 2019). Various approaches—as described below—to assessing food security have been used, but most of the empirical studies have been focused on the poor and low-income countries (Hatlebakk, 2020; Maitra, 2018). To our best knowledge, very few studies have analyzed the issue of food security in a panel of MENA countries, and especially by applying recent econometric methods for panel data (Efron et al., 2018; Keulertz, 2019). There has been little attention to these facts and this research attempts to fill this gap, taking into consideration the particularities of the region as well as its high reliance on food imports for subsistence.

Food security may be considered as the ability to have safe, sufficient, accessible, and affordable food. However, it is very difficult to apply this definition in practice. There is no clear meaning of 'ability', 'hunger', and 'undernutrition', as well as no specification of the time horizon and population. This ambiguity has led to often confuse between food security and self-sufficiency. The two are often wrongly equated in many countries. In fact, there is hardly no country in the world not reliant on other countries for at least some food items (FAO, 2013, 2016). On one hand, a country with a low share of food imports as a share of total exports can ensure much easier its macro-level food security, but this does not mean it is achieving the same for the micro-level food security in vulnerable households. On the other hand, some food-importing countries, such as Singapore, UAE, and Kuwait can be generally food secure, without being self-sufficient. However, their reliance on food imports, even though it can provide security, constitutes a source of vulnerability.

In this regard, food security is often perceived as a multidimensional and flexible concept that gained prominence since the World Food Conference in 1974 (Fig. 1). In fact, the concept of 'food security' was originated in the discussions of international food problems at a time of global crisis (mid-1970s), where the initial focus was primarily on food supply, food availability, and price stability.

For example, according to The State of Food Insecurity 2001, food security is defined as a situation that exists when all people, at all times, have physical, social and economic access to sufficient, safe and nutritious food which meets their dietary needs and food preferences for an active and healthy life. In the same context, the Committee on World Food Security identifies four main dimensions of food security (Fig. 2). The first one represents the *availability* which is ensured if adequate amounts of food are produced and are ready to meet people disposal. The second one is related to *access*, which is assured when all households and all individuals within those households have sufficient resources to obtain appropriate foods. The third one is about the *utilization,* which represents a situation when the human body is able to ingest and metabolize food. The fourth pillar represents *stability,* which is ensured when the three other pillars are maintained over time. However, the recent FAO reports (2013) have noted the need of an additional pillar related to *environmental sustainability*, which is the ability of agricultural systems to provide sufficient food for future generations without depleting natural resources.

For the aforementioned reasons, and in order to apprehend the complexity of the food security issue, a multi-dimensional approach should be utilized taking into account various perspectives such as socio-economic, political, and environmental

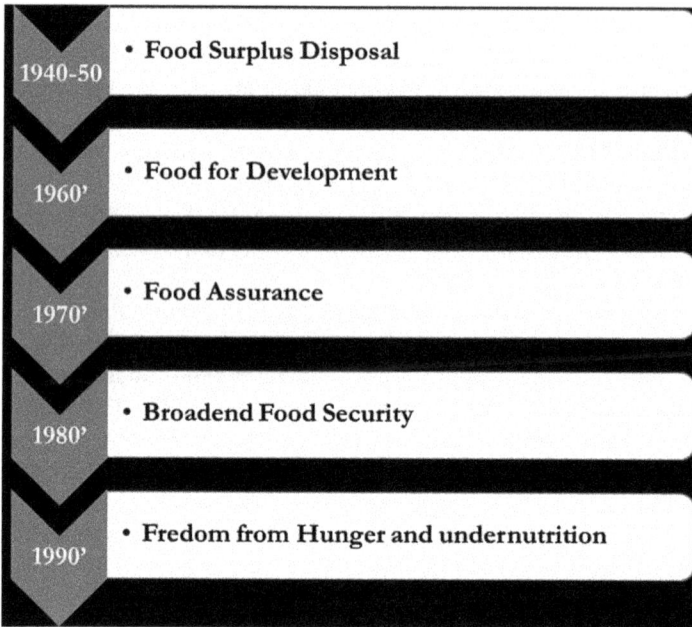

Fig. 1 The evolution of food security concerns. *Source* CFS (2014)

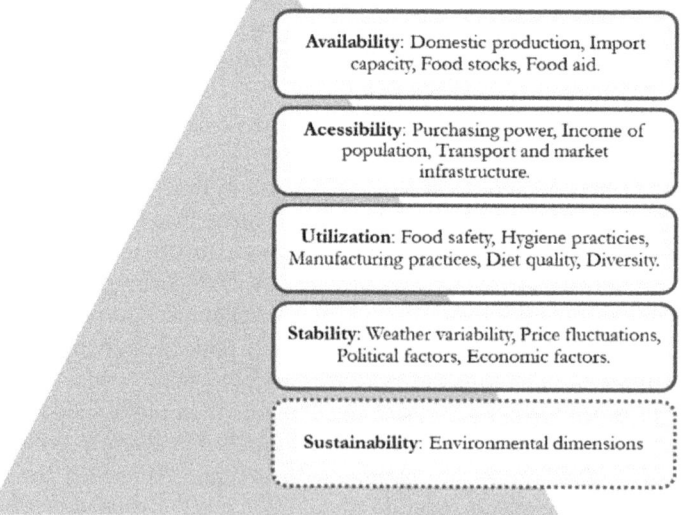

Fig. 2 Pillars of the food security. *Source* FAO (2013)

dimensions. The socio-economic lens confirms that the major cause of malnutrition, in relation to food security, is poverty. To be eradicated, poverty needs sustainable development mechanisms and a more equitable distribution of wealth, especially in the agricultural sector. Many other variables can be mentioned and which are related to the socio-economic context such as education, living conditions, food prices, and the overall conditions of health and hygiene. The political lens is important to approach food security in the sense that the lack of agreements, the complexity in relations and management of the various positions often force institutions to suspend the search and implementation of solutions and initiatives. Finally, the environmental dimension is important to be considered, in the sense that natural resources are fundamental assets to the production of food, rural development, sustainable growth, and population wellbeing. However, the pressure exerted on natural resources and the competition over their control often degenerate into conflicts, violence, and impoverishment of natural capital.

In this regard, a vast and growing literature analyses the relationships between food security and the above dimensions, which can be considered as potential determinants. Table 1 summarizes the literature related to the main determinants of food security as well as the results of the selected research. For example, Aker and Lemtouni (1999) present a framework for assessing the food security at a national level. In this work, six variables were used to capture the domestic and global supply and demand mechanism, namely domestic food production, average annual rainfall, world food prices, gross domestic product, Gini coefficient of income distribution, and exports of goods and services. The findings showed that there is inter-variables correlation between rainfall, domestic food production, income, and domestic production. Cereal production, used as a proxy of food production, was not significant to food availability. Meanwhile, income (GDP) was significant and had a positive effect.

Arshad and Abdel Hameed (2010) examined the factors leading to the increase in food commodities prices and its implications for food security in Malaysia. The identified factors include the decline in agricultural production growth, and by the extension food supply, the decline in global cereal stocks, and an increased food demand from emerging economies. As stated by the authors, the second cause is related to the systemic factors that include underinvestment in agriculture and a lopsided policy towards export crops at the expense of domestic food security. The third and fourth causes are pertained to the increase in biofuel demand and technical factors, respectively.

Morrissey et al. (2013) estimated the extent to which local food prices influence the weight outcomes, food insecurity, and food consumption patterns of children from infancy to 5 years of age. It was found that, compared to children living in areas with lower-priced fruits and vegetables, children living in areas with higher-priced fruits and vegetables averaged higher measures of standardized Body Mass Index (BMI) scores. It was also found that there is a significant association between food prices, child and family weight, and food security outcomes.

In terms of concepts, different definitions of food security can arise depending on the number of factors involved. These include the scope of the analysis, whether the examined situation is real or potential and whether the analysis is quantitative or

Table 1 Summarized literature review related to food insecurity determinants

Author (year)	Research title	Research location	Main determinants and results
Abbasi et al. (2016)	Assessment of household food insecurity through use of a USDA questionnaire	Ghana	Household food insecurity was associated with education and household income level
Ijarotimi (2013)	Determinants of childhood malnutrition and consequences in developing countries	Developing countries	Poverty, lack of nutrition knowledge, poor child-feeding practices and lack of care all-cause malnutrition
Abo et al. (2015)	Determinants of the food security status of female-headed households	Ghana	Age of household head, educational level of household head and the size of the family were predictors of household food insecurity
Aker and Lemtouni (1999)	Framework for assessing food security in the face of globalization	Morocco	Rainfall is highly negatively correlated with domestic food production, while, income (GDP) was significant and had a positive effect
Applanaidu et al. (2013)	An econometric analysis of food security and related macroeconomic variables in Malaysia	Malaysia	Biodiesel production, exchange rate and government expenditure on rural development will give the highest shock to food security in the long term
Kombi et al. (2017)	Stunting, wasting and underweight in Sub-Saharan Africa	Sub-Saharan Africa	Low education of mother-father, child's age, sex of child (male) and poverty were associated with all forms of malnutrition
Arene (2010)	Determinants of food security among households	Nigeria	Income and age of the household head were associated with food insecurity

(continued)

Table 1 (continued)

Author (year)	Research title	Research location	Main determinants and results
Tamiru et al. (2017)	Food insecurity and its association with school absenteeism	Ethiopia	Gender, household size and school absenteeism were associated with household food insecurity
Tembo et al. (2009)	The effects of market accessibility on household food security	Malawi	Households without access to the market were more food insecure, as compared to those with access
Dessalegn Tamir et al. (2016)	Household food insecurity and its association with school absenteeism	Ethiopia	Household food insecurity was associated with maternal education, school absenteeism and income
Mbwana et al. (2016)	Determinants of household dietary practices in Rural Tanzania	Tanzania	The literacy status of the mother and the distance to a water source impact household dietary diversity
Ihab (2015)	Assessment of food insecurity and nutritional status of children	Malaysia	The prevalence of underweight, stunting and wasting was high among food-insecure households. Income, household size and the number of children were associated with household food insecurity
Aidoo et al. (2013)	Determinants of household food security	Ghana	Household size, farm size, off-farm income, credit access and marital status were found to influence household food insecurity significantly
Babatunde (2010)	Impact of off-farm income on food security	Nigeria	Income was associated with household food insecurity

(continued)

Table 1 (continued)

Author (year)	Research title	Research location	Main determinants and results
Frelat (2016)	Drivers of household food availability in sub-Saharan Africa based on big data from small farms	Sub-Saharan Africa	Higher food prices and the lack of market access were associated with household food insecurity
Demissie et al. (2012)	Magnitude and factors associated with malnutrition in children 6–59	The Somali region, Ethiopia	Boys were more malnourished than girls; income and illness were associated with malnutrition
Psaki (2012)	Household food access and child malnutrition	Eight country study	Food insecurity was associated with wasting, underweight and stunting
Belachew (2011)	Food insecurity among adolescents	Ethiopia	Food insecurity was associated with illness and place of residence
UNICEF (1990)	Strategy for improved nutrition of children and women	Developing countries	Food insecurity is associated with child and maternal malnutrition
Owusu (2011)	Non-farm work and food security	Northern Ghana	Household food insecurity was associated with income
Obadiah (2014)	Determinants of household food security	Kenya	Income is associated with food insecurity
Titus (2007)	An analysis of the food security situation among Nigerian urban households	Nigeria	Household food insecurity was associated with residence

Source Compiled by the authors

qualitative. Therefore, in terms of analysis, different approaches and indicators can be utilized to apprehend the complexity of the food security concept: as example, anthropometric measures, such as those discussed in the Scientific Symposium on Measurement and Assessment of Food Deprivation and Undernutrition, held by the FAO in 2002. The vulnerability analysis (Lovendal et al., 2005), which comes from the dynamic, forward-looking characteristic, and stochastic framework. The Vector autoregression (VAR) is also a widely used approach for the management of the specific risks associated with the critical threshold level of the nutritional outcome, over a given time horizon.

Finally, we can evoke the food availability approach (Malthus, 1798; Sen, 1981; Dreze & Sen, 1989), which is focused on the balance between population and food. However, Amartya Sen's entitlement approach contributed to challenge this perspective and shifted the focus from national food availability to people's access to food. Thus, the analysis of food security through the capability approach allows a more comprehensive assessment. While the income-based approach would take income as a focal variable, the entitlement/capability approach provides information on how income is used to ultimately reach the capability to be food secure depending on personal and external conversion factors, food choices, and behaviors.

The main objective of this chapter is to show that food security is by no means a tightly-defined concept. Using available data from the MENA region, we try to confirm that food security is not only an agricultural issue, but various factors may directly or indirectly impact on the country's food security position. The adopted methodology is based on a Panel Data approach, which will be deeply detailed in Sect. 4.

3   Brief Overview of Food Security in MENA Region

A rapidly changing global environment and mounting domestic challenges are pushing many MENA countries to rethink their development models. The conventional model has been based on oil wealth, a preference of state over markets, import substitution industrialization, and often untargeted redistribution mechanism. In addition, the development challenges had been complicated by global and regional issues such as climate change and food insecurity. MENA countries become increasingly dependent on international markets to secure physical access to food. Indeed, official statistics show that the MENA region is the most food import-dependent region in the world, and net food imports are projected to rise even further in the future (Breisinger et al., 2010). This fact can be explained by demand factors such as rising population and changing consumption patterns and also by supply-side factors, which include limited natural resources.

However, even if a higher ratio of total exports to food imports may indicate a higher level of food security in some countries, macro-level food security is not equal to food self-sufficiency, which is a fact of particular relevance for the MENA region. Thus, the proportion of undernourished in the MENA region may now even be higher because of the recent food crisis and ongoing global recession (Ivanic & Martin, 2008). Indeed, the MENA region has witnessed an increase in the prevalence of undernourishment from 1990 to 2016 (Fig. 3), mainly as a result of conflicts and crises. Similarly, the prevalence of severe food insecurity in the MENA adult population (15 years and older) reached 9.2% in 2016–2017 (Fig. 4), which is above the world average (7.4%), but below the average for all developing countries (11.9%). Nonetheless, according to the FAO overview on the Near East and North Africa Regional 2016, levels and trends of food insecurity differ widely from one country to another, as well as across its sub-regions. The lowest prevalence of severe food

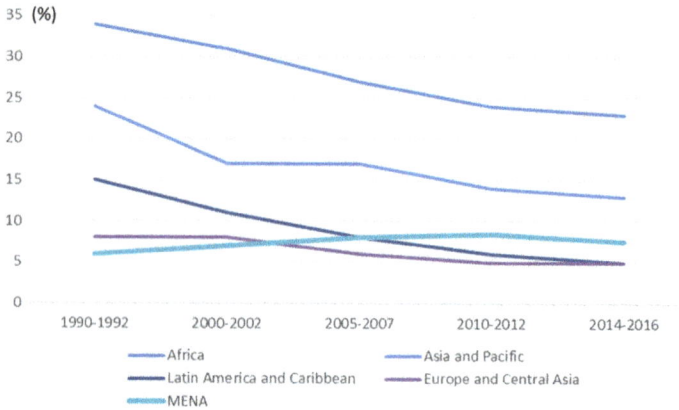

Fig. 3 Prevalence of undernourishment. *Source* FAOSTAT

Fig. 4 Prevalence of severe food insecurity, 2016–17. *Source* FAOSTAT

insecurity in MENA was estimated for the Maghreb at 4.6%, while the Golf Council Countries (GCC) reported a prevalence of 7.2%. Concerning the Mashreq, the most populous sub-region in MENA, it has witnessed a dramatic increase in the prevalence of undernourishment and recorded a prevalence of 11.7% during the period 2016–2017, affected especially by protracted conflicts.

These considerable differences reflect significant dissimilarities in terms of economic structure and adopted trade policies. In fact, many countries, especially those around the Mediterranean Sea (except Algeria), are highly dependent on agriculture, which plays an important role in their economies, while about half of the countries are net oil exporters, with a weak agricultural sector. However, even if some MENA countries have a relatively developed agricultural sector, agriculture is not enough to ensure the macro-level food security of this region. In fact, on one hand, the performance of agriculture, measured by agricultural value added in the percentage of GDP, is highly correlated with the average value of food production

(Fig. 5), one of the important indicators of food availability pillar, and explain well the food production position across the MENA countries.

But, on the other hand, it seems that the performance of agriculture is not able to influence the other food security pillars. In fact, Figs. 6, 7 and 8 shows that there is no link between the agricultural value added and the most important indicators of the three other pillars. This means that other factors may affect the country's food security position, regardless of its dependency on agriculture.

Finally, MENA countries are diverse, and thus policy options on how to improve food security may differ. To have an idea on diversity related to food security, we can extend the commonly used typology for MENA countries based on mineral resource wealth, by using several food security related indicators, to capture both the macroeconomic and household-level dimensions of food security. Despite the diversities, we can identify some general features when analyzing the MENA region. Similarities concern growth stories and structural change, the role of agriculture and the rural economy, poverty reduction and job creation, vulnerabilities, and climate change impacts.

Fig. 5 Average value of food production, 2017. *Source* FAOSTAT (2017)

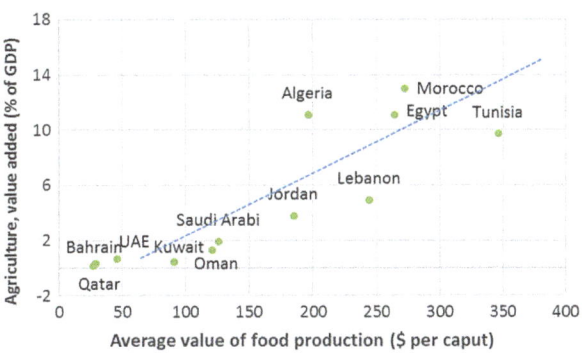

Fig. 6 Domestic food price volatility, 2017. *Source* FAOSTAT (2017)

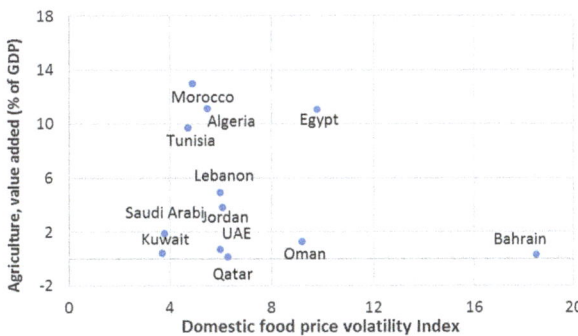

Fig. 7 Depth of the food
deficit, 2017. *Source*
FAOSTAT

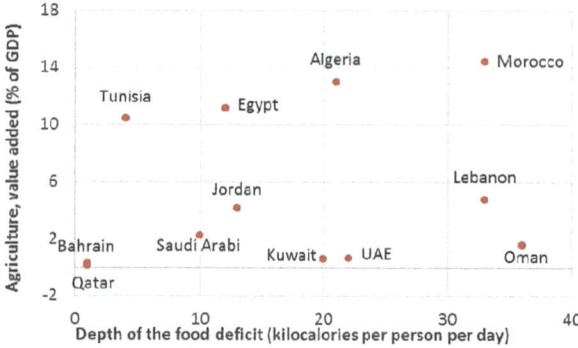

Fig. 8 Prevalence of anemia
among children, 2017.
Source FAOSTAT

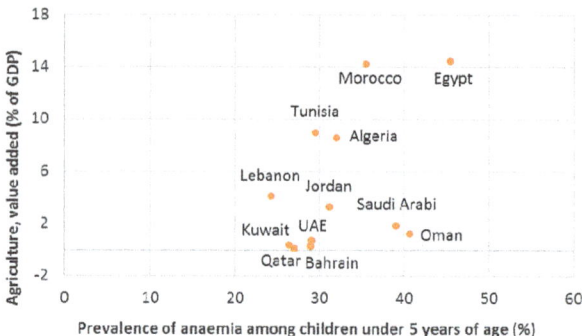

4   Methodology

This section presents our adopted approach to analyze the effects of macroeconomic
factors on food security. We describe the selected variables and the data sources,
taking into consideration the data limitation and the intended objectives. We also
specify the appropriate econometric model, which tackles all statistical problems and
reflects the interconnectedness of the different MENA countries and their specific
characteristics.

4.1   Data Description

Given that it is hard to find a single or a global indicator that takes all dimensions
of food security into account, we consider four indicators as dependent variables
in our models. Each indicator reflects a pillar of food security. The choice of the
four variables was mainly guided by their reliability and availability for the selected
countries during the covered period. The selected indicators are sourced from the FAO
database over the period 1990–2017. Table 2 describes all the dependent variables
used in this chapter.

Table 2 Definitions and data sources of the dependent variables

Pillar	Indicator	Code	Brief definition	Source
Food availability	Average value of food production	**value_food**	This indicator represents the pillar of food availability. It expresses the food net production value (in constant 2004–06 international dollars), as estimated by FAO and published by FAOSTAT, in per capita terms	Food and Agriculture Organization (FAO)
Food accessibility	Depth of the food deficit (kilocalories per person per day)	**food_deficit**	The depth of the food deficit represents the pillar of food accessibility. It indicates how many calories would be needed to lift the undernourished from their status, everything else being constant	FAO
Food utilization	Domestic food price volatility	**food_price_volatility**	The domestic food price volatility index represents the pillar of food utilization. It measures the variability in the relative price of food in a country. The indicator is calculated from the monthly domestic food price level index using monthly consumer and general food price indices and purchasing power parity data from the International Comparison Program conducted by the World Bank	FAO/World Bank

(continued)

Table 2 (continued)

Pillar	Indicator	Code	Brief definition	Source
Food systems stability	Prevalence of anemia among children under 5 years of age	**anemia_5y**	The Prevalence of anemia, children under age 5, represents the pillar of food systems stability. It is the percentage of children under age 5 whose hemoglobin level is less than 110 g per liter at sea level	FAO/World Bank

Regarding the macroeconomic determinants, we try to explain the selected dependent variables by using a few numbers of key factors, according to the existing literature and taking into account the specificities of the MENA region, in line with a fairly parsimonious specification. After several statistical tests to verify the validity of the economic theory, we selected five variables, which could explain the four dimensions of food security in the region, namely: GDP per capita, agriculture value added, trade, inflation, and food prices. Table 3 presents definitions and sources of all variables used in our empirical analysis. Except for the agriculture value-added and trade that are measured in terms of their ratio to nominal GDP, the other variables are expressed in growth terms.

According to the literature, a number of other macroeconomic variables could have been relevant for this study. For example, the unemployment rate, household incomes, industrial production, and other demographic and political indicators are also considered as the key factors in one of the four equations of our adopted methodology, but they are not available or not reliable for our purposes. Fortunately, ignoring these variables might not be problematic, since our adopted approach describes well the important channels through which main macroeconomic drivers can affect food security in the MENA region.

At the econometric analysis level, we construct four models (Table 4) using the same factors to examine the determinants of food security in the MENA region, according to the four described pillars. The purpose of using the same explanatory variables for the four adopted model is to understand how the selected determinants affect each pillar of the food security in the MENA region, and which pillar is more influenced by the macroeconomic factors.

For estimation purposes, we reduce the number of countries to ten in models 2 and 3, under data constraint environments, in order to obtain reliable and adequate results. Finally, stationarity of the variables was tested using the Augmented Dickey-Fuller test (ADF), which indicates that all the series selected in this model are stationary.

Table 3 Definitions and data sources of the independent variables

Indicator	Code	Brief definition	Source
GDP per capita, PPP (constant 2011 international $)	**gdp_capita**	GDP per capita based on purchasing power parity (PPP). PPP GDP is gross domestic product converted to international dollars using purchasing power parity rates	World Bank/IMF
Trade (% of GDP)	**trade_gdp**	Trade is the sum of exports and imports of goods and services measured as a share of the gross domestic product	World Bank
Agriculture, value added (% of GDP)	**Agr_va_gdp**	It measures the Contribution of Agriculture sector in the economy	World Bank
Annual consumer prices Index	**CPI**	A consumer price index (CPI) measures changes in the prices of goods and services that households consume	World Bank/IMF
Food price index, 2000 = 100, constant 2000 $	**food_price**	Food price index includes fats and oils, grains and other food items	World Bank
Oil Exporter	**oil_Exporter**	Dummy variable = 1 if it is an oil producer country and = 0 elsewhere	Author's estimation

4.2 Model Specification

In this subsection, we examine the determinants of food security in the MENA region, by conducting econometric methods for panel data analysis, as well as some diagnostic tests to make sure that the underlying assumptions for a good model are fulfilled. Thus, our dynamic model specified in the equation below is characterized by the presence of a lagged dependent variable among the other explanatory variables. This empirical model could be expressed as follows:

$$Y_{it} = \alpha + \lambda Y_{it-1} + \beta X_{it} + \tau_i + \varepsilon_{it}$$

where Y_{it} is the dependent variable of the country i for the period t, which could represent one of the four selected food security indicators described above. Moreover, X_{it} is the vector of explanatory variables, capturing macroeconomic characteristics of the countries, such as the economic performance, the agriculture sector contribution,

Table 4 Adopted approach

	Model 1	Model 2	Model 3	Model 4
Pillars	Availability	Access	Stability	Utilization
Dependent variable	Average value of food production	Depth of the food deficit	Domestic food price volatility index	Prevalence of anemia among children under 5 years of age
Independent variables	GDP per capita Trade, % of GDP Agriculture, VA Food price index Oil exporter	GDP per capita Trade, % of GDP Agriculture, VA Food price index Oil exporter	GDP per capita Trade, % of GDP Agriculture, VA Food price index Oil exporter	GDP per capita Trade, % of GDP Agriculture, VA Food price index Oil exporter
Number of countries	12	10	10	12
Selected countries	Algeria Bahrain Egypt Jordan Kuwait Lebanon Morocco Oman Qatar Saudi Arabia Tunisia UAE	Algeria Egypt Jordan Kuwait Lebanon Morocco Oman Saudi Arabia Tunisia UAE	Algeria Bahrain Egypt Jordan Kuwait Morocco Oman Qatar Saudi Arabia Tunisia	Algeria Bahrain Egypt Jordan Kuwait Lebanon Morocco Oman Qatar Saudi Arabia Tunisia UAE
Time coverage	1990–2016	1990–2017	2000–2017	1990–2016
Number of observation	324	280	180	324

the trade openness, and the international food prices. λ and β are the coefficients, while τ_i and ε_{it} are the unobserved country-specific fixed effect and error terms.

However, since the variables may be endogenous, the estimations of this equation by the ordinary least squares (OLS) could generate biased and inconsistent estimator. To tackle the heteroscedasticity and autocorrelation problems as well as the endogeneity problem of the lagged dependent variable, we use the Generalized Method of Moments (GMM) estimation technique, which employs orthogonality moment conditions to obtain valid instruments. Thus, we estimate our model using the System GMM estimator proposed by Arellano and Bond (1991)[2] as well as Blundell and Bond (1998) which combines, within a system, the regression in levels and the regression in differences. For the regression in levels, the instruments used are the lagged differences of the endogenous and exogenous variables. The instruments for the regression in differences are lagged levels of the endogenous and exogenous variables previous or equal to (t − 2). Thus, in order to eliminate the country-specific

---

[2] For more details, see Arellano and Bond (1991), Bond (2002), Baltagi (2001, pp. 131–135) and Blundell and Bond (1998).

effect that might cause the biases of estimators, we estimate first-differences of our equation.

$$\Delta Y_{it} = \lambda \Delta Y_{it-1} + \beta \Delta X_{it} + \Delta \varepsilon_{it}$$

It should be noted that the validity of the System GMM estimator depends on two key assumptions. The error terms are not serially correlated and the instruments used in the regression in levels and in differences are valid. In order to test both hypotheses, we run two specification tests proposed by Arellano and Bond (1991) and Arellano and Bover (1995). The first test examines the null hypothesis that the differenced error term $\Delta \varepsilon_{it}$ has no second order serial autocorrelation, which means $E(\Delta \varepsilon_{it} \Delta \varepsilon_{it-2}) = 0$. The non-rejection of the null hypothesis provides support to our model estimations. The second is Hansen test of over-identifying restrictions, which tests the overall validity of the instruments by analyzing the sample analog of the moment conditions used in the estimation procedure. The hypothesis tested is that the instrumental variables are uncorrelated to some set of residuals, and therefore they are acceptable instruments. Thus, our model specification is valid if we cannot reject the null hypothesis of over-identifying restrictions.

Blundell and Bond (1998) show that the standard errors of the two-step System GMM estimator are biased downward in finite samples. To overcome this problem, we employ a lower number of instruments than the number of sample countries, in order to mitigate the over-fitting problem of the endogenous variable and improve the efficiency of the two-step estimator.[3] Finally, failure to reject the null hypotheses of both tests gives support to our estimation procedure.

5   Model Estimation and Obtained Results

After presenting our theoretical approach, we started this section by estimating the model for each of the four dimensions of food security over the period 1990–2017, under data availability constraints. Each equation of the four adopted models is expressed by the same four explanatory variables described above. This is to understand how the complex concept of food security is dependent on some macroeconomic factors such as the standard of economic development, the trade openness and the international food prices. The estimated results allow us to examine the determinants of food security in the MENA region and to confirm the importance of some economic policies in the country's food security position, regardless of its dependency on agriculture.

---

[3] For more details, see Beck and Levine (2004) and Roodman (2009).

5.1 Model Estimation Results

Table 5 shows the results of the regression estimated using the GMM procedure for our sample of 12 MENA countries over the period 1990–2017. Each equation of the four adopted models is expressed by the same four explanatory variables. The obtained results are generally satisfactory and in line with the hypothesis of the study.

As discussed above, we rely on the Hansen Test for the overall validity of our instruments, as well as on the Arellano and Bond Test (1991) for the presence of second-order autocorrelation in the differenced residuals. These results reveal that for the four adopted models, the test of Hansen cannot reject, at the 1% level, the null hypothesis of the overall validity of the used instruments. Furthermore, the Arellano and Bond Test cannot reject, at the 1% level, the null hypothesis of absence of autocorrelation of the second order in the residuals. Therefore, these results provide support for our use of dynamic panel models to assess the described determinants of food security in the MENA region.

Initially, all the specifications have been run with inflation, but given its high correlation with the food prices, this variable was dropped from our model to avoid any autocorrelation problem or statistically insignificant estimates. As shown in Table 6 (column 2) results, the food net production value depends mainly on the standard of economic development, as measured by GDP per capita, and the agriculture sector contribution. This is not surprising since the countries with strong agricultural sectors are characterized by significant food production. Moreover, the estimation results of Table 6 (column 3) indicate a negative relationship between the depth of the food deficit and the agriculture sector performance as well as the ratio of trade to GDP. This suggests that the increase in the agriculture contribution and improving the trade openness policy will reduce the food deficit and, therefore, raise the access to food in the MENA region. Table 6 (column 4) shows that the international price volatility has a significant impact on domestic price volatility, while the economic performance, the agriculture sector's contribution, and the trade openness don't affect the food prices in the domestic market. Finally, according to Table 6 (column 5), the prevalence of anemia among children under 5 years of age is affected positively by the trade openness, and negatively by the standard of economic development and the agriculture sector contribution.

In other words, except the first pillar related to the availability of food, which depends on the agricultural sector's performance, the three other pillars depend on other macroeconomic factors, such as the standard of economic development, the trade openness, and the international food prices.

Table 5 Estimation results of adopted models

Variables	Equation (1): VALUE_FOOD	Equation (2): FOOD_DEFICIT	Equation (3): FOOD_PRICE_VOLATILITY	Equation (4): ANEMIA_5Y
GDP_CAPITA	**0.856*****	−1.025	−1.158	**−1.105*****
	(0.000)	(0.175)	(0.405)	**(0.000)**
Log (AGR_VA_GDP)	**0.468*****	**−0.529***	0.699	**−0.183****
	(0.000)	**(0.087)**	(0.635)	**(0.041)**
Log (TRADE_GDP)	0.031	**−0.784*****	1.716	**0.372*****
	(0.522)	**(0.014)**	(0.372)	**(0.000)**
FOOD_PRICE	0.072	0.128	**1.127***	−0.045
	(0.447)	(0.533)	**(0.064)**	(0.308)
Hansen test p-values	0.525	0.654	0.818	0.465
Second-order Serial correlation p-values	0.691	0.367	0.318	0.148

Numbers in parentheses are P-value
***significant at 1% error level, **significant at 5% error level, *significant at 10% error level

5.2 Robustness Checks

Given that seven of the twelve countries selected in our panel data are net oil exporters,[4] our results could reflect mainly the food security in the oil-rich economies. In fact, since many of these countries have a weak agricultural sector, characterized by low productivity and limited resources, especially water, they are often highly dependent on food imports. For this reason, in order to investigate the robustness of our results, we include in our models the dummy variable 'Oil_Exporter' that takes the value 1 to indicate that the country is a net oil exporter and 0 otherwise. Thus, we estimate our regressions using GMM procedure and compare both results, in order to examine if this new variable will be significant in explaining the food security in the MENA region, and if it will influence the first obtained results.

Thereby, Table 6 reveals that the dummy variable is not statistically significant at the 10% level in the four adopted models, which means that the food security issue is not necessarily related to the oil-rich economies. This shows that macroeconomic factors, such as the trade openness and the international food prices, play a major role in achieving and maintaining food security in the MENA region, even in those countries highly dependent on agriculture.

Moreover, concerning the other independent variables, the results reported in Table 6 are mostly similar to those presented in Table 5. The only difference is that the coefficient of international food price, described in Eq. (3), is not any more significant in explaining the domestic food price volatility, while the trade openness became statistically significant at the 10% level. This could be accepted since the domestic prices in a country with an open trading regime are more sensitive to international food prices.

These findings clearly confirmed our assumption that food security depends not only on agricultural factors, but on various macroeconomic factors as well, which may affect directly or indirectly the country's food security position, regardless of its dependency on agriculture or oil sector.

6 Conclusion and Policy Recommendations

Food security is still a growing challenge for the MENA region, despite the continued efforts in developing its agricultural production and in improving access to food. In this context, this chapter has tried to show the impact of certain determinants of food security in this region, by constructing a panel data model and using annual series over the period 1990–2017. This adopted approach takes into consideration the interconnectedness of the different MENA countries and their specific characteristics, in order to obtain plausible and reliable results. As results, we found that the determinants of food security in this region vary according to the adopted pillars. Each

---

[4] The net oil exporters in our panel data are Algeria, Bahrain, Kuwait, Oman, Qatar, Saudi Arabia and United Arab Emirates.

Table 6 Estimation results of the adopted Models, including an oil exportation effect

Variables	Explained variable: VALUE_FOOD	Explained variable: FOOD_DEFICIT	Explained variable: FOOD_PRICE_VOLATILITY	Explained variable: ANEMIA_5Y
GDP_CAPITA	**0.852***** **(0.000)**	−1.100 (0.603)	0.355 (0.754)	**−1.075***** **(0.000)**
Log(AGR_VA_GDP)	**0.529***** **(0.000)**	−0.778 (0.266)	0.379 (0.463)	**−0.341***** **(0.057)**
Log(TRADE_GDP)	−0.211 (0.348)	**−0.654**** **(0.049)**	**0.794*** **(0.096)**	**0.251***** **(0.010)**
FOOD_PRICE	0.056 (0.399)	0.322 (0.302)	0.271 (0.472)	−0.027 (0.592)
OIL_EXPORTER	0.200 (0.657)	0.659 (0.640)	0.240 (0.242)	−0.579 (0.368)
Hansen Test p-values	0.559	0.710	0.819	0.369
Second-order Serial correlation p-values	0.493	0.953	0.313	0.151

Numbers in parentheses are t-statistics
***significant at 1% error level, **significant at 5% error level, *significant at 10% error level

pillar is affected differently by some macroeconomic factors. In fact, the agricultural sector plays an important role in securing food availability, food utilization, and food stability, while it is not significant in explaining food accessibility, which depends mainly on trade openness and international food prices.

These findings clearly confirmed our assumption that food security depends not only on agricultural factors, but on various macroeconomic factors as well, which may affect directly or indirectly the country's food security position, regardless of its dependency on agriculture or oil sector.

Based on the results of this study, several recommendations can be implemented to increase the food security of MENA countries. First, these countries should reduce their dependency on imported food, in order to decrease their exposure to volatility and uncertainty in the international market. Second, they should build on the vast accumulated experience to accelerate the implementation of current strategies and to support the long-term transition to a more efficient and sustainable water use and agriculture sector. Third, they need to increase regional economic integration and establishing an intra-MENA food security policy to provide emergency food or water supplies that could reduce the potential risk of supply disruptions or during shocks. Finally, these should continue to promote inclusive economic growth and diversify their economic structure, with the capacity to generate productive employment and reduce poverty. It is the most important measure for safeguarding food accessibility for vulnerable people, whether that food is imported or comes from domestic production.

References

Abbasi, N., et al. (2016). Assessment of household food insecurity through use of a USDA questionnaire. *Plants and Agriculture Research, 4*(5), 379–386

Abo, T., et al., (2015). Determinants of the food security status of female-headed households.

Aidoo, R., et al. (2013). Determinants of household food security. *European Scientific Journal.*

Aker, J., & Lemtouni, A. (1999). A framework for assessing food security in face of globalization: The case of Morocco. *Agroalimentaria Junio, 8*, 13–26

Alam, M. M., Siwar, C., Murad, M. W., & Toriman, M. E. (2011). Farm level achievement of climate change, agriculture and food security issues in Malaysia. *World Applied Science Journal, 14*, 431–442

Amaza, P.S., Adejobi, A. O., & Fregrene, T. (2008). Measurement and determinants of food insecurity in Northeast Nigeria: Some empirical policy guideline. *Journal of Food Agriculture and Environment, 6*, 92–96.

Applanaidu, et al. (2013). An econometric analysis of food security and related macroeconomic variables in Malaysia.

Arellano, M., & Bond, S. R. (1991). Some tests of specification for panel data: Monte Carlo evidence and an application to employment equations. *Review of Economic Studies, 58*, 277–297

Arellano, M., Bover, O. (1995). Another look at the instrumental variable estimation of error-component models. *Journal of Econometrics, 68*(1), 29–51.

Arene, C. J., et al. (2010). Determinants of food security among households. *Pakistan Journal of Social Sciences, 30*(1), 9–16

Arshad, & Abdel, H. (2010). Global food prices: Implications to food security in Malaysia. *CRRC Consumer Review, 1*(1), 21–38.

Babatunde, R. (2010). Impact of off-farm income on food security. *Food Policy, 35*(4), 303–311.

Baltagi, B. (2001). *Econometric analysis of panel data.* (2nd ed.). Wiley.

Barilla Centre for Food & Nutrition. (2011). Food security: Challenges and outlook. Barilla Centre for Food & Nutrition.

Barrett, C. B., & Lentz, E. C. (2009). *Food insecurity.* The International Studies Compendium Project. Wiley-Blackwell Publishing.

Belachew, T., et al. (2011). Food insecurity among adolescents. *Nutrition Journal, 10*(1), 29

Blundell, R., & Bond, S. (1998). Initial conditions and moment restrictions in dynamic panel data models. *Journal of Econometrics, 87*(1), 115–143

Breisinger, et al. (2010). Food security and economic development in the Middle East and North Africa. ifpri discussion paper.

Carter, M. A., Dubois, L., & Tremblay, M. S. (2013). Place and food insecurity: Aa critical review and synthesis of the literature. *Public Health Nutrition, 17*(1), 94–112.

Committee on World Food Security (CFS). (2014). Global strategic framework for food security & nutrition (GSF), Third Version, https://www.fao.org/fileadmin/templates/cfs/Docs1314/GSF/GSF_Version_3_EN.pdf

Demissie, S., et al. (2012). Magnitude and factors associated with malnutrition in children. *Science Journal of Public Health, 1*(4), 175–183

Dreze, J., & Sen, A. (1989). *Hunger and public action.* Oxford University Press.

Efron, S., et al. (2018). Food security in the Gulf cooperation council. Emerge85 and the RAND Corporation.

Elbehri, A, Elliott, J., & Wheeler, T. (2015). Climate change, food security and trade: an overview of global assessments and policy insights.

FAO. (1996a). *Declaration on world food security.* World Food Summit; FAO.

FAO. (1996b). Food, agriculture and food security: Developments since the world food conference and prospects. Technical Background Document No 1 for the World Food Summit, Rome.

FAO. (2000). *Multilateral trade negotiations on agriculture: A resource manual.* FAO.

FAO. (2001). *The state of food insecurity 2001.* FAO.

FAO. (2003). *Trade reforms and food security: Conceptualizing the linkages.* FAO.

FAO. (2006b). Trade reforms and food security: Country case studies and synthesis. FAO.

FAO. (2006a). World agriculture: Towards 2030/2050. Interim Report FAO, Rome

FAO. (2008). *Methodology for the measurement of food deprivation: Updating the minimum dietary energy requirements.* FAO Statistics Division.

FAO. (2011a). The state of food insecurity in the world, Rome, Italy.

FAO. (2011b). The state of the world's land and water resources for food and agriculture. Rome, Italy.

FAO. (2013). *The state of food insecurity in the world* 2013. FAO.

FAO. (2016). *Near East and North Africa regional overview of food insecurity.* Cairo, pp. 35

FAO. (2017). *The state of food security and nutrition in the world.* FAO.

FAO. (2018). *Assessing the contribution of bio economy to countries' economy.* Rome, Italy.

Frelat, et al. (2016). Drivers of household food availability in sub-Saharan Africa based on big data from small farms. *Proceedings of the National Academy of Sciences., 113*(2), 458–463

Gebre, G. G. (2012). *Determinants of food insecurity among households in Addis Ababa city, Ethiopia.*

Hatlebakk, M. (2020). *Corona and food security in poor countries.* Bergen: Chr. Michelsen Institute (CMI Brief no. 2020:04), p. 6.

Ihab, N., et al. (2015). Assessment of food insecurity and nutritional status of Children. *International Medical Journal., 22*(6), 509–516

Ijarotimi, O. S. (2013). Determinants of childhood malnutrition and consequences in developing countries. *Current Nutrition Reports, 2,* 129–133

Ivanic, M., & Martin, W. (2008). Implications of higher global food prices for poverty in low-income countries. *Agricultural Economics, 39*, 405–416

Keulertz, M. (2019). *Water and food security strategies in the MENA region*" Rome, IAI, p. 7.

Kombi, B. J. A., et al. (2017). *Stunting, wasting and underweight in Sub-Saharan Africa*, Vol. 142017.

Le Mouël, et al. (2018). Food dependency in the Middle East and North Africa region: retrospective analysis and projections to 2050, at French National Institute for Agriculture, Food, and Environment (INRAE).

Lovendal, et al. (2005). *Tomorrow's hunger: A framework for analyzing vulnerability to food insecurity*. ESA Working Paper No.05–07.

Maitra, C. (2018). *A review of studies examining the link between food insecurity and malnutrition*. Technical Paper. FAO, Rome, p. 70. License: CC BY-NC-SA 3.0 IGO.

Malthus, T. R. (1798). *An essay the principle of population*. Royal Economic Society.

Mbwana, H. A., et al. (2016). *Determinants of household dietary practices in rural Tanzania*.

Mitchell, L. (2003). *Economic theory and conceptual relationships between food safety and international trade*. International Trade and Food Safety/AER-828.

Morrissey, et al. (2013). Local food prices: Effects on child eating patterns, food insecurity. IRP No. 16–2013.

Obadiah, K. (2014). Determinants of household food security. *Journal of Macro-Economies, 22*(7), 105–128

Owusu, V., et al. (2011). Non-farm work and food security. *Food Policy, 36*(2), 108–118

Psaki, S. (2012). Household food access and child malnutrition. *Population Health Metrics, 10*(1), 24

Santos, N. (2015). *Egypt, Jordan, Morocco and Tunisia key trends in the agrifood sector*. Investment Centre Division, FAO and EBRD.

Schmidhuber, J., & Tubiello, F. N. (2007). Global food security under climate change. *PNAS, 104*, 19703–19708

Seed, B. (2019). Food security indicators: Review of literature. BC Centre for Disease Control, Population & Public Health. Vancouver, BC July 15, 2019.

Sen, A. K. (1981). Ingredients of famine analysis: Availability and entitlements. *The Quarterly Journal of Economics, 96*, 433–464

Shah, M., Fischer, G., & van Velthuizen, H. (2008). Food security and sustainable agriculture: The challenges of climate change in Sub-Saharan Africa. *International Institute for Applied Systems Analysis*.

Tamir, D., et al. (2016). Household food insecurity and its association with school absenteeism.

Tamiru, D., et al. (2017). Food insecurity and its association with school absenteeism. *BMC Public Health, 16*(1), 802

Tembo, D., et al. (2009). The effects of market accessibility on household food security. Natural Resource Management and Rural Development 2009.

Timmer, C. P. (2010). Reflections on food crises past. *Food Policy, 35*, 1–11

Titus, B., et al. (2007). An analysis of the food security situation among Nigerian urban households. *Journal of Central European Agriculture., 8*(3), 397–406

UNICEF. (1990). Strategy for improved nutrition of children and women.

United State Department of Agriculture. (2013). *World agriculture supply and demand estimated*.

Wooldridge, J. M. (2002). *Econometric analysis of cross section and panel data*. MIT Press.

Assil El Mahmah is currently a Senior Economist at the National Bank of Kuwait (NBK) since July 2018. Before joining NBK, he worked in many public and private institutions in several countries, gaining significant international experience. He published many papers in scientific indexed journals related mainly to macroeconomic research, fiscal and monetary policies, as well as forecasting models. El Mahmah was an Economic Advisor at the Gulf Monetary Council in Saudi Arabia for one year and a Senior Economist at the Central Bank of UAE for two years. Prior to that, he was a Senior Macroeconomist at an investment bank in Morocco for two years. From 2006 to 2013, El Mahmah worked at the Central Bank of Morocco in different positions and different departments.

Amine Amar is a statistician engineer from the National Institute of Statistics and Applied Economics (INSEA), Rabat, Morocco. He holds a Ph.D. from Mohamed V University of Rabat and has expertise in conducting surveys, statistical methodologies, economic analysis, and public policy evaluation (High Commission of Planning, Morocco). Currently, he is a Senior at the Moroccan Agency for Sustainable Energy (MASEN) and a Fellow Researcher at the Center for Research on Environment, Human Security and Governance (CERES, Morocco) and the Economic Research Forum (ERF, Egypt). He published papers in scientific indexed journals, related to elaborating and implementing economic and statistical methodologies for some research areas such as environment, sustainable energy, finance, economy, and public policies.

Chapter 5
Emerging Threats to Food Security in Nigeria: Way Forward

Burhan Ozkan and Wasiu Olayinka Fawole

Abstract Nigeria is a country of over 200 Million people where agriculture plays an important role in maintaining food security and employs about 75% of the population (who are mostly in the rural areas). Like other developing countries, Nigeria faces many challenges affecting its food production potential. This is due, among others, to: the abandonment of agricultural sector because of oil boom in late 1960s; the lack of modern farming implements; the low value addition due to an ineffective value chain that places low priority on value addition which is the fastest growing phase of agricultural value chains in developing countries; and lately the insurgency in the North Eastern part of the country. Among these factors, the emerging effects of climate change have not only affected crop production, but also have contributed significantly to the emerging threats to livestock due to temperature increases, as is evident with the shrinking of Lake Chad which has caused millions of people to lose their major source of livelihood (fishing). This chapter examines the impacts of these factors on the Nigerian agricultural sector and makes some recommendations to policy makers within the perspective of revamping the agricultural sector for a better agricultural delivery and achievement of Sustainable Development Goals (SDGs) by 2030. Finally, this chapter reviews some of the governmental efforts to revamp the agricultural sector and rural development programmes in Nigeria to make the sector attractive to the teeming number of youths.

Keywords Food security · Assessment · MDGs · Hunger · Nigeria

1 Introduction

Nigeria, like other African developing countries, had looked with optimism toward the achievement of Millennium Development Goal (MDG) 1 that aimed to half hunger in the world by 2015. The achievement of that laudable MDG 1 in the world by 2015 remains a big policy question to answer across all countries, particularly the

B. Ozkan (✉) · W. O. Fawole
Faculty of Agriculture, Department of Agricultural Economics, Akdeniz University, Antalya, Turkey

© The Author(s), under exclusive license to Springer Nature Switzerland AG 2021
M. Behnassi et al. (eds.), *Emerging Challenges to Food Production and Security in Asia, Middle East, and Africa*, https://doi.org/10.1007/978-3-030-72987-5_5

127

developing ones. This goal was specifically met in some countries through carefully implemented development policies and they are now working towards eradicating hunger by 2030 as envisaged by the Sustainable Development Goals (SDGs), which is the completing phase of MDGs. Other countries, mostly developing ones particularly in sub-Saharan Africa, have remained food insecure with the situation deteriorating in some cases, thereby threatening the achievement of zero hunger by 2030.

This chapter aims to investigate and evaluate those factors that have constituted threats to food security in Nigeria. It also assesses the socioeconomic drivers of food security among the sampled households. Food security in recent times has been at the forefront of development policies worldwide owing to the place food occupies in the assurance of human survival through healthy and active living. Food security has been defined by various authors, but there has not been one universally accepted definition for food security. This makes the task of assessing food security cumbersome to undertake, particularly where accurate data are often unreliable or unavailable.

Food security started to gain prominence in the early 1970s when it was defined by the World Food Summit of 1974 held in Rome as "availability at all times of adequate world food supplies of basic foodstuffs to sustain a steady expansion of food consumption and to offset fluctuations in production and prices" (UN, 1975). Though there have been arguments that food security predated this era, the consensus among development researchers is that the term as currently known was the fallout of the World Food Summit of 1974. To affirm the complexity of food security as a concept, Shetty (2015) stated that as of 1997, food security had well over 200 definitions and 450 indicators for its assessment, thereby confirming the complexity of food security in terms of scope and assessment. Just as in the case of other sub-Sahara African countries, feeding a fast-growing population in Nigeria has become a herculean task for the agricultural sector.

During the period 2007–2017, Nigeria's agriculture sector underwent major reforms (FAO reference). The introduction of the Agricultural Transformation Agenda (ATA) reformed the input delivery system, strengthened famers' resilience to shocks, and enhanced agricultural credit. In 2016, the government launched the Social Investment Programme, which strengthened school feeding, conditional cash transfers and food bio-fortification programmes which was aimed at building a strong value chain for agricultural commodities.

The Nigerian population annual growth rate currently stands at 3.5% while the growth rate in agriculture is 2.7%. So Nigerian agriculture cannot provide the food requirements of its population (NBS, 2017). Several actions have been undertaken to address the rising food insecurity in Nigeria and these programmes at various points in time paid off with food insecurity declining until 2009, when it began to rise due to many factors, including the incursion of Boko Haram insurgency of late (Fawole & Ozkan, 2017).

Furthermore, food insecurity in Nigeria is very peculiar compared with other developing countries because Nigeria can fund the existing food deficit by importation. These imports can be paid by the large foreign exchange earnings from crude oil sales, which accounted for about 70% of foreign earnings. The recent oil price slump has depleted foreign reserves and forced the Central Bank of Nigeria (CBN) to remove some food items from its priority import list. The resulting price increases exacerbated food insecurity and hurt the capability for households to meet their dietary needs for active and healthy living.

Food insecurity in Nigeria can been attributed to decline in the fortunes of agriculture due to the emphasis on petroleum production and its foreign exchange earnings (CBN, 2016). Nigeria, unlike many countries, has a vast area of cultivable lands that are still underutilized. For instance, Nigeria has 71.2 million hectares of cultivable land, but only 34.2 million hectares (or 34.8%) is used (Daramola, 2004; Fasoranti, 2008). The cut in food production lately has been largely due to the Boko Haram insurgency in the North Eastern part of Nigeria. North Eastern Nigeria traditionally plays a key role in food production (Fawole, 2017). Several factors have been identified to influence food insecurity at the household level. This includes widening inequality which in turn hinders access to resources of production and economic opportunities (Sanusi et al., 2006; Babatunde et al., 2007a). Nigeria, on the other hand, has great human and material resources that could be harnessed to create productive opportunities for its growing population, particularly the youths who are mostly unemployed (Canagarajah & Thomas, 2001).

This chapter assesses food security and some of the socio-economic variables that affect it. It was conducted post-MDGs with a view to coming up with appropriate policies that will ensure the effective implementation of SDGs taking into consideration the unaccomplished goals of MDGs. Nigeria is currently doing a lot to address food insecurity and hunger in the country through ongoing economic diversification programmes of the government which has the agricultural sector at the forefront. This is not targeted only at halting the trend of food insecurity, but also address the rising unemployment among youths through active participation in agriculture and improving the value chains through investment in manufacturing phase of the agricultural chains.

This chapter provides a comparative assessment of its findings with previous studies to draw conclusion on the food security situation in Nigeria. The study identified some of the socio-economic variables that influenced food security in Nigeria. This study will contribute to the body of knowledge by assisting researchers working on food security and its socioeconomic drivers in the future. This is expected to guarantee sustainable development through continuous monitoring and evaluation of food security to guide policy makers and other stakeholders in the sector in coming up with strategies that could solve the problem in a sustainable manner. For instance, Lawrence (2012) traced the growing concerns about food insecurity to the fast depletion of food-producing resources like water, oil and phosphorus due to pressure mounted by growing population. In Nigeria's case, the current population growth rate of 3.5% and growing demands for food without corresponding growth in food supply has put more pressures on available resources.

1.1 The MDG Agenda on Food Security in the World

The World Food Summit of 1996 aimed to reduce the number of hungry people by 20 million annually, thereby halving the number by 2015. However, since that memorable and important summit, the average annual decrease has been 2.5 million which is still far below the set target (Clover, 2003). According to the chairperson of World Food Summit held in 2002, as cited in Clover (2003), "together with terrorism, hunger is one of the greatest problems the international community is facing". Corroborating this assertion, James Morris, then Executive Director of the World Food Programme (WFP) in his address to the United Nations (UN) Security Council in December 2002, stated, in addition the problem of AIDS/HIV, the danger posed by food insecurity where about 38 million people in Africa alone faced imminent threat to their peace, security, and stability (WFP, 2002). The UN recently hinted that the world was facing the largest humanitarian crisis since 1945 by warning that more than 20 million people from four countries might die as a result of starvation and famine (The UK Telegraph, 2017).

1.2 Review of Previous Studies on Food Security in Nigeria and Selected African Countries

Table 1 shows the findings of similar studies on food security in Nigeria and some selected African countries. Fakayode et al. (2009), Obayelu (2010, 2012), Ndhleve et al. (2013) and Oni et al. (2013) corroborated the position of the Food and Agriculture Organization (FAO) (2017) on the State of Food and Nutrition Insecurity in the world. They all agreed in unequivocal terms about the high prevalence of food insecurity in Nigeria and sub-Saharan Africa as a whole.

According to the reports published by the FAO, food insecurity in the world had risen from 777 million in 2015 to 815 million in 2016 (FAO, 2017). This corroborated the findings of those studies as presented on Table 1 affirming the frightening level the food insecurity. Table 1 showed that even where the findings showed contrary results by reporting some level of food security in households, there is still a low ratio of food secure to food insecure households, as in the case of Oyebanjo et al. (2013) which reported that 59.2% of households were food secure while 40.8% were food insecure. Irohibe and Agwu (2014) estimated that 74% of the households in Kano state (Nigeria) were food secure, while 26% were not. Kano is in northern Nigeria which is an important agricultural zone renowned for its staple crops and livestock production with semi-developed irrigation system. There are lots of economic opportunities around Kano for people to guarantee their economic access to food that meet their dietary needs for healthy and active living. It is pertinent to note that despite the importance of food security, there have been few studies that have provided details among Nigerian people. This is due to the absence of centrally processed and reliable official data. This has made planning very difficult if not impossible (Olanike et al., 2007; Sanusi et al., 2006).

Table 1 Review of food security status in selected African countries in previous studies

Authors	Food secure (%)	Food insecure (%)	Location
Adetunji (2016)	36.5	63.5	Nigeria
Amaza et al. (2008)	42	58	Nigeria
Arene and Anyaeji (2010)	40	60	Nigeria
Babatunde et al. (2007a)	36	64	Nigeria
Babatunde et al. (2007b)	37.2	62.8	Nigeria
De Cock et al. (2013)	47	53	South Africa
Fakayode et al. (2009)	12.2	87.8	Nigeria
Gebre (2012)	41.84	58.16	Ethiopia
Ibok et al. (2014)	12.4	87.6	Nigeria
Ibrahim (2009)	70	30	Nigeria
Iorlamen et al. (2013)	67.3	32.7	Nigeria
Irohibe and Agwu (2014)	74	26	Nigeria
Kuwenyi et al. (2014)	48.3	51.7	Swaziland
Magana-Lemus et al. (2016)	60	40	Mexico
Mitiku et al. (2012)	50.5	49.5	Ethiopia
Muche et al. (2014)	42.9	57.1	Ethiopia
Ndhleve et al. (2013)	13	87	South Africa
Obamiro et al. (2003)	52	48	Nigeria
Obayelu (2010)	23.7	76.3	Nigeria
Obayelu (2012)	16	84	Nigeria
Okwoche and Asogwa (2012)	68	32	Nigeria
Omonona and Agoi (2007)	49	51	Nigeria
Omotesho and Muh'd-Lawal (2013)	34.6	65.4	Nigeria
Omotesho et al. (2006)	34.2	65.8	Nigeria
Oni et al. (2013)	12	88	South Africa
Oyebanjo et al. (2013)	59.2	40.8	Nigeria
Sanusi et al. (2006)	30	70	Nigeria

Source Adapted from previous studies on food security in Nigeria and selected countries

2 Materials and Methods

2.1 Study Area

The study discussed in this chapter was conducted in Nigeria. According to the Nigerian Bureau of Statistics (NBS), the population is 200 Million with an annual growth rate of 3.5%. It is expected to hit 210 million by 2021, which will put more pressure

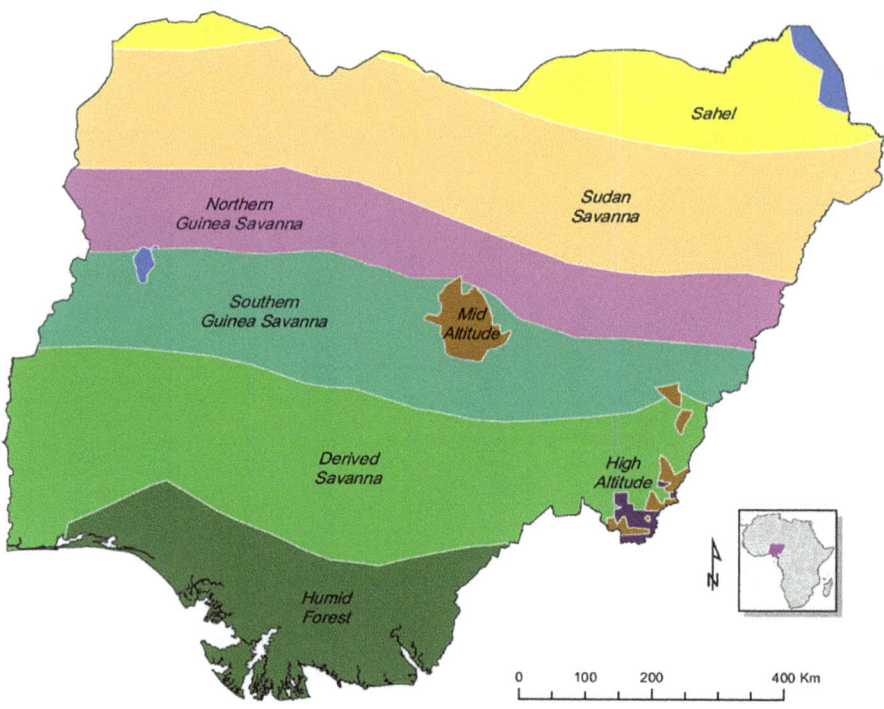

Fig. 1 Agroecological zones in Nigeria. *Source* Ojanuga (2006)

on resources required for food production (land and water), therefore aggravating in turn the already difficult food security situation (NBS, 2017).

Figure 1 shows the Agroecological Zones in Nigeria. The study area (South West), highlighted in deep green, comprises six states and falls within the humid forest. the states of Oyo and Osun served as the representative samples for the Zone due to their culture, tradition, language, socioeconomic characteristics, and the kind of foods consumed by households.

2.2 Sampling and Analytical Techniques

The research used multistage sampling techniques involving three stages. The first stage involved purposive selection of Osun and Oyo states due to time and resource constraints. The second stage was the random selection of Nine Local Government Areas from the two states (five from Oyo and four from Osun) according to their population. In Oyo, the local governments of Ibadan North East, Saki West, Surulere, Oyo East and Ibarapa East were selected with one local government each from the five zones of the State. For Osun, the local governments of Osogbo, Obokun, Ayedaade and Ife East were selected with at least one from each of the three zones in Osun State (two were from Osun East due to its large population size). The third and last stage was a random sampling of 161 households from the selected towns and villages in those local governments. For the food consumption data, a recall period of 7 days was used following the works of Muche et al. (2014), Fawole et al. (2016), and Mitiku et al. (2012). This implies that questions were asked on the type and quantities of foods consumed by households in the last 7 days on the assumption that consumption patterns are the same throughout the month. The sampling of 161 households was arrived at using the sampling formula stated below:

$$n = \frac{Z^2 p(1-p)}{d^2}$$

n   Sample size,
Z   1.96 (95% confidence interval),
p   0.29 represents the existing rate of the planned study (Determined by a pilot survey),
d   0.05. (Malhotra, 2004).

For the data analysis, both descriptive statistics and econometric methods (binomial Logit) were used. Four indicators were used to assess the food security situation among the households. This was done to lend more credence to the results of this study going by the identified weaknesses of various known food security indicators including those used in this study (Fawole & Ozkan, 2017). The results are presented using descriptive statistics that show the distribution of households according to their food security status in number and percentage; while the socioeconomic drivers of food security were identified using binary logistic regression models.

2.3 Food Security Assessment

2.3.1 Recommended Minimum Daily Per Capita Calorie Requirements/Recommended Daily Allowance (RDA)

This is one of the most commonly used indicators for assessing food security at the household level (Mitiku et al., 2012; Muche et al., 2014; Omotesho et al., 2006). It is described as gold standard by Mekonnen and Gerber (2017) who compare household per capita calorie intake (PCCI) with food security recommended daily allowance (RDA) of 2700 kcal/AE/day. Households were subsequently categorized into four food security groups following Arene and Anyaeji (2010), Devereux (2006), and Sanusi et al. (2006):

Food secure: $PCCI_i \geq RDA$,
Food insecure without hunger: $2300 \leq PCCI_i \leq 2699.99$ kcal/AE/day,

Food insecure with moderate $1900 \leq PCCI_i \leq 2299.99$ kcal/AE/day,
hunger:

Food insecure with severe hunger: $PCCI_i \leq 1900$ kcal/AE/day.

2.3.2 Food Security Assessment Based on Perception (Self-Report)

This self-reporting method of assessing food security at a household level is new and based on the perception of household head or other appointed representative who is familiar with food supplies and consumption. Self-reporting is relatively cheap compared with other traditional methods of data collection and it has proven to be efficient. Hossain et al. (2016) and Magana-Lemus et al. (2016) used a similar technique to assess food security while recommending its use with traditional indicators like recommended daily per capita calorie requirement (RDA) to serve as a control indicator in ascertaining the prevailing food security.

2.3.3 Food Security Assessment Based on United States Households' Food Security Scale (US-HHFSS)

The Food Security Scale is another famous technique in the assessment of food security at the household level that was developed in the United States in the mid-90s when food security started gaining grounds. It was revised in 2000 and used to investigate the severity of food insecurity and hunger among the US households (FANTA, 2003). It had since been modified and domesticated by various authors such as Obayelu (2010), Sanusi et al. (2006), and Fakayode et al. (2009) in their respective studies (all conducted in Nigeria) to suit the peculiarities in terms of number and types of foods consumed. There are 18 standard questions used in this method: 3 are asked on food experiences of the entire households; 7 on the experiences of the adult members of the households; and the remaining 8 questions on the food experiences of the children. The households were subsequently classified into four discrete food security groups based on the number of affirmative answers to the questions:

Food secure, FS (0–2 affirmative responses),
Food insecure without hunger, FIWH (3–5 affirmative responses),
Food insecure with moderate hunger, FIMH (6–8 affirmative responses),
Food insecure with severe hunger, FISH (9–10 affirmative responses).

2.3.4 Food Security Assessment Based on Households' Food Expenditure Share

Food security index of households was assessed by comparing per capita food expenditure share of each household with two-thirds of the mean per capita food expenditure share for all households:

$$FS_i = \frac{\text{Per Capita Food Expenditure of the } i\text{th Household}}{2/3 \text{ Mean Per Capita Food Expenditure of all Households}} \tag{1}$$

FSi   Food security index of the ith household,
FSi ≥ 1 Food secure households,
FSi < 1 Food Insecure households.

Households whose per capita food expenditures were above the food security threshold were marked 1 (food secure), while those below the threshold marked 0 (food insecure).

2.4 Binary Logistic Regression Model

This model was used to identify those factors that influence the food security of households either positively or negatively.

2.4.1 The Model Specification

$$L_i = \ln\left(\frac{P_i}{1 - P_i}\right) = Z_i \tag{2}$$

From the general model as specified in (2) above,

$$Z_i = \beta_0 + \beta_i X_i + \mu_i \tag{3}$$

where i = 1, 2, 3 ... n. The Eq. (2) above can thus be rewritten as;

$$L_i = \ln\left(\frac{P_i}{1 - P_i}\right) = \beta_0 + \beta_i X_i + \mu_i \tag{4}$$

n   number of explanatory variables specified in each model.
Dependent variable   food security (food secure = 1, food insecure = 0).
X_i explanatory variables as listed on Table 2.

Four models were specified from Eq. (4) based on the four indicators used to assess food security of the households in order to capture as much as possible number of explanatory variables to explicitly identify the major drivers of food security in those households investigated particularly those that significantly affect it in order to come up with appropriate recommendations based on the results of the analysis according to the significant effects such variables have on food security. The explanatory variables and the exact models specified are as presented in Table 2. The binary logistic regression specification for the four models (1–4) are as shown in Eqs. (5)–(8).

Table 2 Distribution of explanatory variables according to the number of models specified

Explanatory variables included in the models	Model(s) where specified
[+]Access to credit facilities (access = 1, no access = 0)	3, 4
[+]Access to public health facilities (access = 1, no access = 0)	2, 3, 4
Age of the household head (years)	2, 3, 4
Asset values of the household (Naira)	4
[+]Cooperative membership (member = 1, non-member = 0)	2
Coping strategies utilized by household (0 – 9)	1, 3
Educational status of household head (years spent in school)	1
Food market accessibility to household (Km)	1, 2, 4
[+]Gender status of household head (Male = 1, Female = 0)	3
Household size (No. of people eating from the same source)	1
[+]Income from other members of the household (yes = 1, no = 0)	4
[+]Major occupation of the head (farming = 1, Non-farming = 0)	1, 3, 4
[+]Marital status of household head (married = 1, otherwise = 0)	3
Non-food expenditure shares (Naira)	1, 3
[+]Secondary occupation by the household head (yes = 1, no = 0)	1, 2, 4
[+]State of the location of the household (Osun = 1, Oyo = 0)	1, 2, 3, 4
Total monthly income of household head (Naira)	1, 2, 3

Field survey, 2016 [+] Dummy variables

$$\text{Model 1: } L_i = \ln\left(\frac{P_i}{1 - P_i}\right) = \beta_0 + \beta_1 X_1 + \beta_2 X_2 + \ldots + \beta_9 X_9 + \mu_i \quad (5)$$

$$\text{Model 2: } L_i = \ln\left(\frac{P_i}{1 - P_i}\right) = \beta_0 + \beta_1 X_1 + \beta_2 X_2 + \ldots + \beta_7 X_7 + \mu_i \quad (6)$$

$$\text{Model 3: } L_i = \ln\left(\frac{P_i}{1 - P_i}\right) = \beta_0 + \beta_1 X_1 + \beta_2 X_2 + \ldots + \beta_{10} X_{10} + \mu_i \quad (7)$$

$$\text{Model 4: } L_i = \ln\left(\frac{P_i}{1 - P_i}\right) = \beta_0 + \beta_1 X_1 + \beta_2 X_2 + \ldots + \beta_9 X_9 + \mu_i \quad (8)$$

2.4.2 Estimation of the Parameters of Binary Logistic Regression Models

Parameters for the explanatory variables were estimated through Maximum Likelihood Estimates (MLE) by using STATA 10.0 statistical software package.

P_i = Probability of household being food secure,

$1 - P_i$ = Probability of otherwise (i. e., household being food insecure).

Table 3 Distribution of models according to food security indicators used in specifying them and the number of explanatory variables included in each model

Models	Food security indicators for specification	Number of explanatory variables
1	Recommended daily allowance (RDA)	9
2	Food expenditure share	7
3	Based on perception (Self-report)	10
4	US-HHFSS	9

$$P_i = \frac{1}{1 + e^{Z_i}} \tag{9}$$

$$1 - P_i = \frac{e^{Z_i}}{1 + e^{Z_i}} \tag{10}$$

$Z_i = L_i$ as stated in Eq. (1) above while $Z_i = \beta_0 + \beta_i X_i$ as explicitly stated in Eqs. (3) and (4)

Finally, z-statistics were used to test the statistical significance levels of the estimated coefficients. This chapter uses pseudo R^2, Count R^2, and Hosmer–Lemeshow chi-squared value for goodness of fit. Count R^2 was estimated as:

$$\text{Percentage of correct predictions (Count } R^2) = \frac{\text{Number of correct predictions}}{\text{Total number of observations}} \tag{11}$$

Correct predictions occur when the household is food secure (1) with the estimated Logit greater than or equal to 0.5 and when the household is food insecure (0) with the estimated Logit less than 0.5. The mathematical expressions of the above are:

$L_i \geq 0.5 = 1$ and $L_i < 0.5 = 0 \approx$ correct predictions,

$L_i \geq 0.5 = 0$ and $L_i < 0.5 = 1 \approx$ incorrect predictions.

$L_i =$ estimated Logit from Eq. (4).

(Gujarati & Porter, 2009; Abbas et al., 2017).

3   Results and Discussions

3.1   Results

The findings as presented in Table 4 reveal that the majority of the households manifest varying degrees of food insecurity and hunger (moderate and severe). Some households manifest moderate hunger with US-HHFSS and self-report (perception) based assessments while severe hunger is prominent when the households' food security was assessed based on US-HHFSS. The later account for as high as 12.4%,

Table 4 Food security status of households according to four indicators

FS Status	RDA		Self-Report		US-HHFSS		Food expenditure share		
	N	%	N	%	N	%	Status	N	%
FS	86	53.4	64	39.8	53	32.9	FS	73	45.3
FIWH	47	29.2	43	26.7	53	32.9	FI	88	54.7
FIMH	20	12.4	51	31.7	35	21.7	**Total**	**161**	**100**
FISH	8	5.0	3	1.9	20	12.4			
Total	**161**	**100**	**161**	**100**	**161**	**100**			

Field survey, 2016

which implies that there is still high prevalence of hunger in the study area. Furthermore, the findings reveal that at least one of the seventeen explanatory variables specified in the four models influences food security in the households.

As for the drivers of food security among the sampled households, explanatory variables such as access to credit and public health facilities, age of the household head, secondary occupation by household head, income from other members of the household, and total monthly income of household head significantly influence food security of the households positively.

Explanatory variables such as cooperative membership of household head, coping strategies utilized by households during food shortage, household size, and major occupation of the household head significantly influence food security of the households negatively. The explanatory variables that significantly affected the food security of the households in the study area are highlighted in the green background and bold inks in Table 5 subsequently.

3.2 Discussions

The findings on the food security status of the households confirm the persistent and growing food insecurity in Nigeria. These findings are consistent with Okwoche and Asogwa (2012) but contrast with those of Fawole et al. (2016) and Muhammad-Lawal and Omotesho (2010) in terms of per capita calorie intake. The findings based on self-reporting contrast with those of Magana-Lemus et al. (2016) and congruent with those of Obayelu (2012) and Sanusi et al. (2006) based on US-HHFSS.

Finally, the food expenditure share findings of the study are consistent with those of Adetunji (2016), Omonona and Agoi (2007) and Arene and Anyaeji (2010), but contrast with those of Iorlamen et al. (2013).

Muche et al. (2014), Arene and Anyaeji (2010), Mitiku et al. (2012), and Obayelu (2012) all reported a positive relationship between food security and access to credit facilities, while Babatunde et al. (2007) reported contrary findings. A positive relationship with access to credit is quite understandable because household heads with

Table 5 Logistics regression results showing drivers of food security based on four specified models

Explanatory variables	Model 1		Model 2		Model 3		Model 4	
	Coefficient	Odds ratio	Coefficient	Odds ratio	Coefficient	Odds ratio	Coefficient	Odds ratio
+ Access to credit facilities	–	–	–	–	1.4749*	4.3706	0.3496	1.4185
+ Access to public health facilities	–	–	0.9311**	2.5372	1.3112**	3.7106	0.9604**	2.6127
Age of the household head	–	–	0.0425**	1.0434	0.0627**	1.0647	0.0226	1.0228
Assets values of the household	–	–	–	–	–	–	–1.2E-06**	0.9999
+ Cooperative membership	–	–	–1.2837***	0.2770	–1.3117***	0.2694	–	–
Coping strategies utilized by household	–0.1391**	0.8701	–	–	–	–	–	–
Educational status of household head	0.0536	1.0551	–	–	–	–	–	–
Food market accessibility by household (Km)	0.8340	2.3025	0.9579	2.6063	–	–	1.3157	3.7274
+ Gender status of household head	–	–	–	–	2.2734	9.7123	–	–
Household size	–0.3320***	0.7175	–	–	–	–	–	–
+ Income from other members of household	–	–	–	–	–	–	0.7583**	2.1347
+ Major occupation of household head	–0.8131**	0.4435	–	–	–1.0505*	0.3498	–0.6796*	0.5068
+ Marital status of household head	–	–	–	–	–3.6433***	0.0262	–	–
Non-food expenditure shares	4.4E-06	1.0000	–	–	0.0001	1.0000	–	–
+ Secondary occupation by household head	0.9162***	2.4499	0.8159*	2.2612	–	–	0.3714	1.4498
+ State of location of the household	–0.2439	0.7836	1.4880***	4.4284	–1.2800**	0.2780	–0.5321	0.5874
Total monthly income of household head (Naira)	1.3E-05**	1.0000	1.5E-05**	1.0000	5.5E-06***	1.0000	–	–

Source Estimates of the binary Logit analysis of the food data; ***Significant at 1%, **Significant at 5%, *Significant at 10%; (+) Dummy variables enough credit facilities are able to engage in many productive activities that guarantee adequate income to cater for this household's dietary needs.

Obayelu (2012) reported findings contrary to this study regarding the relationship between food security and access to public health. But the implications of the findings of this study is unambiguous as it affirms the existing nexus between food security and public health accessibility, particularly the utilization dimension of food security which requires good personal health and environmental hygiene to guarantee food utilization. These findings are consistent with those of Magana-Lemus et al. (2016) and Mitiku et al. (2012) which reported a positive relationship between food security and age of the household head but in contrast with those of Muche et al. (2014) that reported contrary findings. Okwoche and Asogwa (2012) also reported a positive relationship between food security and income from other members of the household.

Furthermore, the findings of this study are congruent with Amaza et al. (2008) and Babatunde et al. (2007) who reported a negative relationship between food security and cooperative membership of the household head. Again, the findings of this study are consistent with those of Okwoche and Asogwa (2012) in terms of the relationship between food security and coping strategies utilized by households during food shortages.

Amaza et al. (2008), Muche et al. (2014), and Sanusi et al. (2006) reported an inverse relationship between food security at the household level and household size (number of members eating from the same source), but Iorlamen et al. (2013) and De Cock et al. (2013) reported contrary findings. The later findings might be due to other factors such as a larger household having more people who are working and earning income.

Contrary to Babatunde et al. (2007b), the findings of Obayelu (2012) showed that someone whose major occupation was farming were less food secure compared with when the situation was otherwise. This is quite understandable because the majority of Nigerian farmers rely on rain-fed agriculture which makes their production seasonal in nature and their produce in most cases cannot sustain them throughout the year. Those whose major source of livelihoods is non-farming have access to funds that could assist them to procure their dietary needs throughout the year.

Amaza et al. (2008) reported findings contrary to this study while Kuwenyi et al. (2014) reported negative and similar findings for the relationship between household food security and assets values of the household. The findings of this study are unexpected but could have been due to the cultural practices among the people in the study area where assets are somewhat cherished and held in trust thus the assets value in most cases does not depict the actual status particularly the food wellbeing Kuwenyi et al. (2014). Their values may be unconnected with the exact financial position of the household, and therefore have little relationship with food security.

Above all, majority of similar studies conducted on food security, including Sanusi et al. (2006), Muche et al. (2014), and Omotesho et al. (2006), reported a positive relationship between household head income and food security. This implies that income plays a very crucial role in maintaining food security for the household. This is why this study recommends the adoption of poverty alleviation programs which may increase the access of concerned population to production factors, thus enabling them to engage in ventures that would raise their incomes. This would enable them to acquire foods required for their active and healthy living to carry on with their daily life activities (Table 6).

Generally, the results on the description of diagnostic variables on the fitness of the data, presented in Table 7, showed that all the explanatory variables had marginal effects on the food security status of the households considering their relatively low p-value ($p < 0.001$) in the four models except for model 4 with p-value of 0.0096 ($p < 0.01$).

These four models have relatively high values of Pseudo R^2 and percentage of correct predictions (count R^2) as shown in Table 7. The Hosmer–Lemeshow chi^2 (8) for the four models are 8.87, 7.10, 9.81 and 7.66, which are consistent with those of Kuwenyi et al. (2014) and they affirm that the models correctly fit the data analyzed.

Table 6 Distribution of significant variables according to the models

Models	Significant variables
1	Major occupation, secondary occupation, household size, coping strategies, total income of household head
2	Access to public health, secondary occupation, age of household head, total income of household head, state of location of household, cooperative membership of household head
3	Access to credit facilities, access to public health, coping strategies, age of household, marital status, major occupation, total income of household head, state location of household
4	Access to public health, income from other members of household, major occupation of the household head and assets value of households

Table 7 Distribution of diagnostic variables that described the fitness of the model

	Log-likelihood	LR chi^2	Prob > chi^2 p-value	Pseudo–R^2	Hosmer–Lemeshow chi^2	% of correct predictions
Model 1	−90.00	42.42	0.0000	0.193	8.87	72.05
Model 2	−70.33	32.91	0.0000	0.190	7.10	82.6
Model 3	−47.58	121.22	0.0000	0.5602	9.81	86.3
Model 4	−91.12	21.78	0.0096	0.1068	7.66	70.8

3.3   Implications of the Findings on MDGs Accomplishments

The obvious implications of this study are that food insecurity and hunger remain a challenge in the study area and is applicable to the entire Southern Western Nigeria considering the similarities and peculiarities in the characteristics of the households in terms of social, cultural, economic, nature of livelihoods. The populace is often affected by the same socio-political factors and react to government policies in same manner.

Southern Nigeria has always been touted as being food secure, so there has been little or no interventions in terms of food security programmes. This is one of the justifications for this study and the choice of the study area. The general belief has always been that due to high literacy and relatively low poverty levels compared with the North, people are more open to economic opportunities which are expected to reduce the level of food insecurity. Despite the MDG of halving world hunger by the end of 2015, the majority of the sampled households are still battling with food insecurity and hunger in varying degrees. Consequently, it is concluded that the goal of halving hunger in the study area has not been accomplished at least in the study area.

Finally, based on the significant drivers of food security from the econometric analysis and its policy implications, the goal of reaching the SDGs target of ending hunger in the world by 2030, as far as the study area is concerned, is still a work in progress and requires some deliberate efforts to accomplish. Some of these efforts should include programmes to support agricultural production with funds while also discouraging excessive food imports that have hindered the competitiveness of local farmers. If appropriate actions are taken by relevant stakeholders, the lost ground, as far as the MDGs, could be covered by the SDGs. In order to make this a reality, this study suggests some carefully constructed policy recommendations that, if sincerely and properly implemented by stakeholders in the agricultural and food sub-sectors, will go a long way toward achieving the zero-hunger target in Nigeria as being envisaged by the Sustainable Development Agenda 2015–2030.

4   Conclusions and Recommendations

In this chapter, we have assessed the food security situation of the study area in the post-MDGs era. The study used four indicators (qualitative and quantitative) of food security to investigate their effects. In addition to assessing the prevalence of food insecurity in households, we also investigated the drivers of household food security to determine the appropriate intervention measures. Three major conclusions are derived from the findings. First, food insecurity in the study area is still very pronounced and severe, as witnessed by households that manifested moderate and severe hunger. Second, we found that the majority of the households were food insecure under the four indicators used in assessing food security in the study area.

Third, we found that food insecurity in the study area is principally due to the lack of economic access and inadequate income which are among of the major drivers of food security on the household level. Income available is not enough to guarantee food procurement in a sustainable manner for many households. The findings of the study also reveal that majority of households rely on foods available in the market for their food supplies.

This study strongly recommends the formulation of programs and policies that will ensure sustainable flow of income to households on one hand, and boost food production on the other hand to enable households make provision for their dietary needs. These programs may include support to citizens to have secondary occupation such as small and medium-sized enterprises (SMEs). The findings of this study have shown that those who have more than one occupation were more food secure than their counterparts with sole occupation. This income might come as carefully planned social safety nets, such as cash and food transfers for those households that manifested food insecurity with severe hunger to ameliorate their sufferings. Hunger should be a sign of emergency in food security management in developing countries.

There should be adequate provision for improved household access to public health facilities to enhance food utilization, which is crucial to achieving food security apart from availability and accessibility. Government should step up campaigns to educate people on the importance of family planning to ensure people give birth to an economically viable number of children. Overpopulation has been identified as one of the main causes of food insecurity and this was demonstrated by the findings of this study that showed that household size significantly affected food security status of households negatively. The government should also prioritize the formulation of sustainable agricultural policies that would enable food-producing households to produce in quantities that would not only meet their food needs, but generate income that could assist in procuring food items not produced on their farms.

Furthermore, government should work towards ensuring price stability for food items by providing facilities that would enable the farmers to store their produce during the farming season or the government should buy from them to store for the off-season when prices are high. This will enhance food security of the households by ensuring lower prices for foods sold to final consumers.

Finally, a secondary occupation that would guarantee more income to the households should be encouraged by relevant stakeholders. Households headed by those with a secondary occupation were more food secure compared with those headed by household heads with a single occupation. In general, most studies previously conducted asserted that food insecurity in Nigeria is real and growing with figures almost equal between food secure and food insecure households. Food insecure households in some instances reached over 30%. This confirms that the situation requires urgent and affirmative actions from concerned and relevant stakeholders such as government, financial institutions for funding of agriculture, and private sectors with a view to halting the trend.

These findings based on the analysis of food consumption data presented earlier in the results and discussions section with the research question on the accomplishments or otherwise of the MDGs answered by comparing the various findings of the study with findings from previous studies on food security as reported in Table 1 in order to come to conclusion on the status of food security in the study area. To achieve the zero-hunger target, the world should adopt this slogan *"the presence of hunger anywhere is the presence of hunger everywhere in the world"*, and this assertion is only viable when the political dimension of food security is critically examined. Developed countries that are mostly food secure need to assist the developing ones that are mostly food insecure. The developing countries need technologies that can complement their agricultural production and food processing capacities, instead of relying on providing the vulnerable countries with food aid, a practice that is unsustainable and lacks potential for a long-run solution to growing food insecurity and hunger. Food insecurity and hunger have been a recurring issue in developing countries due to limited access to affordable technologies to produce what they eat amidst the resources required for it.

Acknowledgements The authors are extremely thankful to Dr. Michael R. Reed, Emeritus Professor, and former Director—International Programs for Agriculture at the University of Kentucky, USA for reviewing the initial drafts, making helpful comments and offering valuable suggestions. His sincere cooperation highly appreciated.

References

Adetunji, M. O. (2016). Analysis of the food security status of fruit and vegetable marketers in Ibadan north local government, Oyo state. *Journal of Emerging Trends in Economics and Management Sciences, 6*, 252–257.

Amaza, P. S., Adejobi, A. O., & Fregene, T. (2008). Measurement and determinants of food insecurity in North East Nigeria: Some empirical policy guidelines. *Journal of Food, Agriculture and Environment., 6*, 92–96.

Arene, C. J., & Anyaeji, R. C. (2010). Determinants of food security among households in Nsukka metropolis of Enugu state of Nigeria. *Pakistan Journal of Social Sciences (PJSS), 30*, 9–16.

Babatunde, R. O., Omotesho, O. A., & Sholatan, O. S. (2007a). Socio-economics characteristics and food security status of farming households in Kwara State, North Central Nigeria. *Pakistan Journal of Nutrition, 6*, 49–58.

Babatunde, R. O., Omotesho, O. A., & Sholotan, O. S. (2007b). "Factors influencing food security Status of rural farming households in North Central Nigeria. *Agricultural Journal, 2*(3), 351–357.

Canagarajah, S., & Thomas, S. (2001). Poverty in a wealthy economy: The case of Nigeria. *Journal of African Economies, 10*, 143–173.

Central Bank of Nigeria (CBN). (2016). "Nigerian gross domestic product composition". Available online at: www.cenbank.org

Clover, J. (2003). Food security in sub-Saharan Africa. *African Security Review, 12*, 5–15.

Daramola, A. G. (2004). "Competitiveness of Nigeria agriculture in a global economy: Any dividends of democracy? Inaugural lecture series 36 delivered at the Federal University of Technology. Akure. 2004*(1).

De Cock, N., D'Haese, M., Vink, N., Van Rooyen, C. J., Staelens, L., Schönfeldt, H. C., & D'Haese, L. (2013). Food security in rural areas of Limpopo province, South Africa. *Journal of Food Security, 5*, 269–282.

Devereux, S. (January, 2006). *"Chronic or transitory hunger: How do you tell the difference? Desk review prepared by Stephen Devereux, institute of development studies* (p. 40). Available online at: http://documents.wfp.org/stellent/groups/public/documents/ena/wfp100603.pdf

Fakayode, S. B., Rahji, M. A. Y., Oni, O. A., & Adeyemi, M. O. (2009). An assessment of food security situations of farm households in Nigeria: A USDA approach. *The Social Sciences, 4,* 24–29.

FANTA. (2003). *"Food and nutrition technical assistance (Fanta) project, food access indicator review", academy for educational development (AED).* D. C.

FAO. (2017). "The state of food insecurity in the world 2001, Rome". Available online at: www. fao.org

Fasoranti, O. O. (2008). The determinants of agricultural production and profitability in Akoko Land, Ondo-State, Nigeria. *Journal of Social Sciences, 4,* 37–41.

Fawole, W. O. (2017). "Assessment of food security situation among households in South Western Nigeria post-MDGs era", Ph.D. Thesis Unpublished.

Fawole, W. O., & Ozkan, B. (2017a). The systemic review of food security assessment indicators: Understanding the strengths and weaknesses of the indicators. *Journal of Agriculture and Rural Research, 1*(1), 24–31.

Fawole, W. O., & Ozkan, B. (2017b). Identifying the drivers of food security based on perception among households in South Western Nigeria. *European Journal of Interdisciplinary Studies, 9,* 49–55.

Fawole, W. O., Ozkan, B., & Ayanrinde, F. A. (2016). Measuring food security status among households in Osun State, Nigeria. *British Food Journal, 118,* 1554–1567.

Fedoroff, N. V. (2015). Food in the future of 10 billion. *Journal of Agriculture and Food Security, 4,* 1–10.

Gebre, G. G. (2012). Determinants of food insecurity among households in Addis Ababa city, Ethiopia. *Interdisciplinary Description of Complex Systems, 10,* 159–173.

Gujarati, D. N., & Porter, D. C. (2009). *Basic Econometrics.* (5th ed.). McGraw-Hill.

Hossain, M., Mullally, C. & Asadullah, M. N. (6–9 February, 2016). "Measuring household food security in a low-income country: A comparative analysis of self-reported and objective indicators". Selected paper prepared for presentation at the Southern agricultural economics association's 2016 annual meeting, San Antonio, Texas.

Ibok, O. W., Idiong, I. C., Bassey, N. E., & Udoh, E. S. (2014). Food security and productivity of urban food crop farming households in Southern Nigeria. *Science and Education Centre of North America, 8,* 1–12.

Ibrahim, H., Uba-Eze, N. R., Oyewole, S. O., & Onuk, E. G. (2009). Food security among urban households: A case study of Gwagwalada area council of the federal capital territory Abuja, Nigeria. *Pakistan Journal of Nutrition, 8,* 810–813.

Iorlamen, T. R., Abu, G. A., & Lawal, W. L. (2013). Comparative analysis on socio-economic factors between food secure and food insecure households among urban households in Benue State, Nigeria. *Journal of Agricultural Science, 4,* 63–68.

Irohibe, I. J., & Agwu, A. E. (2014). Assessment of food security situation among farming households in rural areas of Kano State, Nigeria. *Journal of Central European Agriculture, 15,* 94–107.

Kuwenyi, S., Kabuya, F. I. & Masuku, M. B. (2014). "Determinants of rural households' food security in Shiselweni region, Swaziland: Implications for agricultural policy". *IOSR Journal of Agriculture and Veterinary Science (IOSR-JAVS), 7,* 44–50.

Lawrence, G. (2012). Food Security. *Australian Journal of International Affairs, 66,* 281–282.

Magana-Lemus, D., Ishdorj, A., Parr Rosson III, C. & Lara-Alvarez, J. (2016). "Determinants of households' food insecurity in Mexico". *Agricultural and Food Economics, 4,* 1–20.

Malhotra, N. K. (2004). *"Marketing research: An applied orientation"* (4 edn). Pearson Prentice Hill.

Mekonnen, D. A., & Gerber, N. (2017). Aspirations and food security in rural Ethiopia. *Food Security, 9,* 371–385.

Mitiku, A., Fufa, B. & Tadese, B. (2012). "Analysis of factors determining households food security in pastoral area Oromia region, Moyale district, in Ethiopia". *International Journal of Agricultural Science, Research and Technology*, 2, 105–110. Available online at: http://journals.iau.ir/article_5 17520_6a416b6674ddf60742bac6e7f10b6167.pdf

Muche, M, Endalew, B. & Koricho, T. (2014). "Determinants of household food security among Southwest ethiopia rural households". *Food Science and Technology*, 2, 93–100. Available online at: http://www.hrpub.org/download/20150101/FST1-11102511.pdf. https://doi.org/10.13189/fst. 2014.020701

Muhammad-Lawal, A., & Omotesho, O. A. (2010). Intensity of food insecurity in rural households of Kwara State, Nigeria. *Journal of Applied Agricultural Research*, 2, 21–30.

National Bureau of Statistics (NBS). (2017). "Nigeria—General household survey-panel wave 3 (Post Planting) 2015–2016", Third round. Available online at: http://www.nigerianstat.gov.ng/ nada/index.php/catalog/51/study-description. (Accessed June 28, 2017).

National Population Commission. (2006). "*Nigerian population census 2006 RESULTS for Oyo and Osun States*. National Population Commission.

National Population Commission. (2017). "Nigerian current population 2017", National Population Commission. Available online at: http://www.population.gov.ng/index.php/80-publications/216-nigeria-s-population-now-182-million-dg-npopc. (Accessed June 25, 2017).

Ndhleve, S., Musemwa, L., & Zhou, L. (2013). Household food security in a coastal rural community of South Africa: Status, causes and coping strategies. *African Journal of Agriculture and Food Security*, 1, 15–20.

Obamiro, E. O., Dopper, W. and Kormawal, P. M. (2003), "Pillars of food security in Rural Areas of Nigeria", Internet Forum of food Africa, March 31 to April 11, available online at: http://foo dafrica.nri.org/security/internetpapers/ObamiroEunice

Obayelu, A. E. (2010). Classification of households into food security status in the North-Central Nigeria: An application of rasch measurement model. *ARPN Journal of Agricultural and Biological Science*, 5, 26–41.

Obayelu, A. E. (2012). "Households' food security status and its determinants in the North-Central Nigeria". *Journal of Food Economics*, 9, 241–256. Available online at: https://doi.org/10.1080/ 2164828X.2013.845559

Ojanuga, A. G. (2006). "Agroecological zones of Nigeria manual". In F. Berding & V. O. Chude (Eds.), *National special programme for food security (NSPFS) and FAO*.

Okwoche, V. A., & Asogwa, B. C. (2012). Analysis of food security situation among Nigerian rural farmers. *International Journal of Biological, Biomolecular, Agricultural, Food and Biotechnological Engineering*, 6, 1–5.

Olanike, D., William, S., & Deji, S. (2007). Socio-cultural factors influencing the feminization of HIV/AIDS in the rural area of Nigeria: Implication of food security. *Research Journal of Social Sciences*, 2, 7–13.

Omonona, B. T., & Agoi, G. A. (2007). "An analysis of food security situation among Nigerian urban households: Evidence from Lagos State, Nigeria. *Journal of Central European Agriculture*, 8, 397–406.

Omotesho, O. A., Adewumi, M. O., Muhammad-Lawal, A., & Ayinde, O. E. (2006). Determinants of food security among the rural farming households in Kwara State, Nigeria. *African Journal of General Agriculture*, 2, 7–15.

Omotesho, O. A., & Muhammad-Lawal, A. (2013). Optimal food plan for rural households' food security in Kwara State, Nigeria: The goal programming approach. *African Journal of Agriculture and Food Security*, 1, 75–81.

Oni, S. A., Maliwichi, L. L., & Obadire, O. S. (2013). Socio-economic factors affecting smallholder farming and household food security: A case of Thulamela local municipality in Vhembe District of Limpopo Province, South Africa. *African Journal of Agriculture and Food Security*, 1, 93–99.

Oyebanjo, O., Ambali, O. I., & Akerele, E. O. (2013). Determinants of food security status and incidence of food insecurity among rural farming households in Ijebu division of Ogun State, Nigeria. *Journal of Agricultural Science and Environment*, 13, 92–103.

Sanusi, R. A., Badejo, C. A., & Yusuf, B. O. (2006). Measuring household food insecurity in selected local government areas of Lagos and Ibadan, Nigeria. *Pakistan Journal of Nutrition, 5*, 62–67.

Shetty, P. (2015). From food security to food and nutrition security: role of agriculture and farming systems for nutrition. *Journal of Sustainable Food and Nutrition Security, 109*, 456–461.

The UK Telegraph. (March 11, 2017). "UN says world faces largest humanitarian crises since 1945 with 20 million people at risk of starvation", UK Telegraphy.

United Nations. (1975). "World food summit 1974, Rome–Italy", 5–16 November 1974. Available online at: www.fao.org

World Food Programme. (2002). World food summit (p. 2). 3 December 2002. Available online at: www.wfp.org

Burhan Ozkan is a Professor at the Agricultural Faculty, Akdeniz University, Antalya, Turkey. He currently chairs the Agribusiness Management Division of the Department of Agricultural Economics in the same faculty. He is the founder of this department chaired it for more than 15 years. He has been working in in the area of agricultural economics for the last 30 years. He has extensive experience in management with expertise in Agricultural Business Management. He has been involved in many international development projects and has authored over 300 publications, including 250 refereed journal articles.

Wasiu Olayinka Fawole is a Nigerian with about 12 years of working experience in academic and development sectors. He obtained his First and Second Degrees from Nigeria in 2008 and 2012, respectively. Dr. Fawole holds a Ph.D. in Agricultural Economics with specialization in development and food economics. He has published articles on various topics in refereed journals. He currently works on a World Bank Assisted Agro-Project in Nigeria and serves as a Visiting Scholar and Researcher under Prof. Dr. Burhan Ozkan at the Department of Agricultural Economics of the Akdeniz University, Antalya, Turkey.

Chapter 6
Food Security in Morocco: Risk Factors and Governance

Mohamed Zahour

Abstract Food security is a major issue and a growing challenge in the Global South. Its absence affects negatively the health and education of people as well as their ability to work and claim their rights. With its commitment to sustainable development, Morocco has implemented several strategies to modernize its agricultural sector, to sustainably manage its natural resources, and to ensure a social and ecological sustainable agricultural transition. The sectoral approaches adopted to improve food security have recently shown their limitations and proved ineffective, as the degradation of biodiversity and unsustainable use of resources are increasing in addition to a noticeable rise of social-ecological inequalities. The main objective of this chapter is to study and assess the food security situation in Morocco by identifying and assessing the various challenges facing food security and response mechanisms adopted by policy makers. The assessment of public policies in the area of food security reveals the lack of appropriate and robust planning and assessment tools of project impacts, which hinders the capacity to meet populations' expectations, particularly in rural areas. In addition, decisive actions are needed to support income-generating initiatives, to mitigate the effects of climate change, and to strengthen knowledge and social-ecological systems. The success of the national food security policy remains as well predicated on the mobilization of all stakeholders.

Keywords Food security · Environmental resources · Climate change · Agriculture development · Governance

1 Introduction

Food security represents a growing challenge in the developing world. It plays a key role in the fight against poverty, and its absence affects negatively people's health, education, ability to work, and their capacity to assert their rights and achieve equality (CESE, 2019). The awareness that the development of each country requires good governance of food security elicits strategic interests that are growingly supported

M. Zahour (✉)
College of Law, Economics, and Social Sciences, Ibn Zohr University, Agadir, Morocco

by public authorities, development partners, civil society organizations, the private sector, and professionals (FAO, 1974a, b). The aim of these sectors is to develop a reference framework for the actors who fight poverty and malnutrition and strive to achieve sustainable food security.

The concept of "food security" dates back to the 1970s; it was formed in a context of food crisis and rising food prices in international markets whose impacts had severely affected several countries in Asia (mainly India and Bangladesh) and in the Sahel. In 1974, the Food and Agriculture Organization (FAO) defined "food security" as the capacity to ensure, at the national level, the "availability at all times of adequate world food supplies of basic foodstuffs to sustain a steady expansion of food consumption and to offset fluctuations in production and prices" (FAO, 1974a, b). Since then, this concept has undergone a significant transformation over the years. In 1996, the FAO (1996) asserted that "food security exists when all people, at all times, have physical, social and economic access to sufficient, safe and nutritious food that meets their dietary needs and food preferences for an active and healthy life".

One of the main challenges that governments have to face is the capacity to provide nutritious food of sufficient quality and quantity to meet increasing demands caused by population growth and changing diets. These needs are increasing while the available natural resources are constantly decreasing. Another element to be taken into account is the impact of climate change on food systems and their natural base. Climate change contributes in increasing the risks of food insecurity especially for the most vulnerable populations (ONU, 2016). To face these emerging challenges, many frameworks had been developed and adopted in 2015, particularly the Sustainable Development Goals (SDGs) and Paris Agreement (ONU, 2016).

In Morocco, the food security situation seems to be a worrying issue due to many risk factors. These include: the increase of water shortage over the years; the over-exploitation of groundwater resources; the decline of biodiversity and soil quality; the adverse effects of climate change; and the fluctuations of energy prices in international markets (IRES, 2005). To meet the aforementioned challenges, Morocco has adopted several strategies with a priority given to the agricultural, social, and resource management sectors in order to achieve sustainable development (IRES, 2005).

With these challenges in view, the present chapter aims to highlight the various risk factors threatening food security in Morocco, the relevant response mechanisms, and the future prospects in this area.

2   Risk Factors Threatening Food Security in Morocco

Food security interacts with several elements pertained to ecological and economic dynamics and social policies. There was a time when the question concerning food security was posed in terms of national self-sufficiency. Currently, the inequality in natural resources endowments, the pressure of population growth, the effects of climate change, the high urbanization rate, the development of the comparative advantages of agriculture, and the liberalization of the economy pose the food security question both in terms of national production and in terms of the improvement of the citizen's social conditions.

In Morocco, food security faces several national and international challenges. Climate change, energy prices, the weaknesses of governance systems, and the spread of poverty have highlighted the challenges experienced by disadvantaged populations in accessing food on continuous basis.

2.1 Drought and Water Scarcity

The climate of Morocco is characterized by its strongly contrasted nature. Its pluvial and hydrological regimes are dominated by a noticeable irregularity in time and space, and characterized by the alternation of rainy and dry periods, which can last several years. The Economic, Social and Environmental Council (CESE) had alerted the government on the overuse of water resources whose quality and quantity are declining. According to the Council, the situation is alarming since water resources are currently estimated at less than 650 m^3/inhabitant/year compared to 2500 m^3 in 1960, and is likely to drop below 500 m^3 by 2030 (CESE, 2014). According to the latest assessments (Otmani, 2019), the resource potential of natural water in Morocco is estimated at nearly 22 billion m^3/year, divided into 18 billion m^3/year of surface waters and 4 billion m^3/year in ground water; that is the equivalent of almost 700 m^3/inhabitant/year.

Water resources represent a major constraint to the sustainability of food security. The Pillar I of the Green Morocco Plan (GMP) is based on an increase in a general production and productivity through the development of production factors including water (Toumi, 2008). However, the already critical state of water resources is likely to severely hinder the economic growth of the country; thus undermining the access of population to safe and quality water (Benmahane, 2018). Figure 1 illustrates the declining trend in water resources, which highlights the seriousness of the issue.

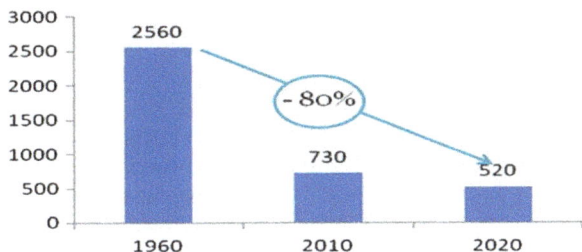

Fig. 1 Evolution of the water resources availability (m^3/inhabitant /year). *Source* Department of Environment, Ministry of Energy, Mines, Water and Environment (2012)

Even if the GMP advocates the use of water-saving irrigation techniques, it remains based on the principle of effective permanent water availability. Indeed, it qualifies the use of water as being "unsustainable on the long run" and observes that the use of water is potentially non-efficient for certain crops. Several assessment studies (Agoumi & Debagh, 2006) have emphasized that five out of eight basins (Bouregreg, Tensift, Oum Er-Rbia, Mouloya, Sud Atlas and Nord-Ouest) are in a situation of water deficit (stress) in the short and medium terms.

To respond to population growth, rising food demands, and economic development support, Morocco should adopt a policy based on the control and mobilization of water resources. This will enable vulnerable people to cope with the effects of drought, thus avoiding massive displacements.

2.2   Biodiversity Degradation

Biodiversity is essential for food security and it is one of the means by which it is possible to reduce poverty and improve livelihoods. Preserving biodiversity implies maintaining a varied diet, specifically a diet rich in nutrients necessary for a good health (UN, 2010). Throughout history, humanity has exploited and developed biodiversity, which is highly crucial for food and agricultural production and supply. Ecosystems were in the past stable and more adapted to the people's needs. This state is undergoing several changes as ecosystems are increasingly fragile because of the introduction of new plant and animal varieties (mainly in the areas of cereals, ovine, and caprine). Aquatic organisms, timber species, microorganisms, invertebrates, and thousands of species and their genetic variability constitute the fabric of ecosystems' biodiversity, and the Moroccan food and agricultural production systems depend on it.

The alarming disappearance of genetic diversity and the depletion of resources interfere with the humanity's potential to adapt to new socio-economic and environmental conditions, such as demographic and climate change (FAO, 2011). Accordingly, like the majority of countries, biodiversity in Morocco is threatened by: the inadequate management and weak conservation of plant and animal genetic resources (terrestrial and aquatic); the degradation of ecosystems which provide services such as pollinating, eliminating pests, recycling nutritious elements, and managing drainage basins; and the lack of ecosystem resilience when confronted to environmental constraints such as climate change (FAO, 2011). Table 1 highlights the consequences of the biodiversity loss and decline in some Moroccan areas.

Morocco is aware of the strategic importance of biodiversity for its sustainable socio-economic development. To honor its commitment to the Convention on Biological Diversity (CBD), the country has adopted a national strategy and action plan in 2004 (Behnassi, 2019) with the aim of promoting the protection of its biological heritage, thus maintaining its conservation and sustainable use. To this end, on the basis of the recommendations suggested in the COP10 of the CBD (UN,

Table 1 Main pressures on biodiversity in the agriculture context

No	Pressures	Consequences	Location
1	Drainage of wetlands for agricultural purposes	Habitat loss (approximately 34,000 ha)	Gharb, Loukkos
2	Transformation of wetlands: dam building	Transformation of rivers and natural lakes (draining) and disappearance of habitat (for migratory birds, disappearance of aquatic flora and fauna, etc.)	Moulouya, Mouloya, Bourgreg, Souss-Massa
3	Overgrazing	Deterioration of the biodiversity	Vulnerable pastoral areas: the oriental, southeast areas of Morocco, and so on
4	Agricultural pollution (fertilizers and pesticides)	Biodiversity loss The appearance of resistant parasites, etc	Irrigated and developed areas of large hydraulics
5	Introduction of exotic/exogenous species	Supplanting native species, or genetic pollution, or genetic erosion	Different regions of agricultural intensification Kenitra, Tangier and Souss-Massa

Source Agricultural Development Agency (Agence de Développement Agricole-ADA) (2012)

2010), Morocco has revised its national biodiversity strategy through a participatory process. This six-fold strategy, which covers the period 2016–2020 (MEMEE, 2016), recommends the incorporation of biodiversity into several policies that are in effect. Indeed, many sectors interact and complete each other in terms of natural resource management; the water sector advances the other sectors, but the forestry and agricultural sectors still represent poles of development comprising interactions and strong influences on biodiversity and its quality (Behnassi, 2019).

2.3   Climate Change

Climate change is currently the subject of deep scientific research and political debate. It represents a major challenge as it negatively impacts the sustainable development of all countries, particularly the developing ones. The current and expected decrease in annual rainfall as well as the increase of spatio-temporal irregularities will worsen the water scarcity situation in many hot-spot regions. The scarcity of water resources will affect particularly agriculture, as it is the socio-economic sector that is most sensitive to climatic variability in terms of agricultural production and income (Mouhot, 2012). The United Nations Framework Convention on Climate Change (UNFCCC) defines climate change in its first article as "a change of climate which is attributed directly or indirectly to human activity that alters the composition of the global atmosphere and which is in addition to natural climate variability observed over comparable time periods" (UNFCCC, 1992).

During the last few decades, Morocco has felt the impacts of climate change on its natural resources (CESE, 2014). Such impacts affect the resilience of both ecosystems and agricultural sector due in particular to the continuous decrease in water availability. Climate change manifests strongly itself in Morocco through the irregularity and unpredictability of rainfall, the higher incidence of storms, and prolonged droughts. These variations in weather conditions favor the appearance of pests and diseases that attack crops and livestock. Cultivated lands, pastures and forests, which cover a large surface of Morocco, are gradually exposed to increased climate variability and change. Furthermore, climate change threatens the progress made in achieving sustainable development goals, particularly those related to the reduction of hunger and poverty and environmental protection (Mahmalat & Bennis, 2012a, 2012b).

Climate change have practically impacted almost all regions in the country. These impacts are partly responsible for the decrease in food and agricultural production and the worsening of the situation of natural environment, as it is noticed through the degradation of many ecosystems like the pastoral areas and the oases. All climatic predictions, including less regular and decreasing precipitations, prolonged heat waves, and shorter cold periods, indicate the advent of a more arid climate in the Mediterranean region (GIZ, 2017). In Morocco, projections show that aridity will gradually increase due to the decrease in rainfall and the increase in temperature. This increase will eventually have negative repercussions on crop yield, especially starting from 2030 (ADA, 2012). Indeed, in the report "Impact of Climate Change on Crop Yield in Morocco," (Benaouda & Balaghi, 2009), the Ministry of Agriculture, Fisheries, Rural Development, Water and Forests and the World Bank, in collaboration with the National Institute of Agronomic Research (NIAR), the Food and Agriculture Organization of the United Nations (FAO) and the Department of National Meteorology (DNM) have jointly undertaken an original study (MAPM and ADA, 2011) to quantify the impacts of climate change on our agriculture by the end of the twenty-first century. The objective of this study consists of determining the economic and political options for adapting our agriculture to climate change. This could offer our country the ability to be prepared for possible critical situations.

Climate change affects both rainfed crops and irrigated crops. An in-depth study of the impact of climate change on fifty agricultural productions in the main agroecological zones carried out by Moroccan national institutions in collaboration with the FAO and the World Bank revealed a number of outcomes caused by climate change (Benaouda & Balaghi, 2009). The drier and warmer the climate becomes in Morocco, the more noticeable the consequences would become: negative effects on rainfed crops, soft wheat yields would experience a deficit of 33% by 2050, and several important irrigated crops would also be affected (FAO, 2011). To cope with this alarming situation, Morocco is required to double its efforts and set a general and conducive framework to manage and mitigate the risks of climate change. Such a framework should consist of establishing mechanisms for strategic monitoring and financing measures of adaptation and mitigation.

2.4 Energy Prices

Energy is essential for the competitiveness and sustainability of agriculture. The fluctuation of energy prices at the global level induces energy insecurity for the majority of countries, especially the energy-importing ones. They become incapable of meeting the energy needs of production sectors, mainly the agricultural sector which is directly linked to food security. The increase in energy prices have a mechanical effect on production costs, especially in countries such as Morocco where mechanization, tillage, crops, transport, and part of transformations are based mainly on fossil fuels. The price of fuel is also important in the areas of sea fishing, which constitutes a source of protein in the diet of Moroccans. The rise of fuel prices would also raise that of fertilizers and raw materials, such as phosphate.

In the current state of events, a state characterized by the globally rapid growth of energy consumption, the costs of producing foodstuffs will increase significantly; such a trend goes against the principle of access to food. The energy challenge for Morocco is substantial, as the country imports more than 95% of its energy needs (Mahmalat & Bennis, 2012a, 2012b). This leads to the necessity of adopting a new approach through investments in renewable energies and other long-lasting resources.

2.5 Issues Hindering Food Governance Systems

Good governance is generally defined as a distribution of national resources that takes into account participation, transparency, accountability, the rule of law, efficiency and equity. It is a process which implies a long-term construction and which involves a relatively slow change in the mental and organizational structures of governance (Behnassi, 2019).

Food insecurity, poverty, and malnutrition are challenges that can be either mitigated or amplified; both directions depend partly on the existing governance systems. The Moroccan governance system pertained to food security and the right to food shows weaknesses at the level of policy implementation and coordination, monitoring, evaluation and participation mechanisms of various stakeholders (Behnassi, 2019). The inefficiency of these mechanisms affects the convergence of policies and programs implemented at national, regional, and local levels. Accordingly, it could be suggested that the development and review of public policies in the food security area does capture the increased participation of civil society in achieving an environment favoring the collaboration and coordination of procedural interventions.

3 Food Security: Procedure and Governance Strategies

In general, governments manage the issue of food security through the implementation of global and sectoral strategies. These last have, in most cases, effects on the dietary, economic, environmental, ecological, and social sectors of the country. With a view of improving the food security situation and ensuring sustainable development, Morocco has implemented strategies affecting several sectors, including the agricultural, fishery, social, and water sectors.

The third section of this chapter consists in examining and evaluating the national food security policy through identifying and analyzing relevant sectoral strategies, particularly those covering the social, agricultural, and water areas.

3.1 Green Morocco Plan

The agriculture sector plays a vital role in the socio-economic development of Morocco since it contributes to almost 19% of the national GDP and the employment of more than 4 million rural individuals. It is unanimously accepted that food security, for more than 34 million inhabitants, is dependent on the development of the agricultural sector (HCP, 2008), even if this perception is questionable given the importance of fishery sector in the country. In order to strengthen the agricultural development process and stimulate the national economic growth, the Ministry of Agriculture and Fisheries has launched in 2008 the GMP as a new strategy (MAPM, 2008). This ambitious strategy, which aims at upgrading agricultural sector as a prominent vector for the socio-economic development of the country, revolves around an integrated global approach whose implementation has been supported by the initiation of numerous structural reforms. At the institutional level, these reforms have focused particularly on reorganizing the Ministry and creating 12 Regional Directions of Agriculture. The objective is to improve the supervision of the sector as well as provide it with the appropriate human and material resources in order to achieve the ambitious objectives of the new strategy.

To meet the food security challenge, the Ministry of Agriculture has invested in the formulation of this new strategy with the objective to: ensure a harmonious and balanced development of agricultural sector; maximize the value captured by farmers; face new challenges while preserving socio-economic balance; and take into account the transformations experienced by the global system of food processing industry. This agricultural strategy is formally articulated around a global approach which covers all stakeholders and their particular objectives and expectations. It is also based on two major pillars: modern intensive agriculture and inclusive smallholder farming. The objective set for modern intensive agriculture is to develop an efficient agriculture adapted to market rules through a new wave of private investments, organized around new models of fair aggregation. As for the inclusive smallholder farming, the objective is to develop an approach oriented towards fighting poverty through a significant increase in the agricultural income of the most vulnerable farmers or operators, especially those located in peripheral areas. Following are the important expected outcomes of this strategy:

- achieving sustainable management of food security;
- a considerable impact on the growth, the upgrading, and the increase of agricultural income, which is set as a robust tool to fight against rural poverty;
- an improvement in agricultural GDP, exports, and private investments;
- a more effective and larger-scale fight against poverty in rural and disadvantaged suburban areas;
- an improvement in purchasing power and in the quality-price ratio for the Moroccan consumer on the domestic market; and
- rebalancing the food deficit over the long term and securing trade as much as possible.

Though the GMP has many weaknesses, which will be presented below, it does succeed in achieving some goals. In terms of agricultural production, an important progress was made from 2008 to 2020, especially in the fields of cereals, sugar, olive crops, red and white meat, dairy and vegetable products. In addition, the plan has contributed to the promotion of local products and the organization of the inter-professions. This achievement was possible through the creation of federations and confederations in each region. Their role is to bring together farmers from each production area to facilitate the implementation of public–private contract programs signed by all concerned professionals. The plan has also made it possible to mobilize multi-stakeholder partnerships and substantial funding.

Despite these achievements, the implementation of GMP during 12 years has revealed many shortcomings and weaknesses, which affected its performance as a national strategy.

First, it should be noted that the authorities have not consulted the McKinsey firm on specific matters even if they initially entrusted it with the formulation of this agricultural development strategy. In addition, the hasty implementation of this plan has overlooked a prior constructive inclusive debate. This has undermined the principle of participation during the formulation of the GMP, thus compromising its inclusiveness. As a consequence, the public legitimacy of the plan was questionable since the public authorities allowed the GMP to be adopted far from any external evaluation, monitoring, and scrutiny or any possibility of improvement.

Second, it should be noted that nothing in the GMP indicates the priority of food security for concerned authorities. Indeed, the plan remained strangely silent on this crucial issue (Alahyane, 2017). Accordingly, the plan does not establish any hierarchy among the sectors. More precisely, the plan calls for the involvement of all sectors while food security requires giving priority to the development of the areas which have the potential to meet growing food demands, such as cereals, oilseeds, sugar, and dairy products. What has been stated so far shows that, in the GMP, food security gives priority to supply at the expense of expressed demands, or demands that could be expressed in the coming years by Moroccans. The GMP has set ambitious objectives in terms of agricultural production, but it has neglected the evolving food consumption needs. This will, as a consequence, hinder the food security and sovereignty of the country. Cereals for instance represent an important component of food security in Morocco. However, because of a limited and declining national production, as shown in Table 2 and Fig. 2 (FAO, 2020), the country uses imports to meet its consumption needs. Therefore, despite the efforts made, the GMP has not been able to rectify the food deficit in Morocco. Agricultural production is still unable to meet the growing dietary demands because the plan has not taken into account the major changes influencing food production and consumption in the country.

Table 2 Cereals production in Morocco

	2015–2019 average	2019	2020 forecast	Change 2020–2019
	000 tonnes			(%)
Wheat	5868	4100	3000	−26.8
Barley	2099	1161	900	−22.5
Maize	101	41	50	22.0
Others	96	95	95	−0.4
Total	8164	5397	4045	−25.1

Note Percentage change calculated from unrounded date
Source FAO/GIEWS Country Cereal Balance Sheet (2020)

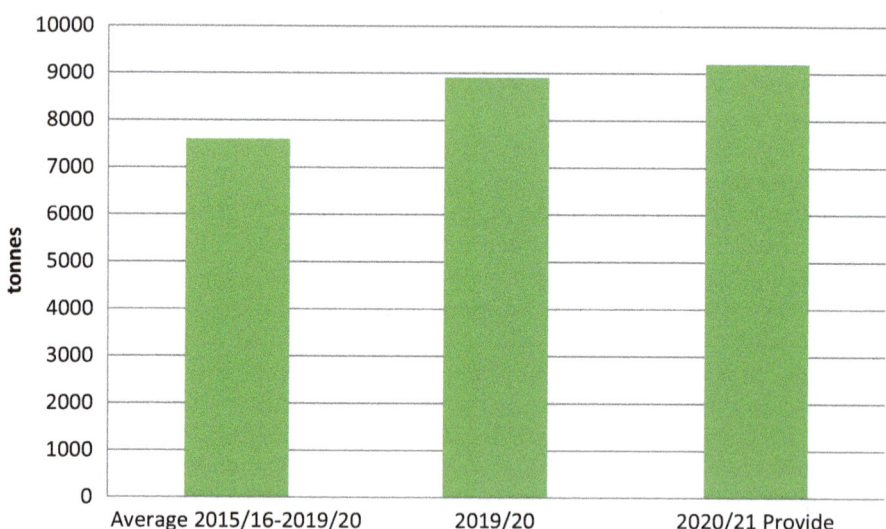

Fig. 2 Cereals imports in Morocco. *Note* Includes milled rice terms split year refers to the individual crop marketing years (for rice, the calendar year of second year shown).
Source FAO/GIEWS Country Cereal Balance Sheet (2020)

Third, the GMP has overlooked many key issues related, among others, to the adaptation to climate change, water and land resources management and governance, the vulnerability of subsistence farming, the poverty and food insecurity of small farmers, and the negative implications of free trade agreements.

All these shortcomings seem pushing decision-makers to rethink the agricultural development and food security challenge by implementing an alternative strategy called "Green Generation 2020–2030".

3.2 The Green Generation 2020–2030

This strategy relies on two keystones: enhancing the human capital; and seeking an effective dynamic for the sector's development through promoting the human and social development. Through this overall objective, the new strategy intends to consolidate the agricultural sector through doubling exports (MAD 50–60 billion) and rising the agricultural GDP (MAD 200–250 billion) by 2030. It also aims at improving the process of product distribution through the renewal of twelve whole-sale and traditional markets. The new strategy also seeks to strengthen the resilience and sustainability of agricultural development as well as improving the quality and capacity for innovation. To accomplish this, it is planned to grant approvals to 120 modern slaughter houses and to strength health and safety control.

The implementation of this strategy will require an annual increase in the sector's budget by almost 2.5% starting from 2020. In view of launching this strategy in time, it targets particularly territorial specificities and assets of each region. The idea is to develop coordination between these regions and all stakeholders, according to the guidelines of good governance in terms of monitoring and evaluating the investors and the indicators of efficiency and performance (MAPMDREF, 2020).

3.3 Land and Soil Management

Efficient land use and good soil maintenance are crucial to the accomplishment of food security. Integrated soil and water management improves agricultural production and enhances the soil's productivity and resistance to desertification and other impacts of climate variability and change. The sustainable management of fertilizers, soil, and water is essential if the country is to benefit from improved crop varieties and practices with the aim of promoting soil fertility (IAEA, 2012). The projects of technical cooperation enacted in the fields of crop improvement, soil preservation, and optimization of fertilizer usage are an opportunity to minimize the excessive use of phytosanitary products and to protect soil and water resources against pollution and degradation.

In Morocco, land issues endanger the profitability of the agricultural sector (Guen-nouni, 2016). They are likely to threaten the food security situation in the medium term. These land issues are not new. They emanate from a lack of responsible governance of land tenure when it comes to strategies applicable to issues related to land, inheritance, and the multiplication of legal status of arable land (Melk, collective, and so on).

To this end, Morocco has been contriving for a long time to put in place political, legal and organizational frameworks capable of promoting a responsible governance of land tenure (FAO, 2012). This was done through the establishment of several national institutions responsible for: protecting the legitimate individual and collective rights to lands; implementing laws, policies, and actions with the aim to protect farmers, small food producers, and vulnerable or marginalized individuals from the effects of climate change, and that in accordance with the obligations established by the country in the framework of agreements relevant to climate change; taking the necessary measures to protect the environment; financially helping farmers within the framework of the Fund for Agricultural Development (FDA) in order to equip and modernize their agricultural operations.

Aggregation constitutes an adequate solution to circumventing the issues related to the fragmentation of arable land and resolving the lack of organization in the agricultural sector (MAPMDREF, 2014). In addition, it is necessary to reform the legal status that regulates land in the country in a way that enhances the access of women to arable land and provides domestic and foreign investors with more opportunities to engage in agricultural investment.

The right to land is closely related to the right to food. Without land we cannot produce our own food products or derive from it a decent income. The constraints weighing on agricultural land call for recognizing the land issues that require urgent and radical solutions. Since land is an essential means of food production, the country has to equally and efficiently distribute it, especially on small-holder farmers who can ensure an efficient use which is likely to meet domestic food demands.

3.4   Strategies for Natural Resources Management: Water and Forests

The management of natural resources represents a major issue for Morocco. For this reason, the country has adopted many sectoral strategies to deal with deforestation and the overall insufficient and spatially heterogeneous rainfall, thus ensuring the sustainable management of the country's water and forests resources.

3.4.1   The National Strategy for Water Resources Management

In Morocco, the water sector has captured the attention of public authorities. It has been placed at the center of economic policy concerns given its decisive role in the water and food security of the country and the support it provides in the development of farming, especially irrigated agriculture. In this context, the country has long adopted a dynamic policy to provide the country with an adequate hydraulic infrastructure in order to improve access to water, meet the needs of industries, and develop large-scale irrigation. The actions taken to achieve this policy's objectives are as follows: establishing a policy of control and mobilization of water resources through the construction of large reservoir dams and water transfer infrastructures; developing technical skills and applied scientific research; bringing forward a long-term planning policy launched in the early 1980s which allowed decision-makers to anticipate water shortages through a long-term visibility (20–30 years); and finally, establishing a regulatory and institutional framework (Law 10–95) to consolidate the integrated, participative, and decentralized management of water resources through the creation of water basin agencies and the introduction of financial mechanisms for the protection and preservation of water resources. This policy has made it possible to provide the country with an important hydraulic infrastructure made up of 139 large dams, totaling a capacity of nearly 17.6 billion m^3. It has also provided the country with thousands of boreholes and wells for extracting groundwater (Mondiale, 2017).

3.4.2 The National Water Plan

To consolidate achievements and meet emerging challenges related to water and food security, Morocco has drafted the National Water Plan (NWP) (MEMEE, 2012). It consists of extending the orientations of the national water strategy (METLE, 2019).[1] The main ideas taken into account to develop the NWP comprise primarily: an integrated and concerted management of water demand and water resources; strengthening the country's water security and adapting to climate change; promoting good governance in the water sector and seeking effective actions from stakeholders; searching for convergence and creating consistency between sectoral programs; and lastly, providing the water sector with the necessary mechanisms and financing means for combining public subsidies and direct cost recovery through the pricing of water services. The NWP is based on three pillars, namely:

- *The management of water demand and water recovery*: this pillar aims at generalizing access to clean water and improving the scope of distribution networks. In the agricultural sector, the plan aims at extending the conversion program to localized irrigation until 2030. To achieve a better development of water resources, it is necessary to accelerate the hydro-agricultural development program downstream from existing or under construction dams.
- *The development of water offer* through: the pursuit of the surface water mobilization by dams; the local development of surface water through small dams; the use of unconventional water resources, including the desalination of seawater, the reuse of treated wastewater; and the possibility of transferring water from the

---

[1] This plan could concretize the philosophy of the national water strategy that is validated by the Moroccan government.

surplus of water basins located in the north-west to the water-deficit basins located in the center-west.

- *The preservation of water resources and natural environment, and adaption to climate change* through: the preservation of the quality of water resources and fighting pollution; sustainable management and protection of groundwater; development and protection of drainage basin; and safeguarding and preservation of sensitive areas, in particular wetlands and oases.

In order to adapt to climate change and enhance the capacity to manage extreme events, the NWP proposes actions with the objective to manage the adverse effects of floods and drought. Additionally, the plan calls for the development of regulatory and institutional reforms along with the revision of the water law (10–95), including an advanced coordination with the different related laws.

3.4.3   A New Strategy Called "Forests of Morocco"

Moroccan forests are in a dilapidated state due particularly to the deterioration of 17,000 ha of forest lands each year, low valuation of cork oak forests, over-harvesting of fuel wood, and over-exploitation of grazing areas (MAPMDREF, 2020).

As shown in Fig. 3, the average value of the Moroccan forest area from 1990 to 2016 was 11.8%, with a minimum of 11.1% in 1990 and a maximum of 12.7% in 2010. The value of 2016, which is 12.6%, remains very low when compared to the world average in 2016 which is 31.4%.

In order to protect the forest ecosystem, whose size is estimated at 9 million hectares, and considering the function that these ecosystems play on environmental,

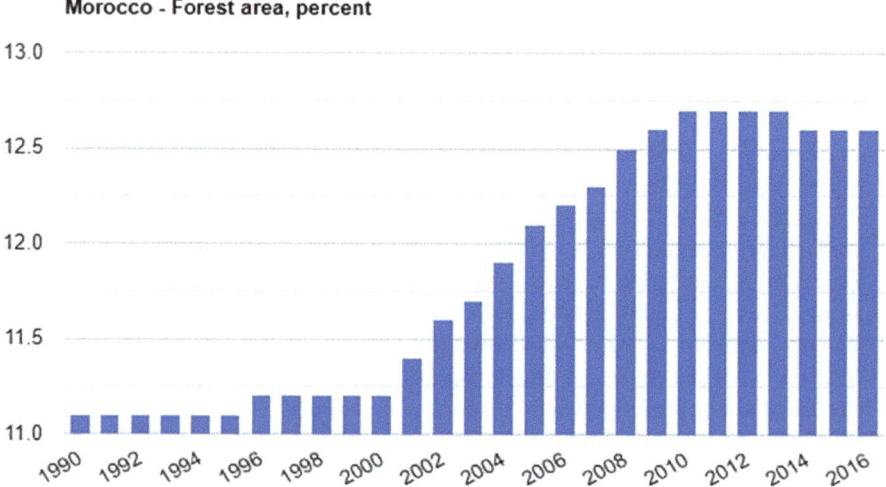

Morocco - Forest area, percent

Fig. 3 Morocco forest area percentage. *Source* FAO (2018)

economic, and social levels, the country is implementing a new strategy called "Forests of Morocco" to strengthen the competitiveness of the sector and ensure its modernization through the use of a management model capable of integration and sustainability and which is likely to generate wealth (MAPMDREF, 2020). This new strategy also aims at transforming forests into a space for development, ensuring the sustainable management of forestry resources, adopting a participative approach involving or connecting users, strengthening the production capacity of forests, and preserving the biodiversity. The strategy aspires to achieve some goals by 2030, including the enrichment of 133,000 ha of forests, the creation of 27,500 additional direct jobs, the improvement of production and ecotourism sectors' incomes to achieve an annual market value of MAD 5 billion (MAPMDREF, 2020).

This new strategy will revolve around four main pillars, namely: the creation of a new model based on a participative approach linking the populations with management; the development of forest areas according to their profusion; the promotion and modernization of forestry professions by creating modern forest nurseries and by adopting digital tools in the management of the sector; and finally the institutional reform of the sector through the upgrading of human resources, the establishment of a training and research center, and the creation of a Water and Forest Agency and a Nature Conservation Agency.

In addition, this strategy plans to develop and enhance 10 national parks to ensure an economic and social growth that respects the natural, cultural, and territorial heritage.

3.5 The Halieutis Plan

In Morocco, the fisheries sector is one of the pillars of development. It contributes to the domestic food security since seafood products are an essential source of food, as they provide an important part of the animal proteins necessary to humans. They are also an essential source of micronutrients such as iron, iodine, zinc, calcium, vitamin A, and vitamin B.

However, despite Morocco's fishing vocation, internal consumption of fresh fish remains low. This due to the fact that access to fishery products is increasingly out of reach of a large part of populations given the rising prices and excessive fish exports and fisheries agreements with many countries, especially the European Union (EU). To overcome this weakness, it is important to organize national commercial circuits of seafood, and impose clear and precise rules in this area. This allows the delivery of quality products to end-consumers at reasonable prices. It is also important to organize awareness campaigns on the quality and nutritional value of seafood.

This situation is still prevailing even if the country has launched in 2009 the Halieutis Plan to vitalize the fisheries sector. The plan is based on three major pillars and sixteen projects which aim at enhancing the sustainable use and management of fishery resources, strengthening the organization and performance of the sector, and encouraging the sector's competitiveness through the promotion of seafood (MAPMDREF, 2013).

To succeed in implementing this plan, the country has set up five key tools: the National Fisheries Committee; a Fund for the Adjustment and Modernization of Fisheries; an Agency for the Development of Aquaculture; a Center for the Valuation of Sea food; and a Monitoring Center for Employment in the Fisheries Sector. To complete the general framework of the sixteen projects of this Plan, other transversal actions have been developed. The aim of these actions is to clarify and complete the legal measures, ensure control effectiveness and traceability throughout the value chain, reinforce competences and improve the appeal toward the fishery trade, organize professional representation and encourage inter-profession, and lastly set up strong public governance to modernize the sector.

3.6   The National Initiative for Human Development (NIDH)

As has been discussed, Morocco has implemented major projects to face the new challenges of globalization (IRES, 2005). With the ambition of endowing Morocco with a more modern economic, infrastructural, and institutional base and strengthening the assets of the national economy, the country has embarked, in parallel with the above-mentioned projects, on a vast process of reforms to achieve an integrated and inclusive human development (Behnassi, 2019). Such a process aims at placing the citizen at the center of development dynamics by focusing on achieving tangible results in the fight against poverty, exclusion, and marginalization. A set of policies and programs have been put in place to this end. The National Initiative for Human Development (NIHD or INDH as commonly used) represents the flagship program in this endeavor.

Launched in May 2005 by the King Mohamed VI, the NIHD has benefited from the support of public authorities and the social development community. Since it has been put in practice, the Initiative has accumulated mixed results, but the positive ones, both on the quantitative and qualitative levels, demonstrate the relevance of its approach.

The NIHD was the vector of a national pact for human development, a framework to boost the development of the country, and a strategic approach of social transformation and governance reform and innovation. The Initiative has introduced a virtuous dynamic of projects for the benefit of targeted disadvantaged areas and populations. It has prompted a process that spreads the values of dignity, trust, participation, good governance. The NIHD has also encouraged the learning of a territorial approach, one based on proximity, strategic planning, synergy, and partnership.

The projects carried out within the framework of the NIHD have contributed to the improvement of social conditions in many rural areas through, among others, the creation of employment, the construction of roads, health centers as well as other infrastructures allowing the citizens to access public health and education.

Despite the weaknesses recorded in the planning, execution, and governance of NIHD programs, especially during the first phase of implementation, the results of this strategy are important, especially in terms of access to essential social services. Nevertheless, to achieve the overall objectives set by the NIHD, public authorities must include the territorial dimension into the national human development policy. They must, in parallel with this, take into account sustainable development indicators as *reporting* tools to measure social, environmental, and economic developments.

4 Analysis and Future Perspectives

Morocco aims to ensure food security, fight against poverty, protect and rationally manage its natural resources, and promote integrated rural development (CESE, 2017). For this reason, the country has adopted several strategies and policies in order to enhance and restructure its missions. The country brings together such frameworks which share the same objectives, while integrating the relevant stakeholders such as the inter-profession, civil society, and the private sector.

The National Food Security Strategy is an instance of the fight against poverty, malnutrition, and hunger. Its purpose consists of facing the risks caused by extreme events and shocks and strengthening the livelihoods of vulnerable households. It makes use of the appropriate tools to create sustainable livelihoods for the vulnerable populations, especially in rural and peri-urban areas. The ultimate goal consists of facilitating access to a healthy and sustainable nutrition as well as building resilience against disruptions. This food policy is a management and coordination tool of various interventions undertaken within the context of food security. Such interventions are carried out by the State in partnership with the concerned stakeholders for the achievement of development.

In the light of the presentation and analysis of risk factors threatening food security in Morocco and in the light of the investigation spotlighting some intervention strategies aimed at improving the general development framework, the following provides an overview of the perspectives that can result from the actions emanating from the overall strategy of the country.

4.1 Capacity Building

The overall strategy of the country's food security has as objective the promotion of an integrated and sustainable development. The agricultural policy, as a component of this overall strategy, encloses the entirety of the orientations, decisions, and actions whose purpose is to strengthen capacities in all dimensions of food security. This is to be done through:

- A sustainable increase in food availability;

- Improving the people's physical and financial affordability to varied and healthy food products;
- Improving the population's nutritional status, especially women, children, and elders;
- Strengthening the resilience of vulnerable populations;
- Strengthening the coordination and governance of food security;
- Strengthening the institutional and technical mechanisms to respond rapidly to food crises;
- Making sure that the objectives related to food security are incorporated into the national strategies concerned with poverty reduction and resource protection;
- Promoting the sustainability of agricultural and rural growth,
- Promoting access to resources and land ownership;
- Encouraging cooperation within the Mediterranean region with the objective of putting an end to financial speculation, and guaranteeing a better functioning of the market that provides services to farmers and consumers.

The success of actions oriented towards managing food security challenges depends on the establishment of a framework of cooperation and exchange connecting all stakeholders (public authorities, NGOs, businesses in the agricultural sector, small and medium-sized- holder farmers, research community, and international organizations).

4.2   Strengthening the Climate Resilience of Agriculture

Climate change has many adverse impacts on deprived individuals, as the socio-environmental context within which they live can be affected by the slightest disruptions. In Morocco, the agricultural sector and the poor are increasingly unable to cope with climate risks and maintain food security. Therefore, to carry out its international commitments, Morocco is taking the necessary measures to foster its adaptive capacity against climate change and to ensure a rapid transition to a low-carbon economy. To this end, the country prioritizes the development of mitigation and adaptation projects. In 2009, the country initiated a national plan aimed at identifying the priority actions to fight against the impacts of global warming. The territorial dynamic initiated by the plan can be subdivided into three types of measures: adaptation measures, mitigation measures, and transversal measures. The main objective of the plan is to mobilize the necessary interdepartmental dynamic and support in order to help implement the adopted actions.

In order to assist the agricultural sector in its struggle against the effects of global warming, Morocco is required to set up a coherent set of approaches to achieve the following goals: structure a stable action framework that can communicate with both the consumers and the producers as regards the costs and benefits of the greenhouse gas (GHG) mitigation/sequestration activities; set a real or implicit carbon price with the aim of encouraging producers and consumers to invest in products, technologies, and processes with low carbon potential; enhance the application of existing technologies and invest in R&D to devise new technologies capable of reducing GHG emissions and increasing productivity; design methods to better understand and measure the carbon potential of agriculture with a view to assess the progress made vis-à-vis national and international objectives relating to climate change; improve the capacity of producers to adapt to climate change and provide indemnity to the most vulnerable categories; and undertake a transformational process that targets integrated approaches, including agro-ecology, agro-forestry, climate-smart agriculture and conservation agriculture, which draw particularly on native and traditional knowledge.

4.3   Ecological Transition

The ecological transition refers to a movement from the current mode of production and consumption to a more ecological mode. This concept refers to the set of ecological practices that must be promoted at the local level to reduce dependence on fossil fuels and face global warming. Ecological transition corresponds to a change in economic and social models, a change which will profoundly transform the methods of consuming, producing, working, and coexisting.

Being aware of the importance of such a kind of transition, Morocco has carried out several actions and programs in this direction by, among others, strengthening scientific research. The areas targeted are related to food security and funding mobilization. The idea is to equip the agricultural sector with modern means to substitute phytosanitary products. The country has also introduced strategies and programs dealing with the issue of food safety to succeed in concretizing this transitional process. The methods used consist of reinforcing the mechanisms and means of evaluation and monitoring (CESE, 2019).

4.4   Evaluating, Monitoring, and Coordinating Strategies

The good governance of food security is essential at the global, national, and local levels to enhance actions against poverty, hunger, and social-ecological vulnerability. In Morocco, priority must be given to the integrated management and financing of the programs and strategies that revolve around food security. As for the interventions of the actors, it is necessary to structure a framework of coordination and cooperation to ensure the harmony of actions and programs whose purpose is to meet the needs of populations.

To achieve the goals that have been set, the country should develop and implement mechanisms of evaluation, monitoring, coordination, and accountability. Such mechanisms would assess and readjust the policies of elected and non-elected authorities at all levels in order to maintain transparency of the whole process. Finally, the participatory approach invites actors from various areas to contribute to the development and application of food security programs and strategies.

4.5 Nutrition Education

Nutrition education is a support and outreach activity that aims at instilling awareness and inviting the population to adopt good nutritional practices. This educational asset could be defined as the set of communication activities that seek to instill voluntary positive modifications of practices that impact the nutritional status of the population (FAO, 1998). Indeed, communication plays a vital role in ensuring the sustainability of food security programs. It could be divided into three categories: social, educational, and institutional communication. The first raises awareness within a society and promotes interactive information between communities. The second manages the transfer of knowledge and techniques. And the third enhances the coordination between different stakeholders. For instance, promoting, supporting, and encouraging breastfeeding (Malkaoui, 2012) is an instance of the actions for which the Ministry of Health has developed an Information, Education, and Communication (IEC) strategy.

5 Conclusion

As has been demonstrated above, food security constitutes a real challenge for Morocco. Achieving food security is essential to secure the right to food. The latter is a universal human right proclaimed by the international law as much as by many domestic legal systems. Promoting this right can also reinforce economic and social rights.

Meeting the food security challenge requires a multisectoral approach that transcends the various actions that are carried out in sectors such as agriculture and energy. This implies the implementation of additional actions in public health, education, industry or other relevant sectors. These actions are necessary to achieve an equitable and sustainable development, as long as it ensures food security for all Moroccans without compromising economic, social, and environmental foundations.

Agricultural systems which allow a considerable consumption of inputs and resources, thus causing massive deforestation, water shortages, soil depletion and excessive GHG emissions, are not suitable for the achievement of a sustainable food and agricultural production. Analyzing the various risk factors threatening food security in Morocco and the relevant response mechanisms shows that the current framework and overall orientation require the reshaping of the whole system. This is because the question of food security is complex and multifaceted and interacts with many areas and dynamics at all governance levels.

References

Agence de Développement Agricole (ADA). (2012). *Etude Environnemental du Plan Maroc Vert, Rapport Définitif* (pp. 40–41). Phénixo.

Agoumi, A., & Debbarh, A. (2006). *Ressources en Eau et Bassins Versants du Maroc: 50 Ans de Développement 1955–2005*. IRES.

Alahyane, S. (2017). *Sécurité Alimentaire et Politique Agricole au Maroc* (pp. 145–160). Unpublished Ph.D. thesis, FSJES.

Behnassi, M. (2019). Coordination, Suivi et Evaluation des Stratégies Multisectorielles Relatives à la Réduction de la Pauvreté Rurale et à la Sécurité Alimentaire et Nutritionnelle, Zoom sur le Cas du Maroc», Project IPC-IG (FAO Country Policy Support to reach SDGs 1 and 2: Capacity development programme to enhance the use of poverty analysis in policy making'), International Policy Center for Inclusive Growth (IPC-IG), Unpublished.

Benaouda, H., & Balaghi, R. (2009). *Les Changements Climatiques: Impacts sur l'Agriculture au Maroc*. AGDUMED, Rabat.

Benmahane, M. (2018). Green economy and sustainable development in Morocco: Assessment and prospects. *Journal d'Economie, de management et de droit, JMED, 1* N°1, Rabat.

Conseil Economique, Social et Environnemental (CESE). (2014). *Rapport annuel de 2014*. Retrieved from: www.cese.ma.

Conseil Economique, Social et Environnemental (CESE). (2017). *Développement du Monde Rural Défis et Perspectives*. CESE report. Retrieved from: www.cese.ma.

Conseil Economique, Social et Environnemental (CESE). (2019). *Le Nouveau Modèle de Développement du Maroc*. CESE.

Department of Environment, Ministry of Energy, Mines, Water and Environment (2012). Développement Durable au Maroc: bilan et perspectives, de Rio à Rio +20, P.4.

FAO. (1974a). *World Food Summit*. Rome.

FAO. (1974b). *La Conférence Mondiale de l'Alimentation*. Rome.

FAO. (1996). *World Food Summit*. Rome.

FAO. (1998). *Les activités nutritionnelles au niveau communautaire: Expérience dans les pays du Sahel*. Etude FAO alimentaire et nutrition 67. Rome.

FAO. (2011). *La Protection de L'environnement Mondial Adapter L'agriculture au Changement Climatique*. Rome. Report retrieved from: www.fao.org.

FAO. (2012). *Directives Volontaires pour une Gouvernance Responsable des Régimes Fonciers Applicables aux Terres, aux Pêches et aux Forets dans le contexte de la Sécurité Alimentaire Nationale*. Rome.

FAO (2018). La situation des forets du monde: les forêts au service du développement durable, disponible sur le lien: http://www.fao.org/3/I9535FR/i9535fr.pdf et http://www.fao.org/faostat/fr/#data/FO, consulté le (15/04/2020).

FAO. (2020). *Global Information and Early Warning Systems*. Retrieved from: www.fao.org/giews/countrybrief/country.jsp?code=MAR

German Technical Cooperation (GIZ). (2017). *Private Sector Adaptation to Climate Change (PSACC)*.

Guennouni, A. (2016). Fruits Rouges: des Problèmes Fonciers Menacent la Rentabilité des Cultures. *Agriculture du Maghreb Journal*. N°93, Marsh 2016, p. 38.

Haut Commissariat au Plan. (2008). *Agriculture 2030: Quels Avenirs pour le Maroc*. HCP, 2008.

Institut Royal des Etudes Stratégiques (IRES). (2005). *50 ans de Développement Humain: Perspectives 2025 Cadre Naturel, Environnement et Téritoires*. Thematic Report. Morocco.

International Atomic Energy Agency (IAEA). (2012). *Agriculture and Food Security*. Contribution of IAEA, Technical Cooperation Program. Retrieved from https://www.iaea.org/sites/default/files/documents/tc/Agri_Eng.pdf

Mahmalat, E., & Bennis, A. (2012a). *Environnement et Changement Climatique au Maroc: Diagnostic et Perspectives*. Konrad-Adenauer-Stiftung E.V.Bureau du Maroc.

Mahmalat, E., & Bennis, A. (2012b). *Environnement et Changement Climatique au Maroc: Diagnostic et Perspectives* (1ière Ed.). Morocco.

Malkaoui, H. (2012). *Projets et Stratégies Adoptés au Maroc par les Différentes Institutions en Matière de Sécurité Alimentaire et Nutritionnelle* (pp. 174–177). Unpublished Engineering Thesis. IAV HASSAN II.

MAPM and ADA (2011). *Projet d'Intégration du Changement Climatique dans la mise en œuvre du Plan Maroc Vert (PICCPMV)*. Etude Cadre de l'Impact Environnemental et Social, Maroc.

Ministère de l'Agriculture, de la Pêche Maritime, de Développement Rural et des Eaux et Forêts (MAPMDREF). (2008). *Plan Maroc Vert*. Morocco.

Ministère de l'Agriculture, de la Pêche Maritime, de Développement Rural et des Eaux et Forêts (MAPMDREF) (2013). *Stratégie Halieutis*. Retrieved from: www.mpm.gov.ma.

Ministère de l'Agriculture, de la Pêche Maritime, de Développement Rural et des Eaux et Forêts (MAPMDREF). (2014). *L'Etat des Ressources Génétiques Animales* (p. 7). Second Report. Rabat.

Ministère de l'Agriculture, de la Pêche Maritime, de Développement Rural et des Eaux et Forêts (MAPMDREF). (2020). *Stratégie Green Génération 2020–2030*. Retrieved from: www.agriculture.gov.ma.

Ministère de l'Energie, des Mines, de l'Eau et de l'Environnement (MEMEE). (2012). *Politique de l'Eau au Maroc* (pp. 2–13). Rabat, Morocco.

Ministère de l'Energie, des Mines, de l'Eau et de l'Environnement (MEMEE). (2016). *Stratégie et plan d'Action National pour la Diversité Biologique du Maroc 2016–2020*. Morocco.

Ministère de l'Equipement, du Transport de la Logistique et de l'Eau (METLE). (2019). *Plan National de l'Eau 2020–2050*. Morocco.

Mondiale, B. (2017). *Pratique Globale de l'Eau: Gestion de la Rareté de l'Eau en Milieu Urbain* (pp. 21–30). Washington.

Mouhot, J. F. (2012). Du Climat au Changement: Chantiers, Leçons et Conflits. *Cultures et Conflits Journal, 88*, 19–42.

ONU. (2016). *La Mise en Œuvre des Objectifs de Développement Durable à L'occasion de la Revue Nationale de la France*. ONU.

Otmani, S. (2019). *Discours Lors de la Séance Mensuelle Consacrée à la Politique Générale*. Chambre des représentants, Rabat. Retrieved from: www.chambredesrepresentants.ma/fr/seances-pleinieres.

Toumi, L. (2008). *La Nouvelle Stratégie Agricole au Maroc Plan Vert: Les Clés de la Réussite* (pp. 7–10). Rabat.

UNFCCC. (1992). *La Convention-Cadre des Nations Unies sur les Changements Climatiques*. Rio de Janeiro.

United Nations. (2010). In *International Conference on Biodiversity of Nagoya-COP 10*. Japan.

Mohamed Zahour is a Ph.D. student at the Faculty of law, Economics, and Social Sciences of Agadir, Ibn Zohr University, Morocco. He graduated from the National School of Administration, Rabat in the area of Partnership and International Cooperation. In 2012, Zahour obtained his first Research Master Degree in Administrative Law and Sciences for Development at Abdelmalek Essaadi University of Tangier. He subsequently obtained a second Research Master Degree in International and Diplomatic Studies at Mohamed V University of Rabat in 2014. Zahour then worked in several ministerial departments. He does his research in the fields of international law, international relations, food security, and human security. Mohamed Zahour has also worked as a temporary lecturer at the University Center of Guelmim and at the Higher School of Technology of Guelmim, Morocco.

Chapter 7
Climate Change, Agricultural Policy and Food Security in Morocco

Saidi Abdelmajid, Ahmed Mukhtar, Mirza Barjees Baig, and Michael R. Reed

Abstract The agricultural sector constitutes a key factor ensuring food security in Morocco, meeting the dietary and nutrition needs through vegetables, fruits, meat, and other products. It employs almost 40% of the working population at the national level and 74% in rural areas. It plays multiple roles in the economy as it provides market and non-market services, contributes to a certain balance between urban/rural, provides a basis to rural people for their social attachment, serves as a cradle for cultural landscapes, and attracts nature lovers by promoting agricultural tourism. However, Moroccan agriculture is still vulnerable to climatic variations and remains strongly constrained by the annual and interannual variability of precipitations. It should be recalled that the Moroccan climate, generally arid to semi-arid, is influenced by global warming with a rise in temperatures and a sharp decrease in rainfall, with prolonged droughts and catastrophic floods. These phenomena are expected to intensify in the coming decades, threatening to further degrade natural resources, especially water scarcity. Furthermore, Morocco is an exporter of agricultural crops that consume a lot of water (such as citrus and tomatoes). It has opted for intensive practices, in particular within the framework of the Green Morocco Plan, in all territories despite their ecological limitations. Intensive agriculture is for instance responsible for the overexploitation of the groundwater in the Souss-Massa region; an area responsible for 60% of the nation's exports of fruit and vegetables. The estimated water deficit of 271 million m^3 for the region was accompanied by a drop in the level of the groundwater and a sea-level rise, with repercussions on the environment

S. Abdelmajid (✉)
Department of Economics, University Moulay Ismail, Meknes, Morocco

A. Mukhtar
Islamic Educational, Scientific and Cultural Organization (ISESCO), Rabat, Morocco

M. B. Baig
Water & Desert Research, Prince Sultan Institute for Environmental, King Saud University, Riyadh, Saudi Arabia
e-mail: mbbaig@ksu.edu.sa

M. R. Reed
Agricultural Economics, University of Kentucky, Lexington, USA
e-mail: Michael.Reed@uky.edu

© The Author(s), under exclusive license to Springer Nature Switzerland AG 2021
M. Behnassi et al. (eds.), *Emerging Challenges to Food Production and Security in Asia, Middle East, and Africa*, https://doi.org/10.1007/978-3-030-72987-5_7

171

and agriculture. Some producers in this region have migrated to Zagora, an extension area of the oasis region, to grow watermelons. In addition, climate change has further aggravated the already water-stressed situation. Indeed, the increasing variability of precipitation and the high frequency of droughts are likely to further reduce the water availability. These constraints will have negative impacts on potential agricultural yields, employment opportunities, and purchasing power of rural people. In such a scenario, it seems imperative to adopt sustainable agricultural practices to adapt and deal with climate change.

Keywords Climate change · Agricultural policy · Natural resources · Sustainable agricultural practices · Food security

1 Introduction

Agriculture is a key sector responsible for the economic, social, and environmental development in Morocco. It contributes 13–14% of the national GDP. In terms of food security, agriculture provides much of the food needs of Moroccans, especially vegetables, fruits, dairy products, and meat. In social terms, agriculture employs almost 40% of the active population at the national level and 74% in rural areas, where nearly 40% of the population still lives. Agriculture also has a significant multiplier effect on the economy, insofar as it provides market and non-market services allowing an urban/rural balance of the population, a social bond in rural areas, a production of cultural landscapes, and attractive locations for tourism (Ministry of Economy, 2019).

Despite the efforts made in irrigation (construction of dams, expansion of the irrigation area, etc.), the agricultural sector is still heavily reliant on rainfall and a good cereal season often depends on timely precipitation. The variability in rainfall is significant for cereal crops (cereal production fluctuated between 3.35 million tonnes[1] in 2016 and 114.7 in 2015). This is explained by the fact that this sector still occupies nearly 60% of the total useful agricultural area, 90% of which is practiced in unfavourable rain-fed "*bour*" areas, thus making it more vulnerable to the harmful effects of climate change (Fallahtrade, 2020).

In Morocco, the impacts of climate change are already tangible because temperatures have increased by +1 to +2 °C on an annual average between 1901 and 2012. Over the past 30 years, Morocco's upward temperature trend has been much higher than the global average, +0.42 °C/decade on average since 1990 for Morocco versus +0.28 °C/decade for all countries (Woillez, 2019). Moreover, the climate in Morocco, which is generally semi-arid, undergoes oceanic influences and is characterized by a great annual and interannual variability, and a marked alternation of wet periods (1995–1996, 2009–2010, 2017–2018) and prolonged droughts (1985–1995, 2004–2006). In recent years, Morocco has experienced catastrophic floods especially in its southern part: more than 100 mm in 24 h, winds from 80 to 120 km/h, and

---

[1] Tonne = 10 Quintal (Qx).

very aggressive late summer storms especially in the mountains, accompanied by lightning and hail (Mastere et al., 2019). According to various reports and studies on climate projections, these phenomena are expected to increase during the coming decades, which risks increasing the degradation of natural resources, particularly the scarcity of water.

So, what will be the impacts of these phenomena on the sustainability of the agricultural sector and, therefore, on the country's food security? To answer this, we will first examine the constraints induced by climate change on natural resources in the Moroccan context. In a second step, we will investigate the extent to which the agricultural sector and public policies (Green Morocco Plan, etc.) can ensure food security for Moroccans under these constraints.

2 Climate Change and Natural Resources in Morocco

Morocco is in the southwest of the Mediterranean region, in the northwest part of the African continent. Morocco has a long coast which extends over more than 3.500 km, including 2.934 km in the Atlantic Ocean, and 512 km in the Mediterranean Sea. Stretching from northeast to southwest, the Atlas Mountains provide Morocco with important freshwater resources on which the country heavily depends for agriculture and irrigation. The climate of Morocco is Mediterranean in the north and semi arid or arid in the south. Figure 1 illustrates the bioclimatic stages: humid, sub humid, semi-arid, arid, and Saharan.

Despite the fact that Africa is the lowest CO_2 emitting continent on the planet (accounting for only 4% of global greenhouse gas emissions), it is considered the most vulnerable to climate change (rise in temperature, droughts, increase in sea levels, floods, etc.) due to its high exposure and low adaptive capacity (Niang et al., 2014). Like other African countries, Morocco is experiencing a significant increase in temperatures and a decrease in precipitation generating disastrous natural phenomena such as drought, floods, salinization, silting up, etc. This increases the pressure on natural resources used by the agricultural sector (especially water and soil).

2.1 Effects of Climate Change on Temperatures and Precipitation

2.1.1 Rising Temperatures

The average annual temperatures have increased in most Moroccan cities between the periods 1971–1980 and 2009–2017 (see Table 1).

Climatic observations over the past few decades clearly show the progression of the semi-arid climate towards the north of the country, close to the Mediterranean,

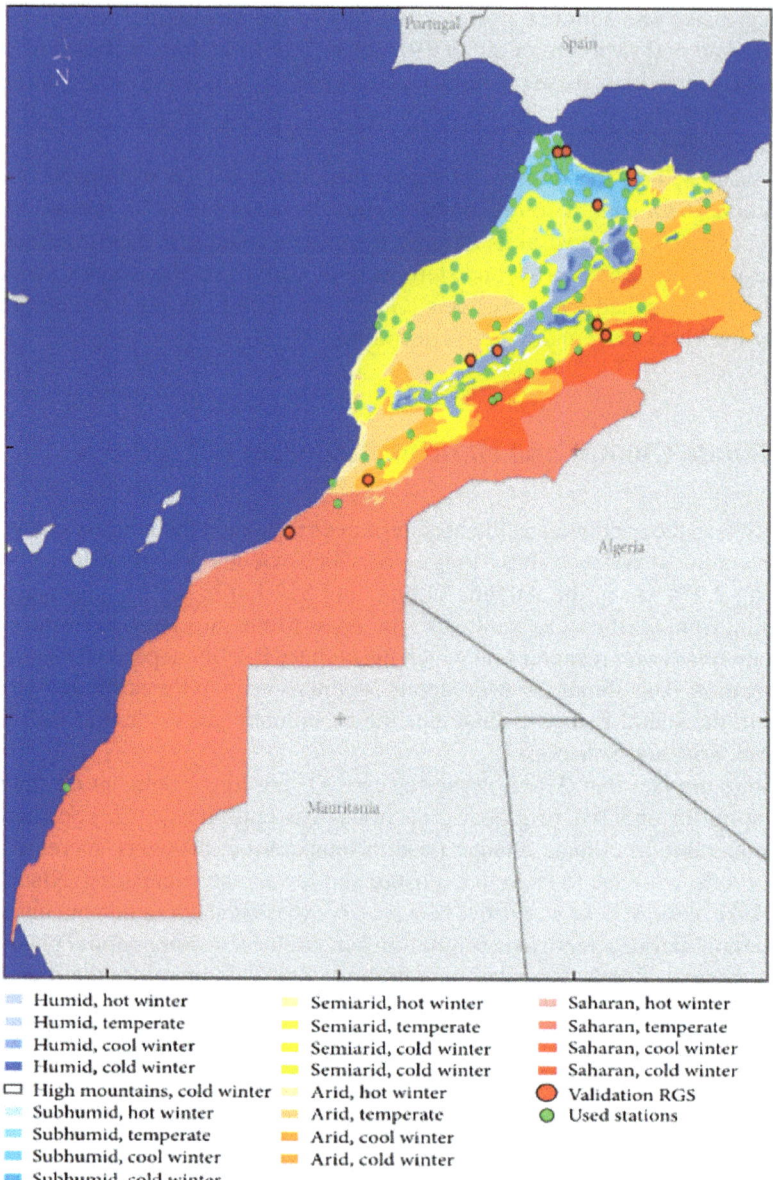

Fig. 1 Bioclimatic stages of Morocco according to Emberger's quotient. *Source* Ezzine et al. (2017)

Table 1 Evolution of average temperatures between the periods 1971–1980 and 2009–2017 in the various stations of morocco weather

| City | Evolution of average temperatures between the periods 1971–1980 and 2009–2017 in the various stations of kingdom of morocco weather (°C) | | | | |
	Period 1971–1980	Period 1998–2007	Variation	Period 2009–2017	Overall variation
Oujda	16.0	17.3	1.3	18.6	2.6
Taza	17.1	19.1	2.0	19.6	2.5
Errachidia	18.5	19.8	1.3	20.8	2.3
Beni Mellal	17.6	19.3	1.7	19.7	2.1
Casablanca	16.8	18.4	1.6	18.6	1.8
Larache	16.7	18.2	1.5	18.5	1.8
Marrakech	19.2	20.1	0.9	21.0	1.8
Meknès	16.6	17.8	1.2	18.4	1.8
Safi	17.6	18.5	0.9	19.4	1.8
Agadir	18.2	19.4	1.2	19.9	1.7
Fès	16.3	17.6	1.3	18.0	1.7
Ouarzazate	18.5	19.8	1.3	20.1	1.6
Tétouan	17.5	17.8	0.3	19.1	1,6
Sidi Ifni	18.4	19.6	1.2	19.6	1.2
Tanger	17.4	18,5	1.1	18.6	1.2
Essaouira	17.0	18.0	1.0	18.2	1.2
Rabat	16.7	18.0	1.3	17.8	1.1
Al Hoceima	17.7	18.4	0.7	18.5	0.8

Source Data of National Meteorology Directorate

whose surface temperature has increased between +0.29 and +0.44 °C per decade according to its different sub-regions (North, South of the Mediterranean) since the early 1980s (MedECC, 2020). The rise in sea level at the Mediterranean coast of Morocco is around 0.6 mm/year for the period 1945–2000. While sea levels on the Atlantic coast have increased between 1.6 and 2 mm/year for the period 1955–2003 (GIZ, 2017). The IPCC predicts an increase of 37–90 cm in sea level by the end of the twenty-first century. Beach surfaces are shrinking and the annual frequency of natural disasters at sea will increase and swells will become more dangerous (Guennouni, 2016). The advantage that Morocco derives from its 3500 km coastline, on which 80% of its industrial and energy infrastructure is installed, will become a serious handicap (Ministry of Environment, 2016).

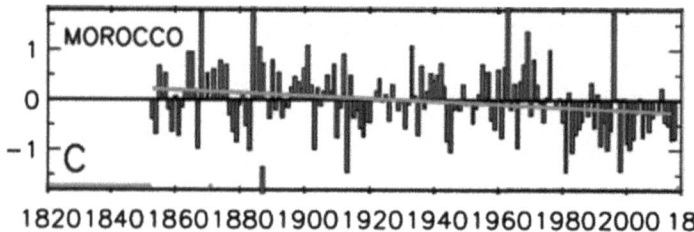

Fig. 2 Average rainfall in Morocco between 1850 and 2018. *Source* Nicholson et al. (2018)

2.1.2 Decreasing Precipitation

Annual precipitation has had a significant decrease over the period 1951–2010, between −10 and −25 mm/year per decade over the north of the country (Woillez, 2019). Considering the past century, the particularly dry character of the last 40 years appears fairly clear (Nicholson et al., 2018), even if some years before 1980 were also marked by drought (see Fig. 2).

El Ajhar et al. (2018) used data from 19 meteorological stations covering the period 1980–2015 to show that the distribution of rainfall in Morocco experienced a change during the period 1980–2015. This change consists of a movement towards drier conditions. The geographic distribution of climate types during the period 1980–2015 shows an increase in semi-arid climates and a reduction in sub-humid climates. In addition, drought appears to become more persistent at the end of the rainy season, with an increase of about 13 days in the maximum period of consecutive dry days (Woillez, 2019).

All the indicators for 2020 show that Morocco is headed for a second year of drought, the third since 2016. The year 2016 is considered the worst drought of the last three decades. The rainfall deficit for 2019 reached 37.6%, or 124.5 mm in 2019–2020, versus 199.5 in 2018–2020. The filling rate of dams for agricultural use fell to 47.6% versus 60.2% a year earlier in the 2018–2019 campaign, which resulted in a sharp decline in cereal production to 5,2 million tonnes (Bank Al-Maghrib, 2020).

The 2020 crop year recorded limited rainfall at 205 mm on April 22, 2020, down 34% from the 30-year average (323.7 mm) and 25% below the previous season (282.1 mm) for the same date. The impact of this low-rainfall volume was exacerbated by its poor and irregular spatiotemporal distribution. There was low rainfall at all stages of grain development, and there were long dry periods (almost 40 days) during tillering and upstream periods. The rainfall deficit affected all cereal regions. In Chaouia and Haouz, the rainfall deficit was 50% on average, while in Saïss, Pré-Rif, and the north, the deficit varied between 30 and 45%.[2]

Many climate change studies (i.e. IPCC, 2014) project an increase in temperature between 2.5 and 3.8 °C by 2050. The rise in temperature is projected to be accompanied by a decrease in precipitation, between −15 and −40% in 2050 depending on the

---

[2] Press release from the Ministry of Agriculture (Morocco) 22/04/2020. http://mapecology.ma/act ualites/campagne-agricole-2019-2020-a-enregistre-pluviometrie-limitee-ministere.

different scenarios. For a warming of +2 °C compared to 1980–2010, the decrease in annual water flows in Morocco would be more than 30% in the whole country (Schewe et al., 2013; Woillez, 2019). The frequency of drought days would also increase by 50% in 2070–2099 compared to 1976–2005 (Prudhomme et al., 2013; Woillez, 2019). Climate change is, therefore, expected to decrease water supply while increasing its demand, leading to increased water shortages (Aqueduct Water Risk Atlas, 2019). Competition for increasingly scarce resources is highly likely: demand for urban water in Morocco is expected to increase by 60–100% in most major cities by 2050, which could seriously threaten the agricultural sector's access to water, disrupting rural incomes and livelihoods (Iceland et al., 2018).

2.2 Impact of Climate Change on Natural Resources

The decline in precipitation and the increase in temperatures for Morocco combine to accentuate the pressure dynamic on natural resources. The availability of water in Morocco has increased from 3500 m^3 per person in 1960, to 730 m^3 per person in 2005 and 645 m^3 per person in 2015 (Dahan, 2017). The situation of the *Al Massira* dam, the second largest reservoir in Morocco, illustrates this phenomenon well. The area of the dam was reduced by more than 60% during the years 2016–2018 (see Fig. 3). The last time *Al Massira* was at this level was between 2005 and 2008, when more than 700,000 Moroccans were affected by the drought and the cereal production fell by 50%. *Al Massira* supplies water to the agricultural sector of the Doukkala region, as well as to many cities, including Casablanca. As reservoir levels continue to decrease, demand for water continues to increase. In addition to the growing demand for urban water and the development of irrigated agriculture, the city of Marrakech plans to draw water from *Al Massira* through a major water transfer project (Iceland et al., 2018).

Fig. 3 Al Massira Dam Morocco surface, water area over time. *Source* Iceland et al. (2018)

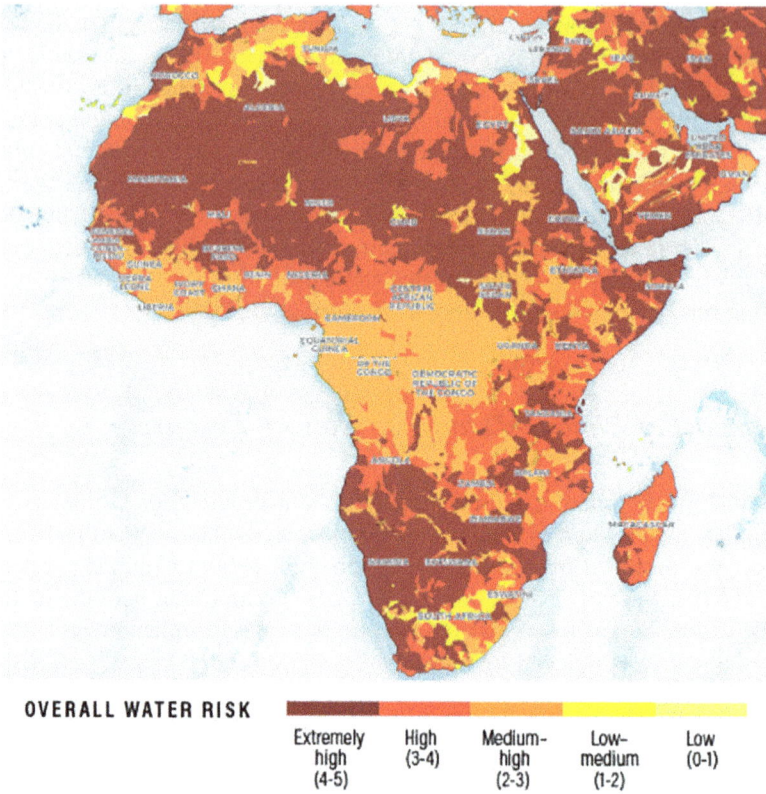

Fig. 4 African countries face some of the highest water risk in the world. *Source* Masson et al. (2019)

According to the World Resources Institute (WRI) report, Morocco is ranked among the countries with high water stress (Mason et al., 2019). In Fig. 4, Morocco is highlighted in red to denote its difficult situation. Morocco is the 22nd most threatened country in the ranking and comes second in North Africa, just after Libya, the 6th most threatened country (Hofste et al., 2019). This is a disturbing place for the country. However, the dry southern most regions are not most at risk, but instead it is the most populated areas where agriculture is most developed that are severely threatened. For instance, the Chaouia-Ouardigha region is most threatened by shortages, with an index of 0.92–1, a threat equivalent to that found on average in Libya. Rabat-Salé-Zemmour region is 0.85. Grand Casablanca is the most threatened region, with an index of 0.78. The Guelmim-Es-Semara region, which is extremely arid, is the least affected in the kingdom (Aqueduct Water Risk Atlas, 2019).

Furthermore, the yield of irrigated crops in Morocco is also expected to decline due to the increased pumping of groundwater and risk of salinization for aquifers. The High Commission for Water and Forests and the Fight against Desertification estimates that the salinization of soils in Morocco affects almost all large irrigated areas; an area of 37,000 ha (out of 414,000 studied) is affected by salinization or alkalization. In the provinces of Zagora and Errachidia alone, it is estimated that 22,000 ha of irrigated land and 5 million ha of rangelands are affected by salinization, which combines its effects with those of wind erosion (Oulhaj, 2013).

According to the Third National Communication of Morocco to the United Nations Convention on Climate Change (Ministry of Environment, 2016), wind erosion is responsible for silting in the southwest part of the country. In the regions of Ouarzazate, Zagora and Errachidia, wind silting causes the loss of around 500 ha/year, and in the Draa valley, it threatens 25% of the irrigation canals, or nearly 65 km. In the southern and eastern regions of the country, silting is one of the main manifestations of desertification. 30,000 ha of palm groves are threatened in Ouarzazate and 250,000 ha in Errachidia. In these two provinces, between 1960 and 1986, the loss of cultivated land from the establishment of sandy dunes was estimated at 155 ha. For all the regions of Ouarzazate, Zagora and Errachidia, silting threatens around 300,000 ha and, therefore, jeopardizes the food security of the local oasis population. These various phenomena have led to increased desertification of agricultural land. Overall, Morocco loses more than 1000 km^2 of productive land per year due to desertification (AKM, 2012). This makes floodplains more floodable, causes destructive flooding downstream, and brings excessive amounts of mud to water tanks, wells, deltas, and river mouths.

The last two decades have been marked by more frequent and more intense floods, like the ones affected the region of Ourika, the province of Guelmim, the province of Sidi Ifni, Grand Casablanca, Tangier, the Gharb valley, and many others. In the Souss Massa region, which accounts for 60% of national fruit and vegetable exports, there has been a sharp increase in the number of floods: between 1982 and 2007 there were four floods in 25 years while between 2008 and 2015 there were seven (OCDE, 2016; GIZ, 2017) The 2010 flood in Agadir Souss-Massa is insightful. It was described as exceptional with precipitation from 18 to 20 February which exceeded 400 mm (480.5 in Issen, 408.4 in the Souss and 430.2 in Tamri). These rains dealt a severe blow to agriculture, causing damage to the 2010–2011 crop year. Hundreds of hectares of cereals, vegetables, and fodder crops were devastated. The bad weather severely damaged the water supply systems of livestock farmers, the existing hydro-agricultural infrastructure, and certain rural roads and tracks in the region. Overall, the assessment of the damage caused by the 2010 floods was $11.38 millions. This has resulted in the displacement of thousands of people in the affected areas (Ezzine et al., 2016).

These different phenomena are leading to an increased variability and reduction of agricultural yields. In fact, the more variability in the climate, the increased frequency and the gravity of extreme events (such as drought and flooding) could cause serious interference in agricultural production, making food availability and rural incomes unstable. This will have a significant impact on food stability and accessibility (Behnassi, 2017).

In the next section, we will identify the agricultural policies carried out in Morocco and the extent to which they have taken into account the above-mentioned constraints which severely affect the country's sustainable food security.

3 Agricultural Policies and Food Security in the Context of Climate Change

Since Morocco's independence, the agricultural sector has been shaped by many public agricultural development programs to improve its performance and its contribution to the economic growth of the country. However, despite these programs from which agriculture has benefited, it has remained subject to multiple constraints (drought, fragmentation and complexity of the legal status of land, overexploitation of water and soil, etc.) and weaknesses in terms of investment capacity, productivity, promotion, professional organization, etc. In addition, cereal production occupied 75% of the Utilised Agricultural Land (UAA), with only 15% of the total crop income generated in 2007. Successive agricultural policies between 1956 and 2007 "were unable to approach agricultural development in its entirety, in its diversity, and in its fundamental relationship with rural development and sustainability" (HCP, 2007). For all these reasons, Morocco has implemented a national strategy called the Green Morocco Plan (GMP) for the period 2008–2020. The main objective of this plan is to improve the productivity of Moroccan agriculture based on modern technologies and to strengthen its position in national and international markets.

However, we will examine the capacity of the GMP to ensure food sufficiency by preserving the natural resources (especially water). A focus will be made on the links between the GMP and intensive practices, the GMP and the retraining policy, and the GMP and agricultural trade.

3.1 Green Morocco Plan (GMP) and Intensive Practices

The GMP has two pillars. Pillar I aims to develop a modern agriculture with high productivity and high added value. It mainly targets medium and large farms through certain practices: the intensification, the aggregation of small farmers, and the drip irrigation. To amplify the public investment effort in the sector and reinforce agricultural production and yields, the country decided to resort to public–private partnerships (PPP)[3] in order to benefit from the assets and capacities of innovation, financing, and private sector management. However, PPP allowed private investment to gain access to large-scale land owned by the state, through long-term land leases. As for Pillar II, the GMP aimed to support agricultural incomes and enhance the productivity of smaller-scale farmers.

The GMP has, according to the Ministry of Agriculture (2019), led to a sustainable increase in agricultural GDP (see Fig. 5). Between 2008 and 2018, the weight of GDPA in GDP varied between 12 and 14% with an average of 12.8%. Thus,

---

[3] The operation of the public-private partnership around agricultural land consists of the long-term rental (17–40 years depending on the type of project) for the benefit of promoters who agree, in a contractual framework with the State, to undertake agricultural investment projects aiming at the development of these lands while creating employment opportunities in rural areas.

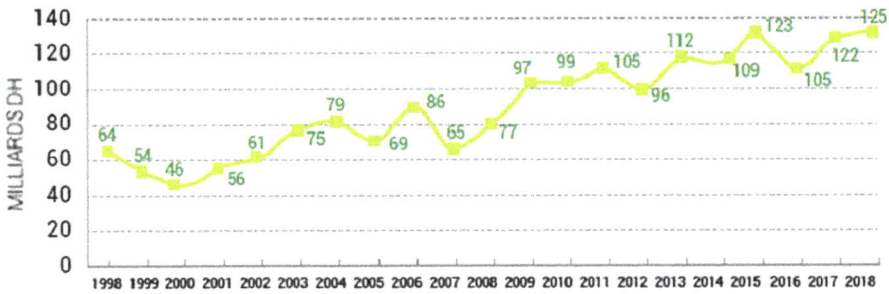

Fig. 5 Gross domestic product of the agricultural branch, Morocco. *Source* Ministry of Agriculture (2019)

the contribution of the agricultural sector to the economic growth has increased significantly, from 7.3% to almost 17.3%.

This growth has happened in many parts of the agricultural sector, notably olive trees, citrus fruits, and red meats whose production recorded an average annual growth rates (CAGR) of 7.8%, 6.3%, and 4.8%, respectively, over the period 2008–2018. It also concerns sectors with high added value, namely fruit trees (27.7% on average over the period 2008–2018), livestock (30.6%), and market gardening (16.1%). It is worth noting that most of these sectors rely on mandatory irrigation uses.

However, these results are from an increase in production which draws on the stock of resources such as land, groundwater, and low-paid hired labor (Saidi & Diouri, 2017). They are also the product of intensive practices (use of chemicals and increased use of water) encouraged by the GMP in all territories, despite their ecological peculiarities. Intensive agriculture is responsible for the groundwater depletion in the Souss-Massa region, which accounts for 60% of Morocco's exports of fruit and vegetables. The estimated water deficit of 271 million m^3 was accompanied by a drop in the level of the water table and an increasing salinization of groundwater with repercussions on the environment and agriculture. Some producers in this region have migrated to Zagora, an extension area of the oasis region, to grow watermelons. It takes 10 million m^3 of water to irrigate 200 ha of watermelons, while the same volume of water can irrigate 1000 ha of palm trees (Agence du bassin hydraulique de Souss-Massa-Draâ 2014, cited by AgriMaroc, 2016).

GMP has also promoted, through subsidies, modern scientific techniques (use of chemical inputs, intensification of mechanization, drip irrigation, etc.) to increase productivity at the expense of ancestral/traditional knowledge and know-how. However, this technoscientific process is not a universal panacea capable of solving all the miseries of peasants and society (Deléage, 2010). Indeed, instead of saving water, drip irrigation is used more to intensify production and extend the area of crops irrigated in summer (olive trees, market gardening, etc.) (Sraïri & Kuper, 2015). High-tech agriculture is often unsuitable for peasant farming systems and the characteristics of their environment. This explains, in part, the failure of Pillar

II (small-holder and family farming) of the GMP, given the incompatibility of the Plan's objectives (generally aimed at improving productivity and promoting specialization through the use of fertilizers and chemicals) with local priorities of farmers such as minimizing the risks of food insecurity, having food on a regular basis, and dividing farm work evenly throughout the year (Saidi, 2013).

Many traditional farming practices are quite effective and safer for the environment (Saidi, 2016). There are very interesting peasant practices in the fight against insects without recourse to chemicals (pesticides). A study conducted by Aurokiatou (2010) showed that the *Cassia nigricans* plant is used by local producers to ward off insects thanks to its bitter taste but also in the conservation and storage of cereals (sorghum, corn, beans, etc.). The growers said at this point: "We know this plant is used to protect the seeds from insects. It settles at the bottom of the attic and is very effective against anything that can damage the grain. We use it in the form of powder that we mix with seeds" (Aurokiatou, 2010: 5). To fight against *orobanches* (parasitic plants), several methods are recommended: crop rotation (unlike specialization), the use of trap plants (Lathyrus sativus, Linum, Flax, Coriander, Mustard, etc.), and biological control (AgriMaroc, 2018):

- Insects: *Phytomyza orobanchia* attacks *orobanches* without damaging the crop. It is found in its natural state in Morocco. 500–1000 insects/ha can reduce the presence of parasitic plants by up to 50%
- Fungi: *Fusarium oxysporum fs porthoceras, Sclerotinia spp, Rhizoctonia solani,* and *Ulocladiumatrum. Fusarium oxysporumfsp o.* is the most used mushroom. It improves tobacco yields by 80.5%. However, it needs high relative humidity and temperatures between 10 and 20 °C
- Bacteria: *orobanches* are sensitive to the *fluorescent Pseudomonas* bacteria; and
- Waste from the crushing of olives: The by-products from the crushing of olives (margines and pomace) are very effective in the fight against *orobanches*. Their use makes it possible to reduce the infestation by 78–97% in the crops of beans (Safour, 2003).

Many of these practices have disappeared due to market laws, the interest of investors, the ignorance of developers, and climatic vagaries. This loss has various societal and environmental consequences.

3.2   GMP and Retraining Policy

To fight against water stress, the GMP has planned the reconversion of cereals—considered by the Ministry of Agriculture as crops that are not very productive and require more water in marginal areas, such as the mountains—for the benefit of the extension of fruit trees, notably the olive tree (see Table 2). In this context, a program to convert cereals to fruit trees on 1 million ha has been launched (ADA, 2012).

Although the GMP succeeded in lowering the area dedicated to cereals from 75 to 60% of the UAA (Ministry of Agriculture, 2019), their contribution to the agricultural

Table 2 The objectives of the GMP by 2020

	Cereals	Citrus	Olive industry	Fruits and vegetables	Milk	Red meats	White meats
Investments (billion MAD)	11	8.9	16.7	24.2	11.4	7.8	5.8
Increase in targeted production (%)	+45	+146	+280	+142	+131	+75	+116
Increase in cultivated areas targeted (%)	−22	+52	+76	+40			

Source ADA (2012)

added value recorded a relative stagnation between 2008 and 2018 (15.6% against 15%). Moreover, the Ministry of Agriculture does not explain whether this reduction in cereal area concerns the irrigated perimeter or the cultivation in the rain-fed *Bour* area. Indeed, if we only take into account the area actually cultivated (sowing), we realise that it hardly changed between 2008 and 2018 (see Fig. 6).

However, the cereal production is still sensitive to rainfall (see Fig. 7). For the 2014–2015 crop years, the production of the three main grains (durum wheat, common wheat, and barley) reached a record production of 1.15 million tonnes. The production of common wheat was at record levels, 5.6 million tonnes, which constitutes a fundamental component for household consumption. Indeed, this level of harvest for common wheat exceeds the record 2012–2013 year which stood at 5.1 million tonnes. Three regions—namely, Doukkala, Chaouia, and El Haouz—accounted for more than 45% of the production.

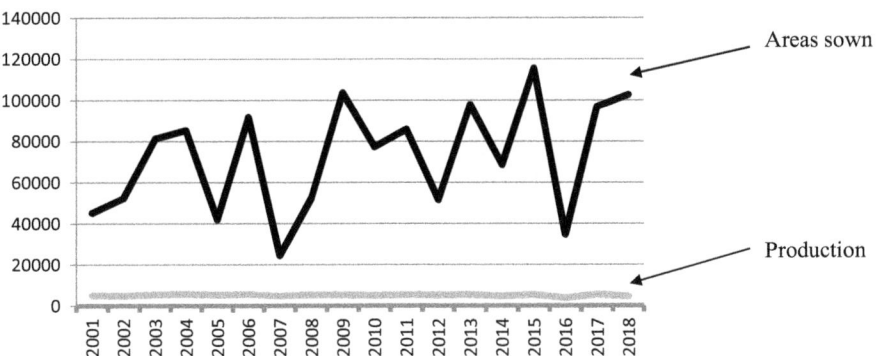

Fig. 6 Area under cereal cultivation and grain production between 2001 and 2018. *Source* Based on data from the ONICL (2020)

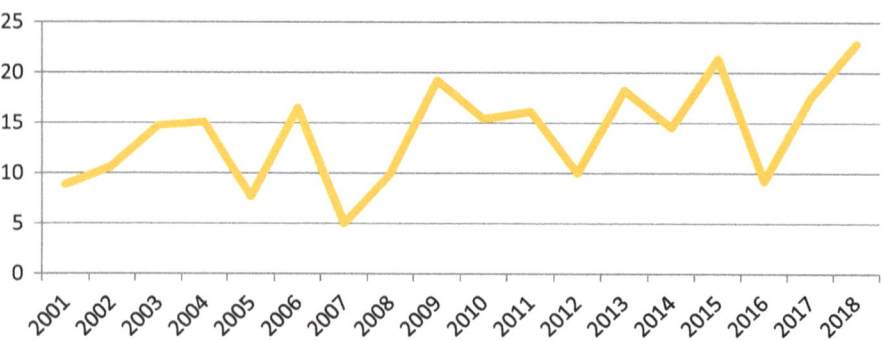

Fig. 7 Yields of cereal crops (Qx/ha) between 2001 and 2018. *Source* Based on data from the ONICL (2020)

According to the Minister of Agriculture (2016), this record production was the result of actions undertaken during the 2014–15 agricultural campaign that involved intensive supervision of farmers, adequate supply of inputs (support for the marketing of selected seeds), mechanization (around 7 tractors per 1000 ha in 2014/15 against 5 in 2007/08), and mobilization of irrigation water throughout the crop cycle. These improved inputs were combined with favorable climatic conditions. In reality, these climatic conditions were the key factor in this record cereal production in 2014/15. The rainfall in 2015/16 was 42.7% below normal, but estimated cereal production was 3.35 million tonnes, a drop of 70% compared to 2014/15. This suggests that rainfall contributed a lot in realizing higher production in the previous year.

In order to make up for this deficit from production, Morocco has imported cereals from world markets. Since 2008, Morocco has not been able to improve its cereal independence, measured by the cereal import dependency ratio[4] (see Fig. 8).

Morocco's cereal imports increased on average from 3.4 million tonnes to 43 million between 2000 and 2007 and 2008–2015 (see Fig. 9). This can be partly explained by the increase in the Moroccan population, which grew from 28.7 to 35.2 million from 2000 to 2018. The country changes the import duty on soft wheat and its derivatives each year based on world market indices. Because of poor production, the import duty dropped from 135 to 35% in 2019. This reduction guaranteed that the import cost of around 2600$[5]/tonne would not change from the last year.

Therefore, the fall in cereal acreage is not consistentwith the goal of food self-sufficiency that could protect the country from dependence on the world market, which is characterized by a large and increasing volatility in food prices. However, cereals, along with sugar, edible oils and oil seeds, constitute the main imported

---

[4] The cereal import dependency ratio indicates how much of the available domestic food supply of cereals has been imported and how much comes from the country's own production. It is computed as ((cereal imports—cereal exports)/(cereal production + cereal imports—cereal exports)) * 100 Given this formula the indicator assumes only values ≤100. Negative values indicate that the country is a net exporter of cereals.

[5] $1 ≈ MDH10.

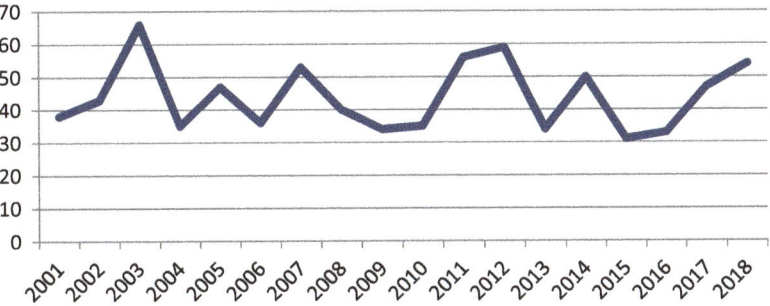

Fig. 8 The cereal imports dependency in %. *Source* Based on data from the ONICL (2020)

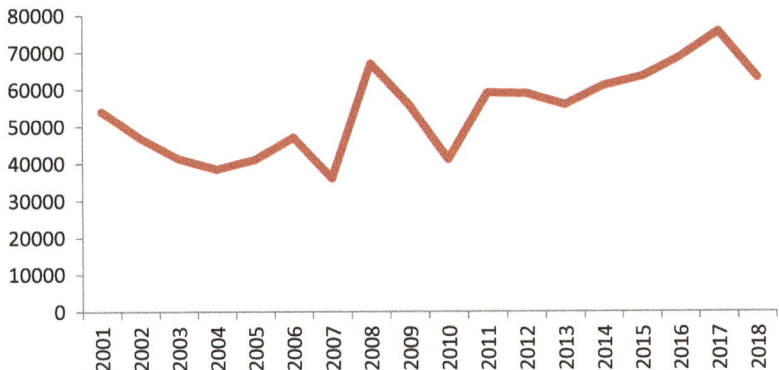

Fig. 9 Grain imports in 1000 Qx. *Source* Based on data from the ONICL (2020)

food products, which aggravate the deficit in Morocco's food trade balance. Cereals represented an average of 36% of the value of food imports between 2008 and 2018. We should recall here that 60% of food energy intake in Morocco is based on cereals, despite growing urbanization and the evolution of dietary patterns (FranceAgriMer, 2015). The average annual consumption of the main cereals is around 200 kg per capita, more than the average world consumption of cereals, which is 152 kg. Soft wheat accounts for almost 70% of the consumption of cereals in urban areas and 66% of that in rural areas. This upward trend will be further accentuated with population growth (33.8 million inhabitants in 2014 compared to 20 million in 1981) and urbanization. Nearly 60% of Moroccans lived in urban areas in 2014. This trend would continue over time with an urbanization rate of 63.4% in 2020 and 67.8 in 2030 (HCP, 2017b). This induces growing food needs for a population following a Mediterranean type diet, based on cereals.

So, we question the relevance of the conversion of cereals to arboriculture, especially in mountainous regions, where warming creates more favourable conditions and wheat yields are increasing. An increase in wheat and barley yields is also being seen in the north of the country, in the Rif region, probably for the same reasons

(Woillez, 2019). However, "alternative" crops (e.g., olive trees) to cereals require irrigation. In addition, the cultivation of cereals produces very important co-products such as straw and stubble, which are vital to the farming system, especially livestock.

Cereal conversion is also planned in the high production regions, such as the Doukkala-Abda region, which recorded the largest cereal production of the 2014/15 campaign. Yet there are plans for this region, as part of the fold 2 projects, to convert 120,000 ha from cereals to fruit arboriculture, including olive, caper, fig, and cactus, and generally go from 847,753 ha to 228,900 ha dedicated to sericulture, by 2020 (Ministry of Agriculture, 2014; ORMVAD, 2012). We, therefore, wonder why the Government would want to move land out of cereal production in a region which contributes 14.5% of national cereal production (or 1.67 million tonnes in 2014/15). It takes 2.09 L to produce one Kcal of energy from fruit, while it takes only 0.51 L to produce one Kcal of energy from cereal (see Table 3). The GMP forecasts that the Doukkala-Abda zone will increase production of beef by 132% and chicken by 89% by 2020. While the production of one kilo of wheat and cereals requires 590 and 1644 m^3 of water, respectively, the production of a kilo of beef requires 15.415 m^3 (26 times more than wheat) and for a kilo of chicken 4325 m^3. In terms of calories from beef and chicken, it takes 10.19 L/Kcal and 3 L/Kcal, respectively (CNRS, 2000; Hoekstra & Mekonnen, 2012).

We also question why Morocco should promote products such as avocado and almonds given their large water footprint per unit of weight, 1,173,000 and 8,047,000 m^3/Kg, respectively (Chouchane et al., 2013).

3.3   GMP and Agricultural Trade

Other sectors that consume less water have been marginalized despite their importance in the diet, namely oilseeds and sugar crops (see Table 3). Indeed, the GMP has not improved the performance of these strategic production chains as shown in Figs. 10 and 11.

Cereals, sugar, and edible oils and oilseeds constitute the main imported food products, which help increase the deficit in Morocco's food trade balance as shown in Table 4.

The dependence rate on sugar imports has worsened since the introduction of the GMP. It has evolved from an annual average of 49% between 1992 and 2008 to 71% between 2009 and 2016, an increase of 22 points in 8 years. The dependence rate on imports of edible oils has evolved from 66% in 1986–1987 to 90% in 2000–2001, and reaching 98.5% in 2009–2010 (Harbouze et al., 2019). In terms of production costs, they do not constitute a major obstacle to the development of oilseeds since the national price of sunflower seeds is 11% lower than its equivalent on the world market (for instance the European Union). Using international prices, farmer margins are high for sunflower and rapeseed (FAO, 2016). This is explained by the significant decline experienced by the Moroccan oilseeds sector since the 1990s. It started with the end of guaranteed minimum prices in 1996. It continued with domestic market

Table 3 The water footprint of some selected food products from vegetable and animal origin

Food item	Water footprint per ton (m³/ton)				Nutritional content			Water footprint per unit of nutritional value		
	Green	Blue	Grey	Tolal	Calorie (kcal/kg)	Protein (g/kg)	Fat (g/kg)	Calorie (L/kal)	Protein (L/g protein)	Fat (L/g fat)
Sugar cops	130	52	15	197	285	0	0	0.69	0	0
Vegetables	194	43	85	322	240	12	2.1	1.34	26	154
Starchy roots	327	16	43	387	827	13	1.7	0.47	31	226
Fruits	726	147	89	962	460	5.3	2.8	2.09	180	348
Cereals	1232	228	184	1644	3208	80	15	0.51	21	112
Oil crops	2023	220	121	2164	2908	146	209	0.81	16	11
Pulses	3180	141	734	4055	3412	215	23	1.19	19	180
Nuts	7016	1367	680	9061	2500	65	193	3.63	139	47
Milk	863	86	72	1020	560	33	31	1.82	31	33
Eggs	2592	244	429	3265	1425	111	100	2.29	29	33
Chicken meat	3545	313	467	4325	1440	127	100	3	34	43
Butter	4695	465	393	5553	7692	0	872	0.72	0	6.4
Pig meat	4907	459	622	5988	2786	105	259	2.15	57	23
Sheep/goat meat	8253	457	53	8763	2059	139	163	4.25	63	54
Beef	14,414	550	451	15,415	1513	138	101	10.19	112	153

Source Hoekstra and Mekonnen (2012)

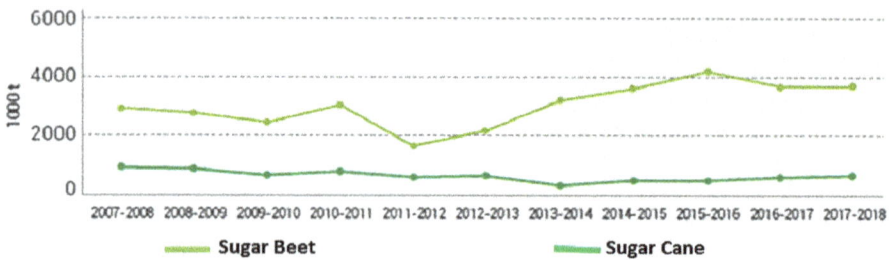

Fig. 10 Sugar production in Morocco, 2007–2018. *Source* Ministry of Agriculture (2019)

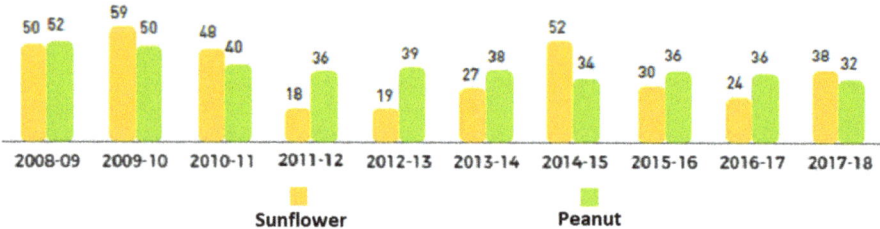

Fig. 11 Oilseed production in Morocco, 2008–2018 (kt). *Source* Ministry of Agriculture (2019)

reforms and bilateral free trade agreements, in particular the agreement signed with the United States in 2006.

The fight against the degradation of water resources is hampered by agricultural policies geared towards exports, which were increased by 67% in value over the 2008–2018 period (Table 4). Morocco exports agricultural products and locally scarce factors despite its water shortage for often disappointing foreign exchange earnings which do not compensate for the value of this virtual water (Akesbi, 2011). Furthermore, their production often requires intensive or even super-intensive practices, using chemical inputs (e.g., dangerous pesticides) which are often harmful to ecosystems. The result is that the rate of import coverage by agri-food exports continues to collapse over the years, from a rate of 200% in the 1970s to 52% over the period 2008–2018 (Ministry of Economy, 2019; Oulhaj, 2013). This dependence is not without risk due to the high variability in world commodity markets, which may be linked to climate change, but also due to increasing concentration of trade and vertical integration. Ninety percent of the cereals produced in the world pass through four large multinational companies (UNCTAD, 2017; Murphy et al., 2012).

A sharp increase in world food prices would decrease poor people's access to food, especially those living in rural areas and suburbs. According to the FAO (2017), food insecurity and poverty are closely linked in Morocco, partly due to the share of income (around 70%) that poor households spend on food. Three of the four million people who live below the national poverty line are in rural areas. Morocco is extremely vulnerable to volatile world prices, largely due to its dependence on

Table 4 Agricultural imports

	2008	2009	2010	2011	2012	2013	2014	2015	2016	2017	2018
Agricultural exports (Billion DH)	15.2	14.8	16.7	19.3	18.0	21.0	22.5	27.0	29.5	33.2	25.8
Agricultural imports in value (Billion DH) of which:	40.4	31.0	35.8	47.3	50.3	42.9	49.0	42.3	51.9	51.2	40.7
Cereals	17.4	8.9	11.9	17.3	19.2	12.8	17.9	13.6	18.5	13.6	10.8
Edible oils and oilseeds	5.6	4.16	4.14	5.11	5.01	4.07	4.6	4.2	5.2	6	4.1
Raw sugar	2.2	3.4	3.3	4.8	5.1	3.7	2.9	3.5	4.6	4.9	2.7

Source Ministry of Agriculture (2019)

imports. However, this aspect alone does not contribute to the country's food inse-curity. Rising food prices pose a significant threat to the health and well-being of Moroccan households, especially the poor. The poor quality of certain products and the change in diet leading to an increase in certain pathologies (diabetes, obesity, etc.) are also underlying factors. In addition, due to the number of Moroccans living slightly above the poverty line (the intensity of deprivation of people without access to basic and necessary needs, is nested at 45.7% according to the UNDP, 2019), any fluctuation, even marginal, in food prices, can have dramatic impacts on national poverty rates. Among the groups most exposed to price volatility are the urban poor, marginalized farmers, and landless rural residents whose income is closely linked to employment in the agricultural sector. This situation is likely to worsen due to the ongoing Covid-19 pandemic, with its repercussions on people's life and livelihoods, global food trade, markets, food supply chains, and livestock.

4   Conclusions

This chapter argues that Moroccan agricultural policy pursued up until now—especially the GMP—aggravates the consequences of climate change on natural resources, especially water, and, by extension, the food security of the country. The issue of water is not only linked to climate change, but it also represents a major public policy issue, inseparable from the future of agriculture and rural society, from the economic viability of many current strategies (ex: the industrial plan), finally from the survival of all Moroccans. The priority issue is to increase agricultural produc-tion in a sustainable manner in order to meet the needs of a growing population with increasingly limited land and water resources in a context of climate change. In 2050, Morocco will have to feed more than 45 million people against around 35.4 million today and 40 around 2040 (HCP, 2017a).

   Morocco must focus on crops improving its food self-sufficiency instead of crops that require a lot of water for export. Morocco should encourage more environ-mentally friendly agriculture based on non-productivist models (local agriculture, organic agriculture, peasant agriculture, etc.). Thus, the production logic of the latter aims to directly improve farmers' incomes based on the development of local crops and to provide consumers with healthy food. These models were born at the end of the 1980s to deal with the multidimensional crisis of the productivist agricultural model: dependence on public subsidies, overproduction, poor sales of agricultural products, lowering of farmers' living standards which causes rural exodus, envi-ronmental degradation, health crises reinforcing consumer mistrust as well as the deterritorialization of agriculture. In this perspective, some practices emerge such as environmentally friendly farming practices (extensive management, exclusion of the use of synthetic products such as pesticides and fertilizers, etc.), the revaluation of family farming, and the promotion of healthy products and those linked to their territorial origin.

In principle, family farming often makes rational use of its resources. This is due to the nature of the 'peasant' lifestyles, which aim to establish a privileged relationship with the environment, proximity to nature, where certain village resource management practices are more the result of family farming than capitalist agriculture. In such cases, family farming can lead to production systems favorable to the preservation of resources, biodiversity, the fight against climate change, etc. (Jackson et al., 2015; Coordination Sud, 2007). It is an agriculture with a low level of inputs or 'low external input sustainable agriculture'; an agriculture which is sustainable and practically uses a minimum of external interventions (chemical fertilizers, pesticides, rental of machines, etc.). We are, therefore, talking about sustainable agriculture, which corresponds to the definition of 'sustainable development' developed by Brundtland (WCED, 1987) and which thus meets the needs of current and future generations in terms of food, basic raw materials and natural resources in economically viable, socially equitable, environmentally sound, and culturally acceptable manner.

However, family farming is no longer the only driving economic activity and sheltered from market pressures which force it, in several parts of the world, to employ practices that have negative impacts on the environment. Moreover, apart from natural constraints (drought, etc.), family farming is largely perceived as a practice based on archaic structures, unproductive, outdated, unable to innovate, and adapt to changes. Such criticism toward family farming is usually advocated by policies in favor of large farms with means of production and a high capital content.

In another way, the relationship of family farmers to nature and their environment is close and respectful if their practices correspond to a peasant way of life and is part of a system that integrates rain-fed crops, irrigated and animal husbandry on farms. Such a system which explains the continuity of cereals and food legumes to represent more than half of the rotations, because they are the crops most adapted to the climate of the country (semi-arid, even arid: precipitation concentrated only in autumn, winter, and early spring). They make the best use of rainwater and generate significant income for farmers (Sraïri & Kuper, 2015). The whole must be embodied by a territorial productive organization which must integrate more dimensions of an economic, social, technical, and natural (pedoclimatic) order. To do this, specific capacity-building programs for smallholder farming must be put in place.

However, food availability can also be improved by tackling food waste and changing food consumption patterns by further integrating other food sources such as fisheries, forest foods, etc.

References

Agricultural Development Agency (ADA). (2012). *Strategic environmental assessment of the Green Morocco Plan* (in French). http://www.ada.gov.ma/etude/download/Etude%20Environnement ale%20et%20sociale%20PMV.pdf

AgriMaroc. (2016). *The cultivation of watermelon prohibited in Zagora?* (in French). https://www.agrimaroc.ma/la-culture-de-la-pasteque-interdite-a-zagora/

AgriMaroc. (2018). *Methods and tips for controlling broomrapes* (in French). https://www.agrimaroc.ma/orobanches/

Ahsan Ullah, A. K. M. (2012). Climate change and climate refugee in Egypt: An overview from policy perspectives. *TMC Academic Journal, 7*(1), 56–60.

Akesbi, N. (2011). La nouvelle stratégie agricole du Maroc annonce-t-elle l'insécurité alimentaire du pays? *Confluences Méditerranée, 78*, 93–105.

Aqueduct Water Risk Atlas. (2019). *Morocco summary results.* https://www.wri.org/applications/aqueduct/country-rankings/?country=MAR&indicator=drr

Aurokiatou, T. (2010). Peasant knowledge: Nature and functions contribution to the debate on the usefulness of local knowledge (in French). In *Symposium: Innovation et développement durable dans l'agriculture et l'agroalimentaire, ISDA*, July, Montpellier, France.

Bank Al-Maghrib. (2020, January). *Monthly review of the economic, monetary and financial situation.* http://www.bkam.ma/Publications-statistiques-et-recherche/Documents-d-analyse-et-de-reference/Revue-de-la-conjoncture-economique/Revue-de-la-conjoncture-economique-2020

Behnassi, M. (2017). Climate security as a framework for climate policy and Governance. In M. Behnassi & K. McGlade (Eds.), *Environmental change and human security in Africa and the middle east* (pp. 3–24). Springer.

Chouchane, H., Hoekstra, A.Y., Krol, M.S., & Mekonnen, M. M. (2013). *Water footprint of Tunisia from an economic perspective value of water*. Research Report Series No. 61. https://www.researchgate.net/publication/271074167_WATER_FOOTPRINT_OF_TUNISIA_FROM_AN_ECONOMIC_PERSPECTIVE_VALUE_OF_WATER_RESEARCH_REPORT_SERIES_NO_61/download

CNRS (Centre national de la recherche scientifique, France). (2000). *Scientific file on water: Uses—crops* (in French). http://www.eaufrance.fr/ressources/groupes-de-chiffres-cles/volume-d-eau-necessaire-pour

Dahan, S. (2017). *Managing rarity of water in the middle urban in Morocco*. World Bank Group, Washington, DC. https://openknowledge.worldbank.org/bitstream/handle/10986/29190/122 698-WP-P157650-Summary-Report-Urban-water-scarcity-in-Morocco-ENG-P157650-2017-12-25-04-12.pdf?sequence=1&isAllowed=y

Deléage, E. (2010). La coproduction des savoirs dans l'agriculture durable. In *Symposium on innovation and sustainable development in agriculture and food [Colloque Innovation et développement durable dans l'agriculture et l'alimentation]*, Jun, Montpellier, France. 8 p. hal-00539813f.

El Ajhar, L., El Khachine, D., El Bakouri, A., El Kharrim, K., & Belghyti, D. (2018). Evolution of the rain from 1960 to 2015 in Morocco (in French). *International Journal of Research Science and Management, 5*(10), 47–56. https://doi.org/10.5281/zenodo.1465732

Ezzine, H., Messouli, M., & Krause, B. (2016). *Analysis and mapping of vulnerability to extreme climate events and estimation of the costs of their impacts in the Souss Massa region (in French)*. https://www.4c.ma/fr/mediatheque/docutheque/analyse-et-cartographie-de-la-vulnerabilite-aux-evenements-climatiques-0

Ezzine, H., Bouziane, A., Ouazar, D., & Hasnaoui, M.-D. (2017). Downscaling of open coarse precipitation data through spatial and statistical analysis, integrating NDVI, NDWI, elevation, and distance from sea. *Hindawi Advances Meteorology* (online). https://doi.org/10.1155/2017/8124962

Fallahtrade. (2020, December 20). Key figures: The cereal sector (in Franch), Le portail agricole, Crédit agricole (online). https://www.fellah-trade.com/fr/filiere-vegetale/chiffres-cles-cerealiculture, consulté le.

FAO. (2016). *Oil crops, oils and meals. Trade and markets division food outlook.* http://www.fao.org/fileadmin/templates/est/COMM_MARKETS_MONITORING/Oilcrops/Documents/Food_outlook_oilseeds/Food_Outlook_June_2016_oilseeds.pdf

FAO. (2017). *A territorial approach to food security and nutrition policy.* http://www.fao.org/3/a-bl336e.pdf

FranceAgriMer. (2015). *Cereals to Morocco (in French).* http://www.franceagrimer.fr/content/download/42240/394761/file/MEP-memoire-logistique-MAROC_20160126.pdf

GIZ (Deutsche Gesellschaft für Internationale Zusammenarbeit GmbH). (2017). *Methodological guide for adapting to climate change in industrial areas (in French).* Eschborn, 100 p.

Guennouni, A. (2016). Global warming: Serious impacts on Moroccan agriculture (in french). *Agriculture Du Maghreb, 98,* 6–7.

Harbouze, R., Pellissier, J.-P., Rolland, J.-P., & Khechimi, W. (2019). *Synthesis report on agriculture in Morocco* (in French). CIHEAM-IAMM, p. 104. hal-02137637f.

Haut Commissariat au Plan (HCP). (2007). *Prospects Morocco 2030* (in French). https://www.hcp.ma/downloads/Maroc-2030_t11885.html

Haut Commissariat au Plan (HCP). (2017a). Population projections for regions and provinces 2014–2050 (in French). Centre d'Etudes et de Recherches Démographiques (CERED). https://www.hcp.ma/region-drta/attachment/861157/

Haut Commissariat au Plan (HCP). (2017b). *Population projections for regions and provinces 2014–2030* (in French). Centre d'Etudes et de Recherches Démographiques (CERED). https://www.hcp.ma/region-drta/attachment/861124/

Hoekstra, A.-Y., & Mekonnen, M.-M. (2012). Water footprint farm animal products. *Ecosystems, 15,* 401–415.

Hofste, R.-W., Reig, P., & Schleifer, L. (2019). Countries, home to one-quarter of the world's population, face extremely high water stress. *World Resources Institute.* https://www.wri.org/blog/2019/08/17-countries-home-one-quarter-world-population-face-extremely-high-water-stress

Iceland, C., Luo, T., & Donchyts, G. (2018). It's not just Cape Town: 4 shrinking reservoirs to watch. *World Resources Institute.* https://www.wri.org/blog/2018/04/its-not-just-cape-town-4-shrinking-reservoirs-watch

Ipcc. (2014). *AR5 climate change 2014: Impacts, adaptation, and vulnerability.* https://www.ipcc.ch/report/ar5/wg2/

Jackson, L., Bernoux, M. & Neufeldt, H. (2015). Family farmers facing the challenges of climate change. In J. M. Sourisseau, R. Kahane, P. Fabre, & B. Hubert (Eds.), *Proceedings of the International Encounters on Family Farming and Research* (Montpellier, 1–3 June 2014) (320 p.). Montpellier: Agropolis International.

Mason, N., Nalamalapu, D., & Corfee-Morlot, J. (2019). Climate change is hurting Africa's water sector, but investing in water can pay off. *World Resources Institute.* https://www.wri.org/blog/2019/10/climate-change-hurting-africa-s-water-sector-investing-water-can-pay

Mastere, M., El Fellah, B., Van Vliet-Lanoe, B., & Maquaire, O. (2019). Assessment of the impacts of climate change on natural risks in northern Morocco (in French). In *The Proceedings of the International Symposium of AIC Thessaloniki—29 May to 1 Juan,* Greece. http://www.climato.be/aic/colloques/actes/Thessaloniki2019_actes.pdf

MedECC. (2020). Summary for policymakers. In W. Cramer, J. Guiot, & K. Marini (Eds.), *Climate and environmental change in the mediterranean basin—Current situation and risks for the future. First mediterranean assessment report* (34 p.). Union for the Mediterranean, Plan Bleu, UNEP/MAP, Marseille, France (in press).

Ministry of Agriculture (Morocco). (2014). *Dynamics of regional agricultural plans* (in French). http://www.agriculture.gov.ma/sites/default/files/doukkala_abda_2014.pdf

Ministry of Agriculture (Morocco). (2016). The agricultural year 2014–15 (in French). Note stratégique 106. https://www.agriculture.gov.ma/sites/default/files/campagne_agricole_2014-2015.pdf

Ministry of Agriculture (Morocco). (2019). *Agriculture in figures 2018* (in French). https://www.agriculture.gov.ma/sites/default/files/19-00145-book_agricultures_en_chiffres_def.pdf

Ministry of Economy (Morocco). (2019). *The Moroccan agricultural sector: Structural trends, challenges and development prospects* (in French). DEPF. https://www.finances.gov.ma/Public ation/depf/2019/Le%20secteur%20agricole%20marocain.pdf

Ministry of Environment (Morocco). (2016). *Third National communication to the United Nations convention on climate change.* https://www.4c.ma/fr/mediatheque/docutheque/troisieme-com munication-nationale-du-maroc-la-convention-cadre-de-nations

Murphy, S., Burch, D., & Clapp, J. (2012). *Cereal secrets: The world's largest grain traders and global agriculture.* Oxfam Research Reports.

Niang, I., Ruppel, O. C., Abdrabo, M. A., Essel, A., Lennard, C., Padgham, J., & Urquhart, P. (2014). Africa. In V. R. Barros, et al. (Eds.), *Climate change 2014: Impacts, adaptation, and vulnerability. Part B: Regional aspects. Contribution of working group II to the fifth assessment report of the intergovernmental panel on climate change* (pp. 1199–1265). Cambridge University Press.

Nicholson, S. E., Funk, C., & Fink, A. H. (2018). Rainfall over the African continent from the 19th through the 21st century. *Global and Planetary Change, 165,* 114–127.

OCDE. (2016).*Study on risk management in Morocco* (in French). Éditions OCDE, Paris. https:// doi.org/10.1787/9789264267145-fr

ONICL (National Interprofessional Office for Cereals and Legumes). (2020). *Development and production of the four main grains* (in French). https://www.onicl.org.ma/portail/situation-du-march%C3%A9/statistiques

ORMVAD (Office Régional de Mise en Valeur Agricole des DOUKKALA). (2012). *Regional agricultural plan* (in French). http://www.ormvad.ma/node/8

Oulhaj, L. (2013, June). *Assessment of Morocco's agricultural strategy (Plan Maroc Vert) using a dynamic general equilibrium model* (in French). FEMISE Research Programme, Research n FEM35-20 Directed.

Prudhomme, C., Giuntoli, I., Robinson, E. L., Clark, D. B., Arnell, N. W., Dankers, R., & Hagemann, S. (2013). Hydrological droughts in the 21st century, hotspots and uncertainties from a global multimodel ensemble experiment. *Proceedings of the National Academy of Sciences, 111*(9), 3262–3267. https://doi.org/10.1073/pnas.1222473110

Safour, K. (2003). *Use of herbicides and olive by-products (pomace and vegetable water: Margines) in the control of bean bean* (in French) (p. 179). Université Mohamed ben Abdallah, Fès, Maroc.

Saidi, A. (2013, July). The Meknes olive growing system, between traditional knowledge and scientific knowledge (in French). In *50e Symposium of the Association des Sciences Régionales de Langue Française*, UCL-Mons, Belgique.

Saidi, A. (2016, July). The Olivier de Meknès Agro-Pole: What actors and what challenges? (in French). *Revue Marocaine De Gestion Et D'economie, 3*(7), 179–198.

Saidi, A., & Diouri, M. (2017). Food self-sufficiency under the Green-Morocco plan. *Journal of Experimental Biology and Agricultural Sciences* (JEBAS), *5.* https://doi.org/10.18006/2017. 5(Spl-1-SAFSAW).S33.S40

Schewe, J., Heinke, J., Gerten, D., Haddeland, I., Arnell, N. W., Clark, D. B, & Gosling, S. N. (2013). Multimodel assessment of water scarcity under climate change. *Proceedings of the National Academy of Sciences, 111*(9), 3245–3250.

Sraïri, M.-T., & Kuper, M. (2015). Les systèmes agricoles basés sur la diversification culturale et l'intégration de l'élevage en vue de favoriser la sécurité alimentaire au Maroc. In M. C. Paciello (Eds.), *Building sustainable agriculture for food security in the Euro-mediterranean area: challenges and policy options* (pp. 139–15). Edizioni Nuova Cultura, Roma.

Coordination Sud. (2007). Défendre les agricultures familiales: lesquelles, pourquoi? Résultats des travaux et du séminaire organisé par la Commission Agriculture et Alimentation de Coordination SUD, 90 p.

UNCTAD. (2017). *Commodity dependence and the sustainable development goals.* https://unctad. org/meetings/en/SessionalDocuments/cimem2d37_en.pdf

UNDP. (2019). *The 2019 global multidimensional poverty index provides the detailed information policy makers need to more effectively target their policies.* http://hdr.undp.org/sites/default/files/2019_mpi_press_release_en.pdf

WCED (World Commission on Environment and Development). (1987). *Report of the world commission on environment and development: Our common future.* UN Documents: Gathering a Body of Global Agreements. https://sustainabledevelopment.un.org/content/documents/5987our-common-future.pdf

Woillez, M.-N. (2019). Review of the literature on climate change in Morocco: Observations, projections and impacts (in French). *Papiers de Recherche AFD, 108*, 1–33.

Dr. Saidi Abdelmajid is a Professor at the Moulay Ismail University, Morocco and a Member of the Economic and Social Studies and Research Laboratory (LERES) at the same Faculty. He undertakes research in liaison with the Public Policy, Political Action, Territories (PACTE) Laboratory, Grenoble Alpes University (UGA), France. His research areas include localized agri-food systems, food safety, territorial development, and sustainable development.

Dr. Ahmed Mukhtar is Former Chairman, HEC, Islamabad, Pakistan and Former Deputy Director General, ICSESCO, Rabat, Morocco. He is an Educational Consultant with a Bachelor and a Master from the University of Agriculture Faisalabad, Pakistan, and a Master of Business Administration, and a Ph.D. from University of California, Riverside USA. He has worked with numerous educational institutions in different capacities. Dr. Mukhtar has over 32 years of educational development and management experience at national and international level, including teaching, research, academic administration, policy development, linking educational research to industry/commercialization, introducing entrepreneurial approaches to education, and initiating diverse range of educational development programs. As Supervisor of the Federation of the Universities of the Islamic World General Secretariat, he was responsible of the coordination with the member universities and other partners for better facilitation and providing platform for knowledge sharing and collaboration among institutions of higher education. As Advisor at ICSESCO, Dr. Mukhtar was responsible for Higher Education, Quality Assurance, Tafahum Program, STI projects and academic maters of Federation of Universities in Islamic world. He was also Responsible of the Cabinet of H.E. DG ICSESCO. As Chairman of HEC, having the status of Federal Minister, and as the Head of the Commission, Dr. Mukhtar was responsible for the implementation of its decisions in addition to the formulation of policies, guiding principles, and priorities for Higher Educational Institutions with the objective to promote the socio-economic development of the country. Being PAO Executive Director, he was responsible for a portfolio of over $500.00 million. As Deputy Director General of ICSESCO, Dr. Mukhtar was responsible for the Directorates of Education, Science, Culture and Communication, ICPSR, CPID (Planning and Strategic Division), and regional centers. Pursuing ICSESCO's agenda in all OIC Conferences, conventions, including Islamic Conference of Foreign Ministers and various ministerial conferences in specified areas of ICSESCO's mandate. He also supervised the preparation of short-term and long-term programs, including three-year action plans for the ICSESCO. Finally, he served as Editor-in-Chief of the ICSESCO Journal of Science and Technology from Feb 2011–Feb 2013.

Dr. Mirza Barjees Baig is a Professor at the Prince Sultan Institute for Environmental, Water and Desert Research, King Saud University, Saudi Arabia. He earned his MS degree in International Agricultural Extension in 1992 from the Utah State University, Logan, Utah, USA and was placed on the 'Roll of Honor'. He completed his Ph.D. in Extension for Natural Resource Management from the University of Idaho, USA and was honored with the '1995 outstanding graduate student award'. Dr. Baig has published extensively on the issues associated with natural resources in the national and international journals. He has also made oral presentations about agriculture and natural resources and role of extension education at various international conferences. Food waste, water management, degradation of natural resources, deteriorating environment and their relationship with society/community are his areas of interest. He has attempted to develop strategies to conserve natural resources, promote environment and develop sustainable communities. Dr. Baig started his scientific career in 1983 as a researcher at the Pakistan Agricultural Research Council, Islamabad, Pakistan. He served at the University of Guelph, Ontario, Canada as the Special Graduate Faculty from 2000–2005. He served as a Foreign Professor at the Allama Iqbal Open University (AIOU), Pakistan through the Higher Education Commission from 2005–2009. He served as a Professor of Agricultural Extension and Rural Society at the King Saud University, Saudi Arabia from 2009–2020. He serves as well on the Editorial Boards of many international journals and the member of many international professional organizations.

Dr. Michael R. Reed is Emeritus Professor of Agricultural Economics at the University of Kentucky, USA. He holds a Ph.D. in economics from Iowa State University (1979); a Doctor Honoris Causa (Honorary Ph.D.) from Bucharest University of Agricultural Sciences and Veterinary Medicine (Romania); and an Honorary Ph.D. from the Faculty of Business Administration, Maejo University (Thailand). Dr. Reed's principal area of research is international trade in agricultural products, including the effects of macroeconomic policies and exchange rates on U.S. food exports, international commodity price dynamics, consumer demand in various countries, and the effects of competition patterns on world agricultural trade patterns.

Chapter 8
Impacts of Climate Change on Livestock and Related Food Security Implications—Overview of the Situation in Pakistan and Policy Recommendations

Hammad Ahmed Hashmi, Azaiez Ouled Belgacem, Mohamed Behnassi, Khalid Javed, and Mirza Barjees Baig

Abstract Pakistan is an agricultural country and livestock plays a pivotal role, with the demand for protein being met primarily by livestock production. Livestock contributes more than 60% to agriculture and 11.22% to the country's GDP. It is estimated that over 35 million people engage in livestock-related activities. Meat production has increased during the past decades, especially poultry. Climate change is severely affecting animal productivity in resource-poor countries like Pakistan directly by heat stress and indirectly by changes in the ecosystem. Pakistan has the world's largest integrated irrigation system; however, water scarcity has pushed farmers to shift cultivation from water-intensive crops like rice, wheat, cotton, and sugarcane to other crops and vegetables, which require less water, thus making increased pressure on the food market. The country still suffers a deficiency in

H. A. Hashmi (✉)
Paws and Claws Animal Consultancy Pakistan (R&D), ADS and OIC DHA Lahore KC, Lahore, Pakistan

A. O. Belgacem
Sustainable Rangeland Management Expert, Food & Agriculture Organization of the United Nations (FAO), Riyadh, Saudi Arabia
e-mail: azaiez.ouledbelgacem@fao.org

M. Behnassi
Faculty of Law, Economics and Social Sciences of Agadir, Ibn Zohr University, Agadir, Morocco

Research Center for Environment, Human Security and Governance (CERES), Agadir, Morocco

M. Behnassi
e-mail: behnassi@gmail.com; m.behnassi@uiz.ac.ma

K. Javed
Department of Livestock Production, University of Veterinary and Animal Sciences, Lahore, Pakistan
e-mail: khalidjaved@uvas.edu.pk

M. B. Baig
Prince Sultan Institute for Environmental, Water and Desert Research, King Saud University, Riyadh, Saudi Arabia
e-mail: mbbaig@ksu.edu.sa

M. Behnassi et al. (eds.), *Emerging Challenges to Food Production and Security in Asia, Middle East, and Africa*, https://doi.org/10.1007/978-3-030-72987-5_8

Crude Protein and Total Digestible Nutrients (TDN) for large and small ruminants in their diet. A detailed understanding is warranted to anticipate the future impacts of climate change on the productivity of ruminants. There is a drastic effect on the performance of water buffaloes/cattle and their calves in terms of growth and production, while a higher Temperature Humidity Index (THI) has the most adverse effects on exotic cattle. The performance of small ruminants (a primary source of animal protein) is also adversely affected by a rise in temperature, becoming more severe with increasing atmospheric humidity along with high temperatures. Moreover, because of the COVID-19 Pandemic, the importance of climate change and the value of food security are expected to increase even more. Mitigation measures are suggested to address productivity losses in the livestock food supply, as the population grows.

Keywords Livestock productivity · Climate change · Food security · COVID-19 · Pakistan

1   Introduction

Pakistan is an agriculture-based country and livestock plays a pivotal role, as the demand per capita protein is met by livestock products (milk, meat, and eggs) predominantly. The availability of animal protein has increased substantially in recent years with a tremendous contribution of the livestock sector. The livestock demand is rapidly growing due to population growth and rising living standards. Livestock, which contributes nearly 60% of value addition in agriculture and nearly 11.22% to the gross domestic product (GDP) in Pakistan, has displayed a healthy trend in the last few years (see Fig. 1). It is currently estimated that over 35 million people are engaged in the livestock sector according to the Economic Survey of Pakistan 2019–2020 (The Express Tribune, 2020). The gross value addition of livestock, increased

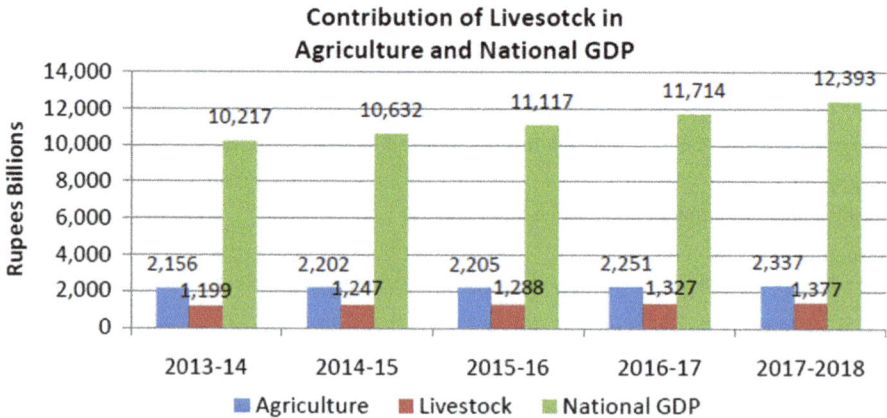

Fig. 1 Contribution of livestock in agriculture in GDP of Pakistan. *Source* year book 2017–18 (Ministry of national food security and research, government of Pakistan)

from Rs. 1430 billion in 2018–19 to Rs. 1466 billion in 2019–20, shows an increase of 2.5% over the same period of last year. Pakistan ranks 7th among the world's countries to be severely affected by climate change (Climate Risk Index (CRI) from 1998 to 2017) (annual averages). The country has experienced the worst drought of its history in the last decade until 2004. Climate change is negatively impacting livestock and its production.

The roles of livestock can be classified in several ways. According to the International Livestock Research Institute (ILRI), two widely used classifications are based depending on the output produced (food, cropping inputs, and raw materials) or production used (household consumption, the supply of inputs, cash income, savings, investment, social and ritual roles) ILRI (1995). Another classification categorizes livestock functions into economic (source of cash income, mean of savings accumulation and investment, economic status), household use (feeding, transportation, fertilizer, and animal draught), socio-cultural (social position, paying bridewealth, providing animals for collective feasts or sacrifices), and leisure (horse racing, chicken fighting, bullfighting, and game hunting).

Livestock species play very important economic, social, and cultural functions in the lives of rural populations in Pakistan. This sector plays multiple roles in the lives of farmers by improving food supply, nutrition, income levels, asset savings, soil yield, transportation, agricultural traction, agricultural diversification and sustainable agricultural production, employment opportunities, ritual purposes, and social status (Moyo & Swanepoel, 2010). The primary purpose of raising livestock is to meet the dietary needs of milk, meat, and eggs for both rural and urban populations. Besides, it helps fulfill the fundamental requirements at a farm (Iqbal & Ahmad, 1999).

Out of the total 79.6 million hectares (Mha) of land in Pakistan, 27% is cultivated (22.79 (Mha)), of which about 19.12 (Mha) area is irrigated and 3.67 (Mha) is rainfed. Ironically, the irrigated area consumes about 80% of all the country's freshwater resources. Figures 2 and 3 show the distribution of land in Pakistan (USAID, 2018; FAO, 2020).

Livestock production adds sizeable revenues to the national economy. Landless farmers and women play a significant role in livestock operations. According to Eckstein et al. (2018) of the Think Tank 'Germanwatch', in Global Climate Risk Index 2019, the agriculture sector is the backbone of Pakistan's economy.

The purpose of this chapter is to identify the impacts of climate change on livestock productivity (mainly livestock health and animal products) and food security, and to propose ways to mitigate losses for livestock farmers, thus keeping them competitive under climate change scenarios. Based on the analysis of the situations, some workable solutions and strategies are discussed.

2 An Overview of Agricultural Resources in Pakistan

Agriculture establishes itself as the major sector of Pakistan's economy, with the majority of the population directly or indirectly dependent on this sector, of which livestock is a major component (Table 1). It contributes nearly 19% of the national

DISTRIBUTION OF LAND IN PAKISTAN

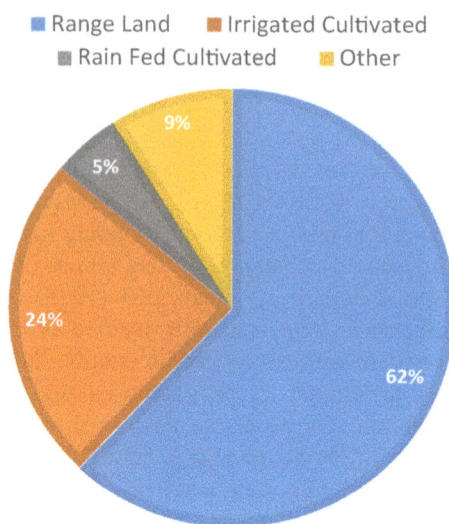

Fig. 2 Distribution of land in Pakistan. *Source* USAID (2018) and FAO (2020)

GDP and about 39% of the labor force. Productive agriculture is key to economic growth and poverty alleviation and is the important source of foreign exchange earnings while feeding Pakistani people, both in rural and urban areas (Pakistan Economic Survey, 2019–2020). The economy of Pakistan depends heavily upon the crop and livestock sectors (Baig et al., 2013). However, the lack of concerted efforts for genetic improvement (selective breeding) is a major issue causing low productivity, in addition to the lack of institutional framework marketing of livestock products and animal health-related issues (National Food Security Policy Pakistan, 2018, Ministry of National Food Security and Research of Pakistan).

Despite these impediments, the agricultural sector recorded a growth of 2.9% in 2019–2020. Livestock contributed 60.5% to the overall agricultural income and 11.2% to the total GDP for 2018–19. The gross value addition of livestock has increased from PKR 1384 billion (2017–18) to PKR 1440 billion (2018–19), showing an increase of 4.0%. According to official statistics, cattle numbers have been increasing in the last three years, with a current estimation of 49.6 millions. Similarly, the population of buffaloes was 41.2 millions, cows 31.2 millions, goats 78.2 millions, camels 1.1 millions, horses 0.4 millions, donkeys 5.5 millions, and mules 0.2 millions. Besides, the fishing sector grew by 0.60% during 2019–20, while the forestry sector increased by 2.29%. Meanwhile, Pakistan is the fourth largest milk producer in the world after India, China, and the USA. Current milk production is estimated at 35 billion liters with 8 million households producing milk and a total herd size of approximately 50 million animals (Economic Survey of Pakistan, Finance Division, Economic Advisor's Wing 2019–2020).

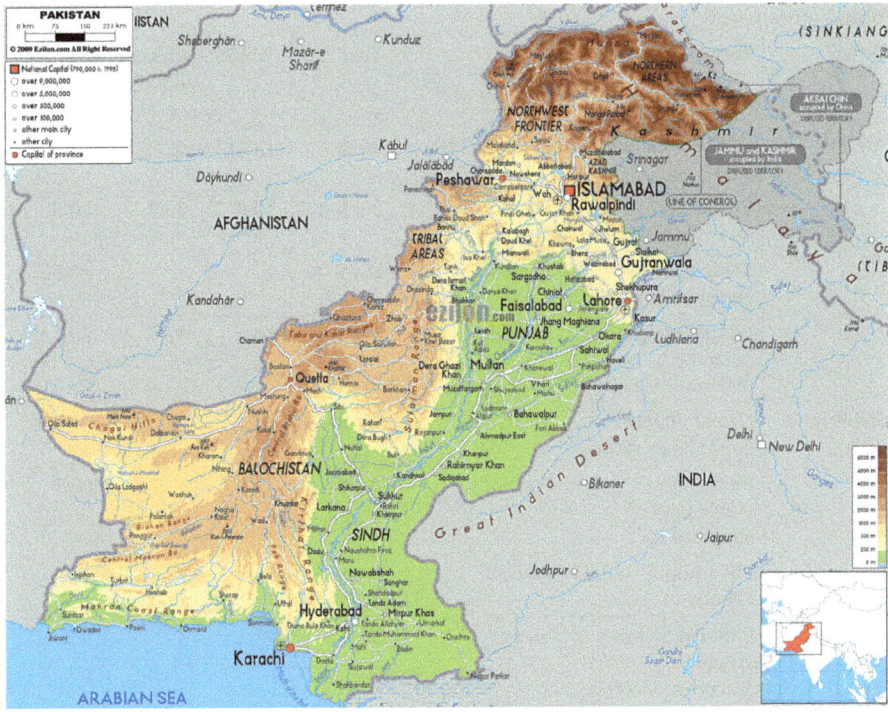

Fig. 3 Map of Pakistan showing the physical distribution of the land. *Source* https://www.ezilon.com/maps/asia/pakistan-physical-maps.html

Table 1 Indicators of food security

Group	Variables	Unit	Year	World	Lower middle income	Pakistan
Availability	Average dietary energy supply adequacy	Percent	2015–17	120	313	108
	The average value of Food production	S per Capita	2014–16	113	210	196
	Average protein supply	gr/caput/day	2011–13	80	55	74
	Average fat supply	gr/caput/day	2011–13	79	64	64

(continued)

Table 1 (continued)

Group	Variables	Unit	Year	World	Lower middle income	Pakistan
Access	GDP per capita (PPP)	Const.2011$	2016	1580.4	6298.5	4857.2
	Prevalence of undernourishment	Percent	2015–17	10.8	13.9	20.5
	Share of food expenditure of poor	Percent	2015–16	–	–	48.52
	Percentage of the population undernourished	Percent	2017	10.6	13.7	20.1
Stability	Cereal import dependency ratio	Percent	2011–13	0.9	−1.5	−17.3
	Percentage of arable land equipped for	Percent	2013–15	23.3	32.8	66.3
	Food Imports/total exports	Percent	2011–13	5	9	16
	Political Stability & absence of Terrorism/violence	Index	2016	–	–	−2.47
	Per capita food production variability	Const.2004–06	2016	2200	3600	2500
	Per capita food supply variability	kcal/per caput/day	2013	6	18	21
Utilization	People using at least basic drinking water services	Percent	2015	88.5	–	88.5
	People using safely managed drinking water services	Percent	2015	71.2	–	35.6
	People using at least basic sanitation	Percent	2015	68.0	–	58.3
	Children under 5 years of age affected by wasting	Percent	2012	–	–	10.5
	Children under 5 years of age are stunted	Percent	2012	24.9	–	45.0
	Prevalence of anemia among women (15–49)	Percent	2016	32.8	–	52.1

Source http://www.fao.org/economic/ess/ess-fs/ess-fadata/en/

3 Effects of Climate Change on Agriculture

According to Germanwatch (Briefing Paper. Global Climate Risk Index 2019), Pakistan is among the top ten countries most affected by climate change in the past 20 years, owing to its geographical location (CRI from 1998 to 2017) (annual averages). According to the Global Climate Risk Index annual report for 2020, Pakistan has lost 0.53% per unit GDP, suffered economic losses worth US$ 3792.52 million, and witnessed 152 extreme weather events, from 1999 to 2018. Lobell et al. (2011) estimated that climate change has already reduced global yields of maize and wheat by 3.8% and 5.5%, respectively.

Agriculture is the backbone of Pakistan's economy, which has also been adversely affected by climate change, as it can disrupt food availability, reduce access to food, and affect food quality. Projected increases in temperatures, changes in precipitation patterns and extreme weather events, and reductions in water availability may all result in reduced agricultural productivity. Crop simulation models-based studies depict significant reductions in wheat, rice, and maize yields in the arid, semi-arid, and rain-fed areas of Pakistan under various Intergovernmental Panel on Climate Change (IPCC) scenarios by the mid and end of the century.

Rising temperatures over the last forty years, as well as changes in rainfall patterns, have resulted in uneven surface water availability, during most crop seasons. Also, the abnormal rise in temperature in the mountainous region of Pakistan, along with the extra melting of glaciers, causes numerous rains annually, displacing millions of people, destroying food, standing crops, necessities, and causing damages to structures estimated in billions of Rupees (Ali et al., 2017). Pakistan's population is growing at a rate of more than 2%. Therefore, important foods such as wheat, rice, corn, sugarcane, and vegetables are out of reach of the poor and this may imply severe social and economic implications for further increasing the risk of the underprivileged sections of society (Asif, 2013). The national Social and Living Standards Measurement (PSLM) survey has reported that more than 84% of households are food secure, while 16% of households are experiencing moderate or severe food insecurity (Dawn, 2020).

Recent developments in the United Nations Framework Convention on Climate Change negotiations (UNFCCC), i.e., the Paris Agreement in 2015 and the Koronivia joint work on agriculture (UNFCCC 2015; 2017; 2018), and the recent IPCC special report (2018) have reinvigorated calls for incentives to reduce GHG emissions, including the pricing of carbon and the levy of a carbon tax. However, the latest analyses on climate-smart agriculture and global food-crop production (Frank et al., 2017; Hasegawa et al., 2018) indicate that a tax on GHG emissions may lead to significant tradeoffs between emissions abatement and food security.

Although the country has the world's largest integrated irrigation system, the growing water scarcity has made farmers compelled to seek less water-intensive crops and vegetables instead of water-intensive crops such as rice, wheat, cotton, etc. Crop yields are also lower in hot summers due to the evaporation and extreme temperature.

The main human-induced factors of environmental degradation are mismanagement and overexploitation of natural resources, deforestation, pollutions, and unsustainable agricultural practices. It has been anticipated that wheat production will decrease by 50% by 2050 in South Asia, which is almost 7% of the global crop production (CGIAR, 2017). Pakistan is sensitive to both increases in temperature and changes in precipitation. Climate change will also cause loss of biodiversity and the shifting of forest areas northwards (to cooler places). If the rains come at the right time, the farmer can harvest reasonable yields. In rain-fed areas, the farmers get approximately about 27% yield of maize, 56% of sorghum and millet, 52% of rape and mustard seeds, 83% of groundnut, and 100% of castor seed, as compared to yields obtained in irrigated agriculture (Supple et al., 1985; Zia & Baig, 1997).

Developing countries are projected to be hard hit by climate change, particularly South Asia and sub-Saharan Africa. Many developing countries are highly dependent on agriculture as a source of food security, economic growth, and livelihood in rural areas (Rosegrant et al., 2005). Numerous desert plants, vegetables, and fruits have much lower water requirements and may be successfully produced in rain-fed areas of the country. The rain-fed areas of the Punjab Province also support some 65% of the livestock population of the country; while 80% of the livestock are sustained by arid lands within the province of Baluchistan. Similarly, rain-fed areas of the provinces of Khyber Pakhtunkhwa (KPK) and Sind sustain a considerable livestock population. Increased variability of monsoon, changes in the availability of irrigation water, severe water-stressed conditions in arid and semi-arid areas, and extreme events such as floods, droughts, heatwaves, cold waves, cyclones, etc. are considered major implications of climate change. Map of Pakistan showing the physical distribution of the land is given in Fig. 3.

Increased evapotranspiration, as a result of high temperatures, will increase the water demand for crops (by 10–30%). These could increase the vulnerability of agriculture, forest, and water resources upon the livelihoods. Increases in temperatures due to climate change could particularly alter bio-physical relationships for crops/livestock/fisheries/forests such as shortening of the growing periods, changing the species patterns, increasing thermal and moisture stresses, changing water requirements, altering soil characteristics, and increasing the risk of pests and diseases in crops (Saif, 2017). The IPCC report (2007) concluded that the rise in global temperature of 2–4.5 °C is almost inevitable during the twenty-first century, whilst the future increase in air temperature is expected in the range of 1.4–5.8 °C for South Asia. The average annual temperatures could rise between 3.5 and 5.5 °C by the end of the twenty-first century.

According to the IPCC Special Report on Climate Change, Desertification, Land Degradation, Sustainable Land Management, Food Security, and Greenhouse Gas Fluxes in Terrestrial Ecosystems (IPCC 2020):

- The land surface air temperature has risen nearly twice as much as the global average temperature (*high confidence*) since the pre-industrial period (1850–1900).

- Climate change, including increases in frequency and intensity of extremes, has adversely impacted food security and terrestrial ecosystems as well as contributed to desertification and land degradation in many regions (*high confidence*).
- Since the pre-industrial period, the observed mean land surface air temperature has risen considerably more than the global mean surface (land and ocean) temperature (GMST) (*high confidence*). From 1850–1900 to 2006–2015, the mean land surface air temperature has increased by 1.53 °C (very likely range from 1.38 °C to 1.68 °C) while GMST increased by 0.87 °C (likely range from 0.75 °C to 0.99 °C). Warming has resulted in an increased frequency, intensity, and duration of heat-related events, including heatwaves in most land regions (*high confidence*).
- Frequency and intensity of droughts have increased in some regions (including the Mediterranean, West Asia, many parts of South America, much of Africa, and north-eastern Asia) (*medium confidence*); and
- The intensity of heavy precipitation events has increased at the global scale (*medium confidence*).

The increasing frequency of flash floods in hilly areas will cause riverbank cuttings and landslides, therefore damaging houses, agricultural lands, roads, and properties.

Climate change impacts livestock production in particular. Other factors include physiological stress on animals, productivity losses (milk and meat), stress on conception and reproduction, reduced productivity of fodder crops, decreased quality and palatability of forages, increased water requirements of animals and fodder crops, and climate-related disease epidemics. Some insect pests and diseases proliferate under high rainfall conditions (e.g. bollworms of cotton, wheat rust, and root rot diseases), some thrive under warm and moist conditions (e.g. thrips and sucking pests) while others under dry conditions (e.g. locusts). At higher elevations, fisheries are likely to be adversely affected by the lower availability of oxygen due to a rise in surface air temperatures. In the plains, the timing and amount of precipitation could affect the migration of fish species from the river to the floodplains for spawning, dispersal, and growth (Saif, 2017).

The Food Insecurity and Climate Change Vulnerability Index, shown in Fig. 4, prepared by the UN World Food Program (WFP) and the Met Office Hadley Centre, illustrates how strong are the adaptation and mitigation efforts given their substantial contribution to the prevention of the worst impacts of climate change on hunger globally and the reduction of vulnerability to food insecurity (Saif, 2017).

International development agencies often consider the settling of a nomadic population (sedentarization) as a solution to food insecurity and poor health care (Fratkin et al., 2004), especially in the African Sahel Zone. The concept of food security is flexible but is widely believed to "exist when all people, at all times, have physical, social and economic access to sufficient, safe and nutritious food which meets their dietary needs and food preferences for an active and healthy life". As per FAO, major scopes, hallmarks include the following;

- *Food availability*: the availability of sufficient quantities and appropriate quality of food, supplied through domestic production or imports (including aid).

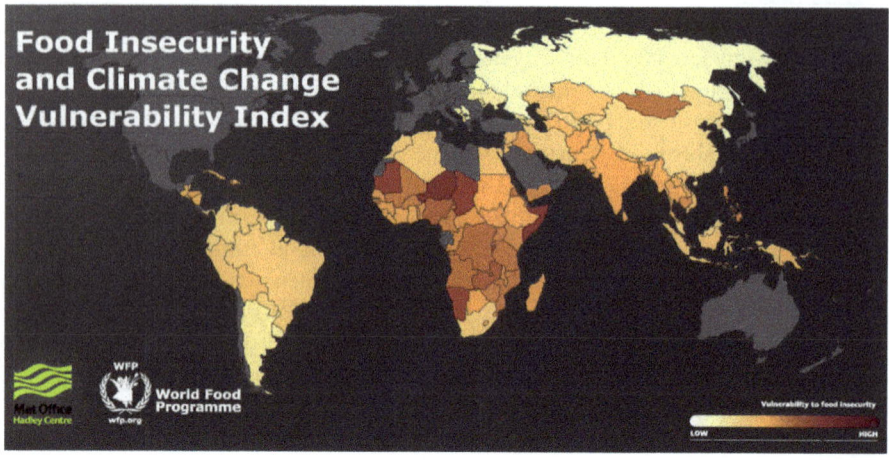

Fig. 4 Food insecurity and climate change vulnerability Index. *Source* https://www.wfp.org/pub
lications/2015-food-insecurity-and-climatechange-map

- *Food access*: access/entitlement by individuals to adequate resources for acquiring
  appropriate foods for a nutritious diet.
- *Stability*: population, household, or individual should not risk losing access to
  food as a consequence of sudden shocks (e.g., climatic crisis) or cyclical events
  (e.g., seasonal food shortages). In short, the stability concept can refer to both the
  availability and access determinants.
- *Utilization*: utilization of food through adequate diet, clean water, sanitation, and
  healthcare to reach a state of nutritional well-being where all physiological needs
  are met (this reflects the importance of non-food inputs in food security).

Despite the various efforts and investments by the government, it is estimated
that nearly 22% of the population in Pakistan is food insecure (FAO, 2014). More
than 20% of Pakistan's population is undernourished, and nearly 45% of children
younger than five years of age are stunted, according to the UN World Food Program
(WFP-Apr 2020). The food security situation of the country is given in Figs. 5 and
6. Pakistan is also prone to extreme weather and disasters (USAID, 2020).

4   Effects of Climate Change on Livestock Productivity and Food Security

Climate change represents a special 'feedback loop' in which livestock production
both contributes to the problem and suffers from the consequences. In the eighteenth
and nineteenth centuries, with the flinch of industrial and agricultural revolutions,
there was a significant increase in the amount of GHGs in the Earth's atmosphere,
causing the Earth's temperature to increase, termed as 'Global Warming' and is

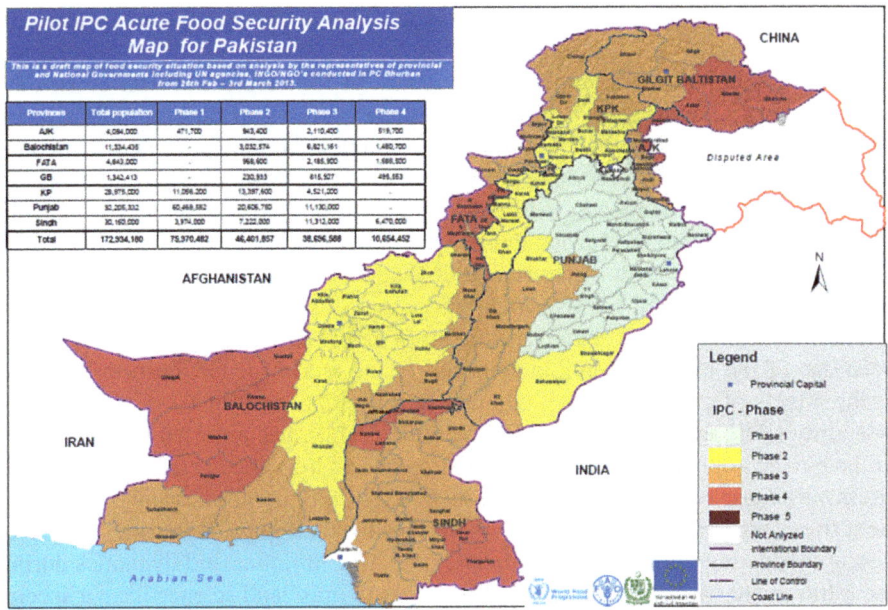

Fig. 5 Acute food security map for Pakistan. *Source* https://www.humanitarianresponse.info/en/operations/pakistan/document/pakistan-ipc-acute-food-security-analysis-maps-jul-dec-2015-draft

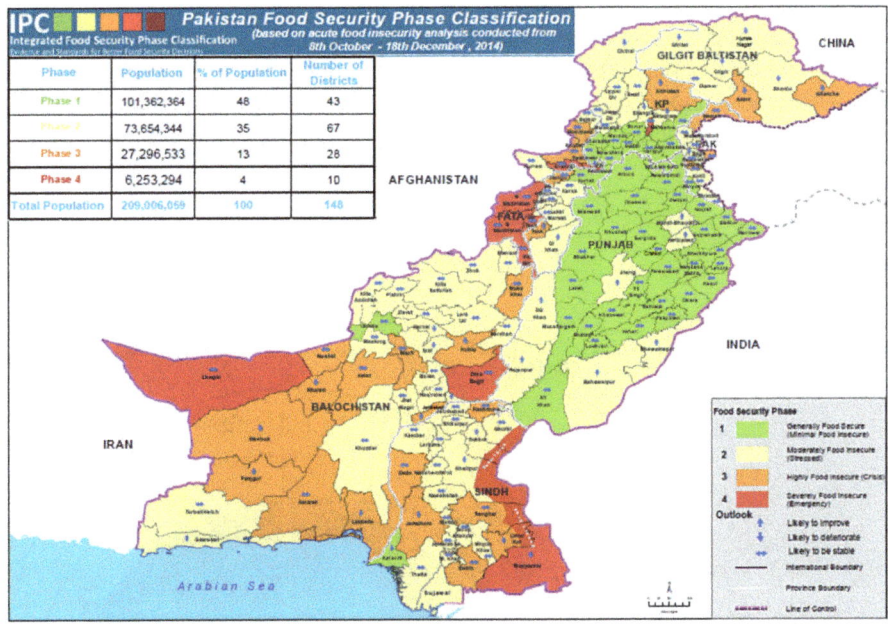

Fig. 6 Pakistan food security phase classification. *Source* Ali et al. (2017)

broadly cited as 'Climate Change'. The types of greenhouse effects are given in Fig. 7.

The livestock sector in Pakistan contributes 14.5% of global GHG emissions, which mainly include gases like CO_2, methane, N_2O, and chlorofluorocarbons. Subsequently, the livestock sector is emerging as among the key players in the reduction of GHG emissions (Jeswani & Mulugetta, 2008). Moreover, climate change poses an additional danger to livestock production in Pakistan, as it is negatively affecting the quality of feed crops and forages, water availability, animal production, milk production, livestock diseases, and biodiversity (Rojas-Downing et al., 2017). Worldwide, the demand for livestock production is predicted to double by 2050, primarily due to development/advancements in global living standards. Pakistan has been one of the most susceptible countries in the world for having been exposed to the effects of climate change, compounded by high rates of food and nutrition insecurity (Baig et al., 2020). Wolfenson et al. (2000), has reported that over 50% of the bovine population is located in the tropics and it has been estimated that heat stress may cause economic losses in about 60% of the dairy farms around the world. Heat stress compromises oocyte growth in cows by altering progesterone, the secretion of luteinizing hormone, and follicle-stimulating hormone and dynamics during the estrus cycle (Ronchi et al., 2001). Advances in grassland science have a special potential to promote the economic development and environmental sustainability of regions, nations, and peoples, especially in some of the world's most resource-limited regions (Delgado, 2005).

There is also a growing awareness of the link between livestock and climate change, affecting the livestock sector. Alternatively, it is also recognized that the livestock sector can play a key role in mitigating climate change by adopting GHG-reducing technologies (FAO, 2009). Therefore, transitioning to sustainable livestock production requires better assessments of causes and impacts, tailored to Pakistan's

Fig. 7 Types of greenhouse effects. *Source* Shahzad (2015)

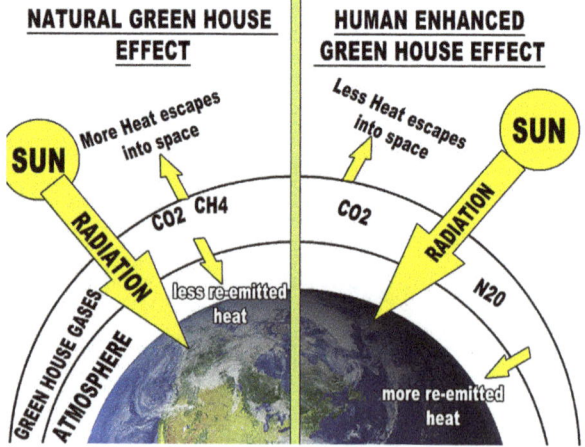

needs, justifying cogent policies that provide sustenance and facilitating the implementation of adaptation and mitigation measures. In the case of meat production, beef cattle with high weights, thick coats, and darker colors are more vulnerable to global warming (Nardone et al., 2010). Global warming may reduce body size, carcass weight, and fat thickness in ruminants (Mitloehner et al., 2001; Nardone, 2000).

The livestock sector is a major segment of agriculture (national and province-wise livestock population till last Livestock Census is given in Table 1 and livestock production of four major livestock species of Pakistan is given in Table 2), as more than half of Pakistan's population's livelihood depends on it. Owing to the traditional ways of production, there has been no significant rise in meat production, and there are no incentives to buy quality livestock due to established traditions. Problems are due to a deficiency of proper services, ancient traditional slaughterhouses that have non-hierarchical distribution systems, and meat distribution with no price structure. Furthermore, animal leather and hides are used to provide income, but due to scant and outdated strategies and marketing regimes, livestock producers are facing problems associated with skin processing and sorting as well.

According to Sulehri and Ramay (2009), the condition of Pakistan was not promising, as the number of food-insecure districts increased from 74 in 2004 to 95 in 2008. As the inflation rate continues to rise, Pakistan's huge population which is already food insecure shall be more threatening, if price fluctuation persists. Keeping in view the fact, the Government has initiated different agricultural programs and marketing strategies under its National Food Policy (NFP) to achieve food security in the country. As per Khan (2000), under NFP, three benchmarks were set to be achieved, which included the following: adequate production of food; the stability of food prices; and access to food by consumers.

Overall, food prices have increased by 75% in dollar terms since 2000, while the foreign exchange earnings and international purchasing power of all countries have decreased. With this situation, food prices continue to increase. This phenomenon is the consequence of rising living standards in neighboring countries like China and India in addition to increased use of food crops for biofuels and animal feeds, increased prices of oil and fertilizers, reduced production capacity in key grain-producing countries due to drought, and failures in the general pricing mechanisms (FAO, 2008:3–4; Headey & Fan, 2008; Simelton, 2011). South Asia has the largest concentration of poor and undernourished people in the world (Kumar et al., 2010:1) and an increase in food prices would further fuel the situation. These countries, which include Pakistan, have very few available options to deal with the challenges being currently faced. Nevertheless, stabilization of food prices may still have a positive role in improving agricultural growth (Cumming et al., 2006) and the overall social welfare including food security (Myers, 2006).

4.1 The Current State of Food Security in Pakistan

Pakistan is presently self-sufficient in major staples—ranked at 8th in producing wheat, 10th in rice, 5th in sugarcane, and 4th in milk production. Despite that, only 63.1% of the country's households are "food secure", according to the Ministry of Health and UNICEF's National Nutritional Survey (2018). The survey incorporates

Table 2 Livestock census in Pakistan–2006 (National and province-wise livestock population in Pakistan) (000 Heads)

Province	1960	1972	1976	1986	1996	2006
Pakistan						
Cattle	16,624	14,674	14,855	17,541	20,424	29,559
Buffaloes	8161	9751	10,611	15,705	20,272	27,336
Sheep	12,378	13,667	18,937	22,655	23,544	26,488
Goats	10,046	15,581	21,693	28,647	41,166	53,787
Camels	490	731	789	958	816	921
Horses	304	391	439	388	334	344
Asses	1474	1901	2157	2998	3559	4269
Mules	43	55	61	69	132	176
Poultry	12,444	17,715	32,033	57,503	63,198	73,648
Punjab						
Cattle	9673	8226	8108	8817	9382	14,412
Buffaloes	6129	7413	7979	11,150	13,101	17,748
Sheep	5583	6280	8037	6686	6142	6362
Goats	2973	5943	7767	10,755	15,301	19,831
Camels	266	365	338	321	187	199
Horses	226	264	286	245	181	163
Asses	897	1063	1139	1657	1948	2232
Mules	23	20	29	36	57	63
Poultry	6440	8688	13,783	27,848	24,511	25,906
Sindh						
Cattle	2936	2800	2854	3874	5464	6925
Buffaloes	1353	1522	1834	3220	5615	7340
Sheep	1590	840	1829	2616	3710	3959
Goats	2201	2275	4237	6755	9734	12,572
Camels	62	80	144	218	225	278
Horses	40	71	94	76	63	45
Asses	159	242	373	500	694	1005

(continued)

Table 2 (continued)

Province	1960	1972	1976	1986	1996	2006
Mules	1	2	3	5	12	20
Poultry	1250	2743	6295	8798	11,549	14,136
KPK						
Cattle	3206	2962	3000	3285	4267	5968
Buffaloes	651	791	762	1271	1395	1928
Sheep	2432	2455	3675	1599	2821	3363
Goats	3035	3737	4686	2899	6764	9599
Camels	76	101	95	70	65	64
Horses	23	31	29	34	47	76
Asses	306	408	381	446	534	560
Mules	19	32	28	23	60	87
Poultry	4190	4939	9708	17,203	22,501	27,695
Baluchistan						
Cattle	643	482	684	1157	1341	2254
Buffaloes	26	22	33	63	161	320
Sheep	2564	3859	5075	11,111	10,841	12,804
Goats	1596	3238	4441	7299	9369	11,785
Camels	86	185	212	349	339	380
Horses	10	19	23	29	43	60
Asses	99	171	244	370	383	472
Mules	(407)	1	1	4	6	6
Poultry	454	1183	1958	3295	4637	5911
Northern Areas						
Cattle	236	204	209	408	–	–
Buffaloes	2	3	3	1	–	–
Sheep	208	233	321	643	–	–
Goats	241	388	562	939	–	–
Camels	(14)	(22)	(24)	–	–	–
Horses	5	6	7	4	–	–
Asses	13	17	20	25	–	–
Mules	(131)	(120)	(188)	1	–	–
Poultry	110	162	288	359	–	–

N.B. = Figures in brackets are actual
Source Livestock census (2006)

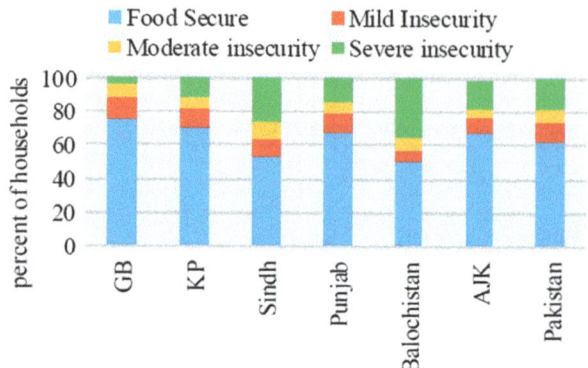

Fig. 8 Household food insecurity in Pakistan by province/region. *Source* Ministry of health and UNICEF (2018)

the Food Insecurity Experience Scale developed by the FAO. The scale trifurcates insecurity along the following dimensions:

- Mild (worrying about the ability to obtain food);
- Moderate (compromising variety/quantity of food and often skipping meals); and
- Severe (experiencing hunger on a chronic basis).

Alarmingly, of the 36.9% of the households in Pakistan labeled as "food insecure", 18.3% face "severe" food insecurity. The remaining 11.1% and 7.6% of the households face mild and moderate food insecurity, respectively. Across the provinces, KP and Gilgit-Baltistan are relatively more food secure than Sindh and Balochistan (Fig. 8). Furthermore, the available estimates of FAO suggested that Pakistan lags behind the progress of lower-middle-income countries in all four dimensions of food security (Table 1).

With a per capita income of US$ 1497, Pakistan is still struggling with issues such as undernourishment, micronutrient (iron, calcium, vitamin-A, etc.) deficiencies, and a deficit of safe drinkable water. Per capita consumption of food products that possess high nutritional value like beef, chicken, fish, milk, vegetables, and fruits is almost 6–10 times lower than that of developed countries. Livestock is rapidly growing in Pakistan (Tables 2 and 3) and is pivotal to the livelihood of its rural people.

The livestock subsector plays an important role in national food security and rural economic uplift. This subsector particularly generates daily cash income for the 8.5 million small farmers and landless families. It also provides a safety net for the poor and self-employment opportunities for women. In most of the rural areas of Pakistan, women are actively involved in livestock production activities (Humera et al., 2010; FAO, 2015). In families with more household members, the livestock responsibilities are divided among different members. In a joint family system having more than one woman, the older woman is generally responsible for milking cows and feeding livestock. Gender distribution of labor in livestock is presented in Table 4.

Pakistan is one of the leading producers of milk with an estimated production of 59.76 million tons (MT) annually. The country produces about 4.48 MT of meat, including 2.23 MT of beef, 0.7311 MT of mutton, and 1.518 MT of poultry meat

Table 3 Livestock production in Pakistan

Livestock population (Million)

Species	2011–12	2012–13	2013–14	2014-15	2015–16	2016–17	2017–18	2018–19	2019–20
Cattle	36.9	38.3	39.7	41.2	42.8	44.4	46.1	47.8	49.6
Buffalo	32.7	33.7	34.6	35.6	36.6	37.7	38.8	40.0	41.2
Sheep	28.4	28.8	29.1	29.4	29.8	30.1	30.5	30.9	31.2
Goat	63.1	64.9	66.6	68.4	70.3	72.2	74.1	76.1	78.2
Total	**161.1**	**165.7**	**170**	**176.6**	**179.5**	**184.4**	**189.5**	**194.8**	**200.2**

Source Pakistan economic surveys (2011–12, 2012–13, 2013–14, 2014–15, 2015–16, 2016–17, 2017–18, 2018–19, and 2019–20)

Table 4 Gender distribution of the livestock in Pakistan

Serial No.	Task	Total			
		F	M	F/M	M/F
1	Animal shed cleaning	6	0	0	0
2	Feeding with summer grass & milking livestock	6	0	0	0
3	Stall feeding	6	0	0	0
4	Make dairy by-products	4	0	2	0
5	Grazing animals in low pastures	2	0	4	0
6	Grazing animals in summer pastures	6	0	0	0
7	Slaughtering animals	0	6	0	0
8	Daily fodder cutting & collection for livestock	6	0	0	0
9	Seling of animals and income possession	0	6	0	0
Total		36	12	6	0
Percentage		67	22	11	0

Source FAO (2015)

(Pakistan Economic Survey, 2018–2019). The organized large and small dairy and fattening units are few; however, commercial dairy and feedlot fattening operations are emerging in the country. Despite a huge population of 88 million cattle, i.e., cows and buffalos, Pakistan imports dry milk and other dairy products. Low productivity per animal and seasonality of milk production are the main root causes behind imports. 90% of the total milk produced enters the marketing channels from subsistence farmers and 5% is processed as dairy pack products. There is a need for decreasing the yield gap in milk production through genetic interventions, improved breeding and feeding programs, the use of local and exotic dairy breeds, and the maximization of fodder and forage production. Growth in population, urbanization, increase in per capita income, and export opportunities are increasing the demand for livestock products. However, the development of this subsector is constrained with lesser profits due to low productivity, poor husbandry practices, and nutrition and health issues (National Food security Policy-Ministry of National Food Security and Research, 2018). The methodological framework of food security assessment in Pakistan is given in Fig. 9.

Despite uncertainty over climate change, the IPCC's Fifth Assessment Report identified a 'possible threshold' on average temperature rise worldwide, by the year 2100, which is between 0.3 °C and 4.8 °C (IPCC, 2013). The potential effects on livestock include changes in production and quality of feed crop and forage (Chapman et al., 2012; IFAD, 2010; Polley et al., 2013; Thornton et al., 2009, water availability (Henry et al., 2012; Nardone et al., 2010; Thornton et al., 2009), animal growth and milk production (Henry et al., 2012; Nardone et al., 2010; Thornton et al., 2009), diseases (Nardone et al., 2010; Thornton et al., 2009), reproduction (Nardone et al., 2010), and biodiversity (Reynolds et al., 2010). These impacts are primarily due to an increase in temperature and atmospheric carbon dioxide (CO_2) concentration,

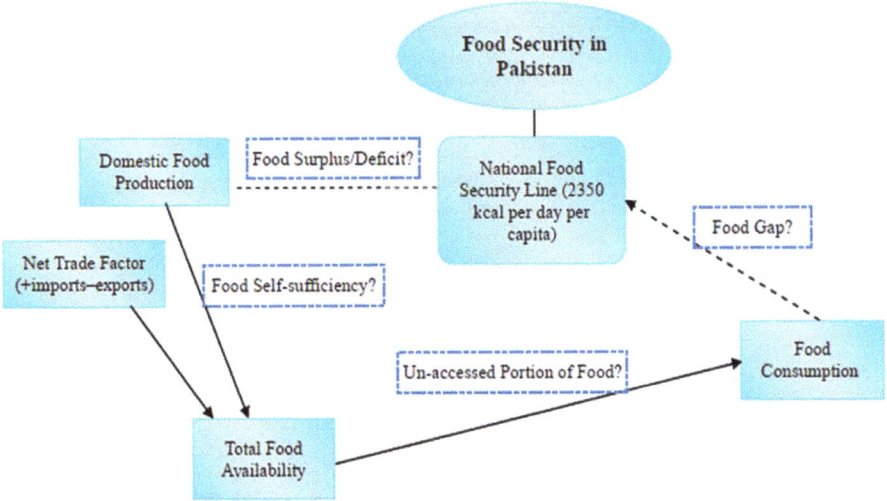

Fig. 9 Methodological framework of food security assessment in Pakistan. *Source* Abid and Jayant (2012)

precipitation variation, and a combination of these factors (Aydinalp & Cresser, 2008; Henry et al., 2012; IFAD, 2010; Nardone et al., 2010; Polley et al., 2013; Reynolds et al., 2010; Thornton et al., 2009). The effects of climate change on livestock are presented in Fig. 10.

5   Effects of Climate Change Adaptation Practices on Livestock

It is quite evident that climate change can adversely affect livestock productivity through changes in the availability of natural resources such as water, changes in the quality and quantity of forage, livestock diseases, and heat stress (Garnett, 2009; Rojas-Downing et al., 2017; Thornton, 2010; Thornton et al., 2009). The increases in temperature and CO_2 levels and changes in the amount and timing of rainfall affect the production of fodder crops and forage, a situation that is deleterious to livestock production. Due to prolonged droughts, the dry period of dairy livestock is extended and, therefore, the volume of milk production is lowered. Heat stress decreases blood progesterone, then the decrease could arise from the effects of heat stress on the follicle which ultimately carries over to the corpus luteum (Roth et al., 2001).

As a result of increases in the demand for water from competing sources and the depletion of water resources resulting from climatic changes, it is estimated that about 64% of the global population will face water stress which will, in turn, reduce the availability of water for livestock sectors. Further, the temperature rise is expected

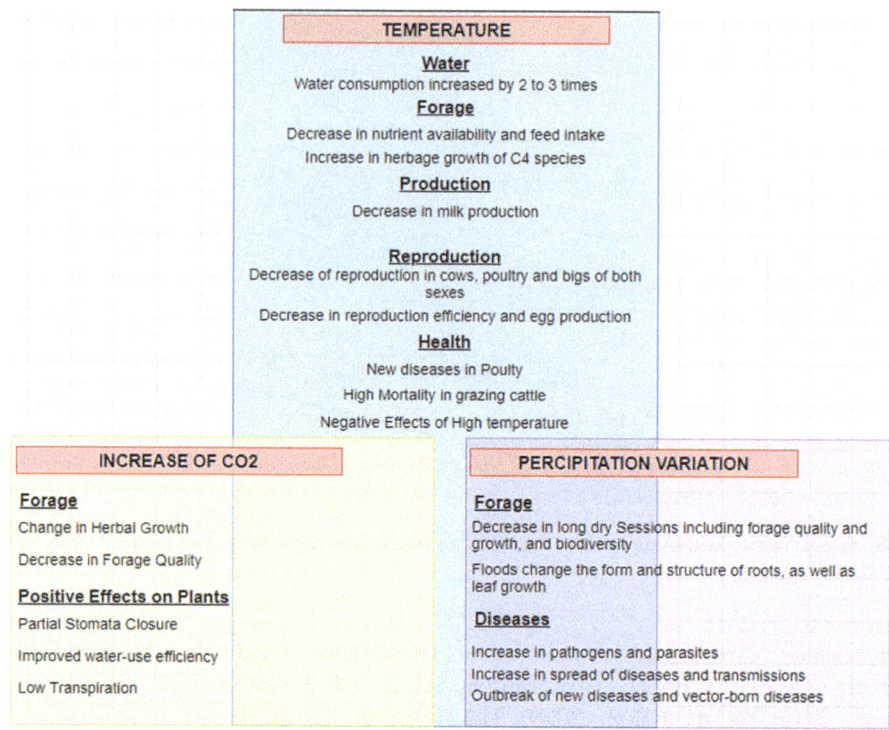

Fig. 10 The effects of climate change on livestock. *Source* Rojas-Downing et al. (2017)

to augment water intake by animals. Prompt policy measures will be needed to address these increasing demands and competition for water. The heat stress induced by climate change also results in decreases in the production of livestock products (Seerapu et al., 2015) owing to alterations in the nutrient content of the forage.

The changing climate also increases the incidence of livestock diseases and mortality (Rojas-Downing et al., 2017), which leads to a drop in livestock production. Climate-induced livestock diseases are manifesting across the globe, threatening the income, livelihood, and food security of the people living in developing countries. The livestock farmers who adopted risk-coping mechanisms generally fared better. Increasing the land area allocated to fodder seems to increase the production of milk and butter, resulting in higher income and lower poverty levels (Rahut & Ali, 2018).

Climate change adversely affects livestock assets, therefore rural households, which are largely dependent on agriculture, should implement coping measures. A wide range of options are available and households choose measures depending on availability and affordability.

Climate change is expected to cause an increase in weather-related hazards, which can severely affect livestock, especially in developing countries. Researchers and policymakers agree that the most significant impact of climate change can be seen in the livestock sector (Naqvi & Sejian, 2011). However, most of the work on climate change has focused on the crop sector, and only a few studies have documented the impact of climate change on livestock productivity. As a result of this adverse effect, farm households adopt several strategies, such as livestock insurance, increasing the area for fodder, selling livestock, and migration to cope with the negative impacts of climate change. Those farm households who adopt strategies to cope with the adverse impacts of climate change on livestock are better off than those who do not.

Climate change is extremely disruptive in poor agricultural communities. In Pakistan enthusiastic young farmers and farmers with higher levels of education use modern adaptation practices, as do farmers that are wealthier with more farmland and joint family systems. Advanced practices are associated with education, female household heads, land size, household size, extension services, access to credit, and wealth. Farmers embracing more adaptation practices had higher food security levels (8–13%) than those who did not, and experienced lower levels of poverty (3–6%). Climate change adaptation practices at the farm level can thereby have significant development outcomes in addition to reducing exposure to weather risks (Ali & Erenstein, 2017). Although food is readily available in Pakistan, the country's overall food security is poor (Table 5). High levels of poverty and inflated food prices have generated an increased number of cases of malnutrition, undernourishment, and childhood stunting in Pakistan.

The key challenges faced by the livestock sector in Pakistan are enumerated below:

Table 5 Area and production of crops by groups

Year	Food crops	Cash crops	Pulses	Edible oilseeds
(Area '000' hectares)				
2000–01	12,358	4078	1329	516
2001–02	11,999	4339	1380	570
2002–03	11,990	4069	1424	564
2003–04	12,657	4291	1447	698
2004–05	12,603	4343	1492	694
5-Years Average	**12,321**	**4224**	**1414**	**608**
2005–06	12,896	4200	1405	729
2006–07	13,066	4320	1472	754
2007–08	13,020	4512	1533	803
2008–09	13,879	4054	1465	748
2009–10	13,758	4295	1395	693
5-Years Average	**13,324**	**4276**	**1454**	**745**
2010–11	13,097	3965	1329	664
2011–12	13,053	4114	1259	713
2012–13	12,761	4272	1221	588
2013–14	13,900	4237	1170	571

(continued)

Table 5 (continued)

Year	Food crops	Cash crops	Pulses	Edible oilseeds
2014–15	13,963	4357	1154	570
5-Years Average	**13,355**	**4189**	**1227**	**621**
2015–16	13,981	4315	1202	473
2016–17	13,832	3952	1228	463
2017–18	13,745	4282	1219	565
(Production '000' tons)				
2000–01	25,986	45,867	621	4091
2001–02	24,311	50,400	594	4080
2002–03	25,890	54,200	930	3948
2003–04	26,854	55,607	871	4155
2004–05	29,906	50,000	1094	5503
5-Years Average	**26,589**	**51,215**	**822**	**4355**
2005–06	30,395	47,185	685	5063
2006–07	32,332	57,236	1089	5106
2007–08	30,689	66,031	755	4870
2008–09	35,120	52,263	992	4767
2009–10	33,974	62,516	763	4940
5-Years Average	**32,502**	**57,046**	**857**	**4949**
2010–11	34,305	67,135	656	4618
2011–12	34,479	72,200	442	5102
2012–13	34,508	66,225	898	5566
2013–14	38,312	80,630	544	5449
2014–15	37,498	76,805	528	5863
5-Years Average	**35,820**	**72,599**	**614**	**5320**
2015–16	38,226	75,716	437	4218
2016–17	40,168	86,564	506	4746
2017–18	38,971	94,604	440	4947

Food Crops = Wheat, Rice, Jowar, Maize, Bajra, and Barley
Cash Crops = Sugarcane, Cotton, Tobacco, Jute, Sugarbeet, Guar seed & Sunhemp
Pulses = Gram, Mung, Masoor, Mash, Mattar, Other Kharif & Other Rabi Pulses
Edible Oilseeds = R&M Seed, Sesamum, Groundnut, Soybean, Sunflower, Safflower, (Production includes Cotton Seed as well)
Source Ministry of National Food Security and Research, Agricultural Statistics of Pakistan (2017–18)

- Expansion of federal and provincial capacity for livestock sector development;
- Promotion of meat as a profitable business for local consumption and exports;
- Low capacity of national control programs on highly infectious and important animal diseases;
- Inadequate compliance with national and international standards for quality and hygiene;
- Prevalence of zoonotic diseases due to the proximity of human and animals;
- Lack of incentives for the generation of quality export surpluses;
- Low quality and contaminated feed; and
- Culling of dry animals and calves under the peri-urban dairy farming system.

6 Initiatives and Strategies by the Government of Pakistan to Safeguard Livestock and Food Security

There is a dire need to foster the regulatory and policy framework, which has the potential to assist the livestock sector in Pakistan to adapt to climate change impacts. Such a framework should be capable of keeping livestock in a serene and better environment, to be resistant against any change. A better environment means no risk to livestock and no threat to food security. Stressed animals mean low production, low output, and animals could be underweight, with low milk production. This also means low output, which is eventually a potent threat to food security. Salient impacts of climate change and recommended coping strategies, are shown in Fig. 11.

Following are the key policy measures, which have been adopted by the Government of Pakistan, with regards to livestock (Ministry of National Food Security and Research, 2018):

- Plans for improvement of local animal breeds, for enriched production of milk and meat.
- Special incentives and encouraging measures for the private sector to invest in dairy production for supplying pure dairy products.
- Promotion of dairy and feedlot fattening through commercial and corporate livestock farming segments.
- Encouragement of the value-added industry for livestock and livestock products to enter into the global Halal food market.
- National programs for risk-based progressive control of trans-boundary animal diseases of trade and economic importance, including Foot and Mouth Disease (FMD) and Peste des Petits Ruminants (PPR).
- Improved legal framework addressing legislative gaps, standards, grades, monitoring, and enforcement to enhance national and international quality compliance.
- Encouragement of provinces and private sectors for improvement of veterinary health services, nomads' movements, disease-free zoning, and livestock markets.
- Enhancement of training opportunities for milk and meat technology to develop a cadre of skilled human resources for the modernization of the sector.

Fig. 11 Impacts of climate change and recommended coping strategies. *Source* Rojas-Downing et al. (2017)

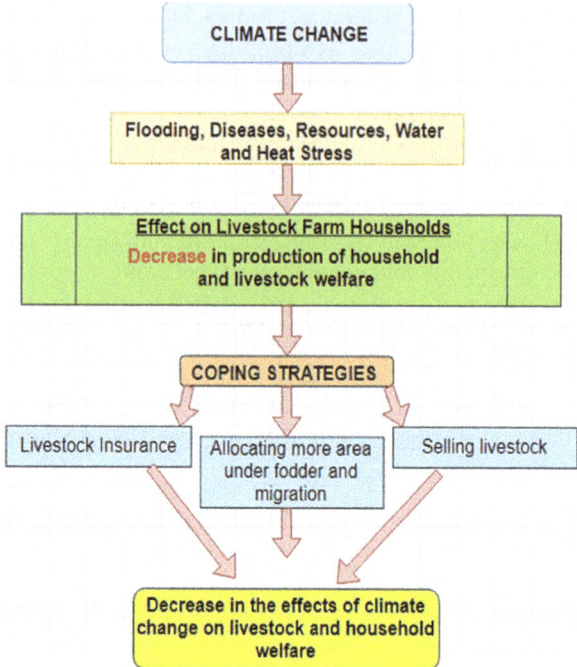

- Up-gradation and capacity building of National Veterinary Laboratory (NVL), National Reference Laboratory for Poultry Diseases (NRLPD), Animal Quarantine Department (AQD), and Livestock and Dairy Development Board (LDDB).
- Coordination for the implementation of 'One Health' programs to manage zoonotic diseases for containment and eradication as well as controlling of deaths and illnesses.
- Establishment of drug regulatory authority for improving the quality of veterinary medicines and vaccines imported or locally produced.
- Strategies to increase fodder area and yield, and to improve rangeland.
- Regulation of the availability of quality feed free from contaminants, including aflatoxins.
- Animals and animal products export facilitation by developing infrastructure on cold chain and traceability aspects.
- Enhancement of duties on import of cheaper dry-milk powder to protect the local dairy industry.
- Developing programs to regulate the culling of adult animals and calves by governments.
- The price of fresh milk may be fixed to encourage dairy producers.
- Development of high producing exotic animal model farms in cooler environments, with the requisite supply of silage and hay.

As mitigation options in the rangelands, reducing livestock numbers will surely reduce emissions but it will negatively affect the net cash income of pastoral communities. However, changing the time of lambing, culling unproductive ewes, and reducing stock in overgrazed areas may lead to a significant reduction in GHG emissions without a substantial effect on net income. Grazing the mix (cattle, sheep, goats, dromedaries) of animals may be both ecologically and economically efficient. Changing animal distribution, the establishment of shaded areas, development of water sources, and fencing can improve carbon sequestration through some increase in plant cover and improved health of the root system through the lighter intensity of grazing. However, the main way to reduce significantly methane emissions is the improvement of the quality of the diet such as by providing protein supplements.

Moreover, with a vision to drive a nationwide cleanliness movement by Pakistani people to combat the effects of climate change, enable a clean and green environment, protect livestock, and ensure food security, the Government launched the Clean Green Pakistan Movement (CGPM) in 2018. A major program under the CGPM has been launched as well in 2019 to ensure the public support and backing of civic bodies for this national effort. It is designed to seek the voluntary participation of citizens for keeping the areas clean, improving civic amenities, and creating in them the spirit and wisdom of holding their habitats and cities unsoiled. This comprises as well greywater recycling and rainwater harvesting, and education of the public on household water treatment options and tree plantation (type and time of plantation for respective areas).

7   Effect of COVID-19 Pandemic on Agriculture Sector in Pakistan

According to the Pakistan Economic Survey 2019–20, the strategy of lockdown has not only affected the domestic demand for livestock but also created supply and demand issues for exports. More than 800,000 rural families are engaged in livestock production and earn more than 35–40% of their income from this sector. In recent years, livestock has been the biggest contributor to value addition in agriculture, leaving crops behind. It currently accounts for 60.56% of agriculture and 11.69% of GDP during 2019–20. The total value of livestock has increased from Rs 1,430 billion in 2018–19 to Rs. 1466 billion in 2019–20, at an increase of 2.58% over the same period in 2018–19. The total milk production during 2019–20 was 61,690,000 tons, while the estimated milk for domestic human consumption was 49,737,000 tons. A healthy trend of consumption of camel's milk is increasing in Pakistan, and its milk production during the financial year 2019–20 was 920,000 tons (Pakistan Economic Survey 2019–20).

Meat production is also increasing. During 2019–20, meat production was reported at 4,708,000 tons, buffalo at 29,805,000 tons, and Poultry at 1,657,000.

Animal Quarantine Department provided critical services and issued 36,853 certificates for the export of live animals, mutton, beef, eggs, and other livestock products, valued at −20,337.162 million (The Daily Dawn, Published on June 14, 2020).

The COVID-19 epidemic has adversely affected the entire value chain of livestock globally, as well as the livelihood of farmers and other players, as they faced challenges in repaying their outstanding loans. In Pakistan, there was no significant impact of COVID-19 on the agriculture sector since it grew by 2.67%, while the livestock sector has shown a growth of 2.58%. Like many other countries across the globe, the COVID-19 pandemic has also spread in Pakistan. However, the COVID-19 emergency came at a time when the economy of the country was already in stabilization mode. According to the FAO, food security will be the second most important area of concern after health if the COVID-19 emergency is extended beyond a certain limit. According to the COVID-19 Emergency in Pakistan, which is already home to around 53 million poor people (a quarter of the country's population) who live below the national poverty line, whereas around 84 million people (around two-fifth of the population) are multi-dimensionally poor. Similarly, food insecurity is also very high and between 20 and 30% population (40 to 62 million people) is food insecure.

In the wake of Coronavirus Pandemic, the Government took the prompt initiative through a masses mobilization drive, that included awareness on self-hygiene, general hygiene of the areas, safe sanitation, liquid/solid waste management, e.g. proper disposal of used masks and gloves for COVID-19 prevention, safe drinking water (drink clean water, purify/treat and ensure the conservation, judicious use of water resources in response to COVID-19). To mitigate the negative impact of COVID-19 on the overall economy, the Government announced many economic relief and stimulus packages which will also strengthen the agriculture sector. Moreover, even in the lockdown, livestock products, food, fruits, and vegetable items were allowed to continue their supply (FAO, 2020).

8 Climate Change Predictions and Impacts on Rangeland Management and Livestock in Pakistan

Farooqi et al. (2005) have analyzed the global climate change perspective in Pakistan and concluded that annual mean surface temperature is on a uniform rising trend, since the start of the twentieth century. Charlotte (2011) added that the major risk from climate change in South Asia has increased summer precipitation, the intensity in temperate regions, increased flash-flood prone areas, and further added that the arid and semi-arid regions would be drier in summer, which could lead to severe droughts. The number of heatwave days per year has grown by almost fivefold in the last three decades (Chaudhry, 2017). The accumulated rainfall map of Pakistan is shown in Fig. 12, and the physiographic features of Pakistan are given in Fig. 13.

Pakistan has a fully operational irrigation system and the rivers flowing through it had abundant water. But until recently, water was scarce enough to meet the needs of

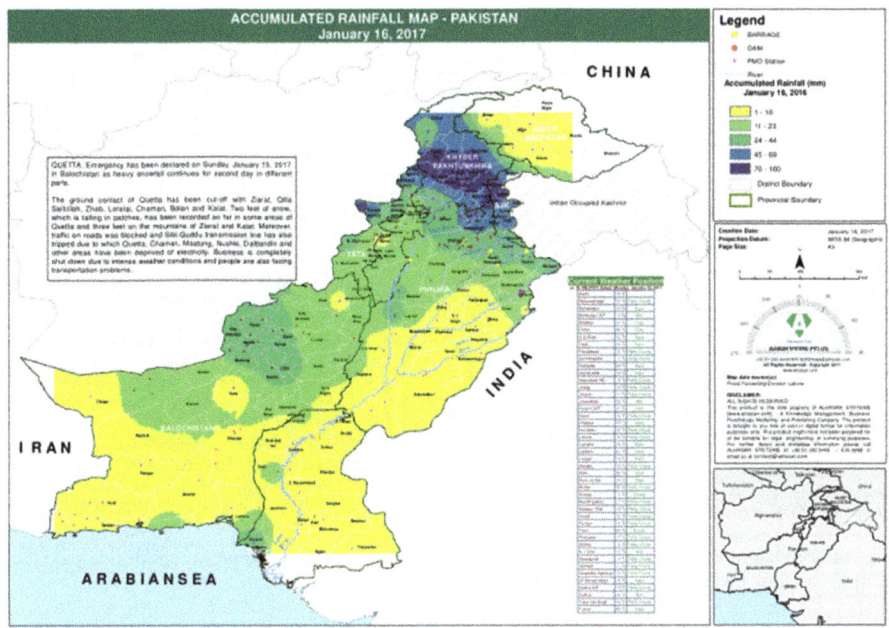

Fig. 12 Accumulated rainfall map of Pakistan. *Source* https://reliefweb.int/map/pakistan/accumulated-rainfall-map-pakistan-january-16-2017

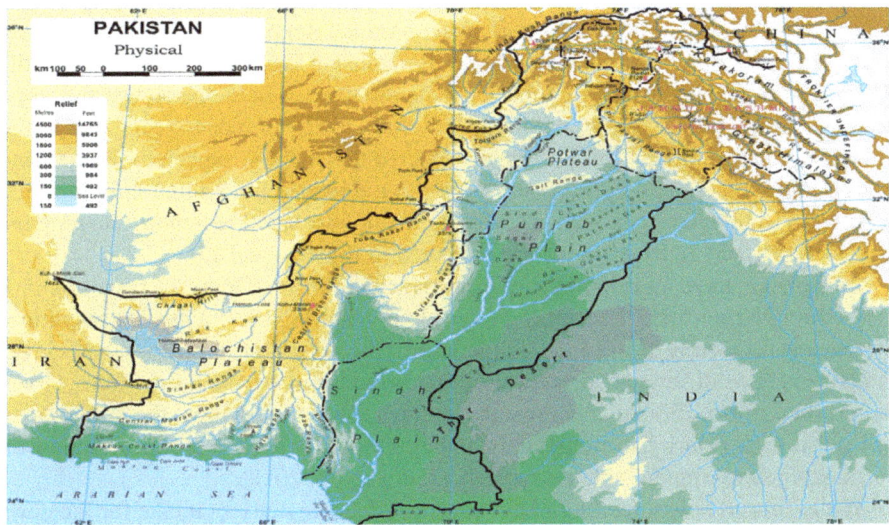

Fig. 13 Physiographic features of Pakistan. *Source* Hussain and Lee (2009)

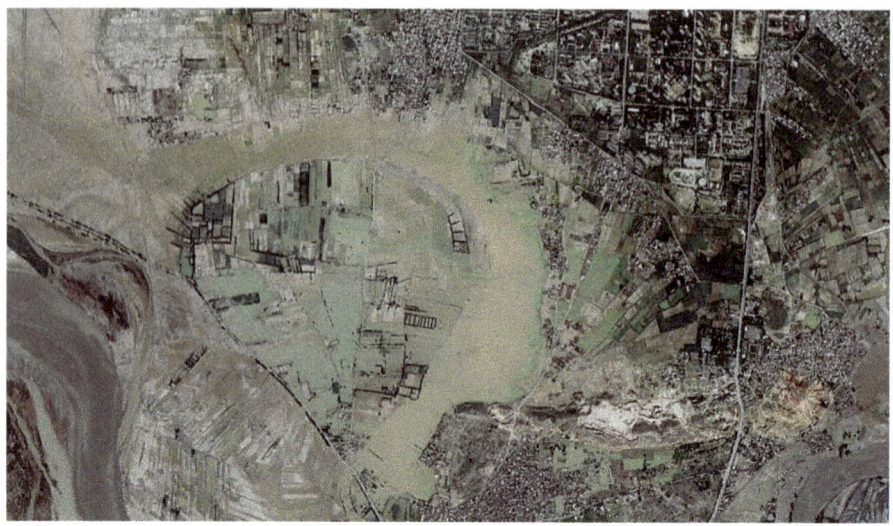

Fig. 14 Floods in Pakistan. *Source* IUCN website (https://www.iucn.org/content/floods-pakistan-satellite-images-provide-and-after-information)

end-users. Population growth began to put unprecedented pressure on the irrigation system over the last decades (UN, 2001) due to a steady increase in demands for food and fodder. Similarly, drought has severely affected the livestock population, especially in 2003–04 (State Bank of Pakistan, Annual Report 2003–04), thus slowing down their development. Researchers in Baluchistan have also reported that the drought has affected the entire country and the province of Baluchistan was the worst affected, resulting in casualties (Shafiq & Kakar, 2007). In recent years, due to the change in rainfall, floods have seriously affected the country, especially Punjab and Sindh. The FAO (2012) report, confirming these trends, reported that only in Sindh province 116,000 livestock were affected by the floods. Satellite images showing the intensity of floods in River Indus are indicated in Figs. 14 and 15.

Rangelands have an important role in improving the economy of rural communities through livestock. They play a key role in restoring the country's agricultural system and protecting biodiversity while improving access rates. Besides that, rangelands provide wildlife and fish, natural habitats in addition to recreational areas. Given the current global climate concerns and in the scenario of change, the government has given high priority to rangelands management, so there is a concentration on sustainable development and ascendency (Jamil et. al., 2018).

While formulating a feeding plan to ensure adequate nutrient intake for the cows facing challenges for heat stress, it depends on optimizing the rumen un-degradable protein to improve milk production response. The forage level must decrease due to the high heat of increment (West, 1999). De Rensin and Scaramuzzi (2003) recommended decreasing appetite and dry matter intake, so in this way, it prolonged the postpartum negative energy phase in dairy cows. Tao et al. (2011) found that cooling

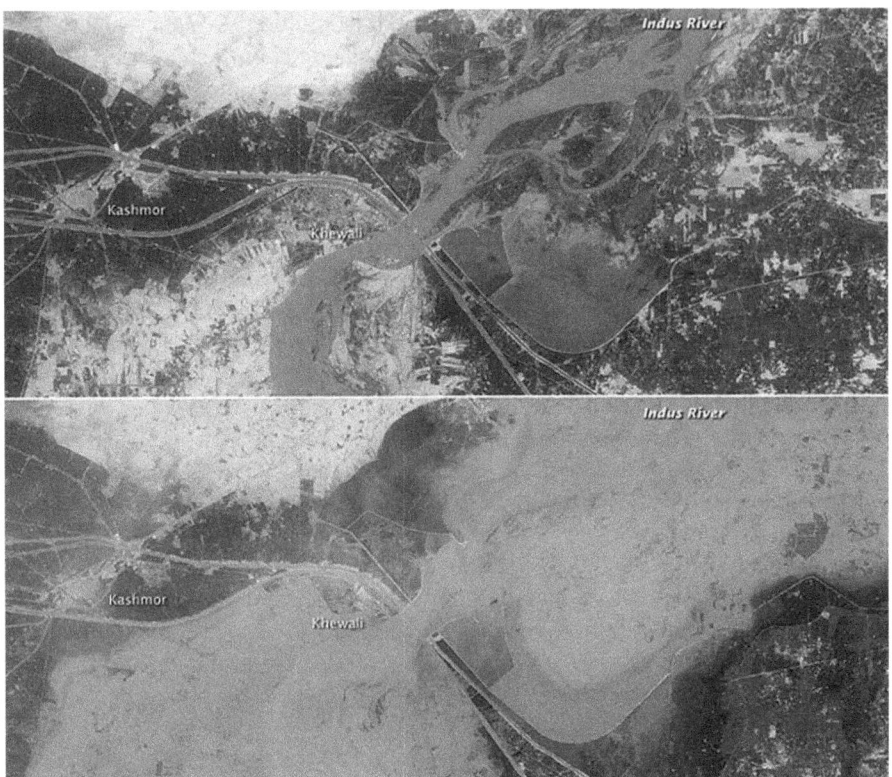

Fig. 15 Floods in Pakistan. *Source* Wired website (https://www.wired.com/2010/08/pakistan-flood-pictures)

of warmth stress can increase milk production (28.9 versus 33.9 kg/day) and lower milk protein (3.01 versus 2.87%). The physiological change regarding milk synthesis during heat stress may be due to hepatic glucose preferentially used for a process other than milk synthesis (Baumgard et al., 2011). An increase in thermal load decreases reproductive efficiency (Fuquay, 1981; Hussain et al., 1992). The Pakistan Agricultural Research Council (PARC) initiated programs that are helping directly or indirectly in the conservation and improvement of purebred animals, which included the establishment of the Embryo Transfer Technology facility, with coordinated Research Project on Improvement of Sheep and Wool (PARC, 1988).

Under current conditions, the interest in agricultural livestock is increasing due to the increase in demand, but because of the advantages of traditional production, the structure has not changed. Similarly, supported growth in the livestock sector encourages poverty reduction and the food supply of small producers has implications for public health and the environment which must be addressed under the supportability implications.

Climatic factors—for example, temperature, precipitation frequency, and severity of extreme events—affect the livestock and crop yield (Thornton et al., 2009). Heat stress has adverse effects on milk production and the reproduction of dairy cattle (Hansen, 2007; Kazdere et al., 2002; West, 2003). The Temperature Humidity Index (THI) has been used as an indicator of heat stress. The critical values of THI are 72 (Igono et al., 1992), while some studies (i.e., Dikmen & Hansen, 2009) showed that dry bulb temperature may be used as a tool to gauge the rectal temperature of lactating Holstein cows during a sub-tropical environment. The increase in thermal load above the thermal neutral zone affects animal performance. The animal fails to dispel excess heat to maintain body temperature (West, 1999). Mukherjee et al. (2011), found that a decrease in lymphocytes in heat stresses cows. Physiologically, there will be an increase in rectal temperature and respiration rate due to an increase in ambient temperature beyond the thermo-neutral zone (Chase, 2006). Gwazdauskas (1985) found that estrous hormones were found lower during heat stress and resulted in a shortening of estrus duration. He further investigated that lower fertility in heat-stressed male cattle is due to the impaired spermatogenesis and testosterone during exposure to hyperthermia.

Baumgard and Rhoads (2007) remarked that the effect of heat stress harmed the selection of dairy parameters, including milk yield and reproduction, which caused a heavy economic encumbrance. An increase in thermal load decreases reproductive efficiency (Hussain et al., 1992). Most of the reproductive problems are decreased duration and intensity of estrus (Her et al., 1988), lower conception rate, and high embryo mortality.

The change in ambient temperature severely affects the animal body. Many scientists have investigated the dynamics of the animal body due to heat stress. During 1973–78 trials, it was concluded that the combined use of showers and fans as a thermal relief measure in dairy buffaloes was found to be significantly useful as compared to the use of either fans or showers. Improved milk yield and the occurrence of estrus were more pronounced when the treatment includes the combined use of fans and showers. Similar observations were recorded within the follow-up studies, Younas et al. (1979) indicated that under the environmental temperatures of 32–47 °C with a mean ratio of 33–75%, the physiological norms of the buffalo calves were significantly affected and their weekly body weight decreased by 43 kg as compared to 46 kg under open-air tree shade than inside a shed with showers plus ceiling fans, body temperature was higher 101.6 °F than 101.0 °F, respiration rate was higher 28 to 26 per minute, and therefore the pulse high which was 53 to 54 per minute under treatment with open-air tree shade as compared to inside with ceiling fans and showers.

The conclusion from these studies was that the declining effect of thermal stress on certain components of milk and blood, could not be solely ascribed to heat stress. In recent studies, it has been observed that daily milk production was significantly higher in cows under shade and fans (SF) (7.9 ± 1 kg) followed by shade+fans+sprinklers (SFS) (6.9 ± 1.2 kg) and Shade (S) (6.1 ± 0.9 kg) treatments. The mean Rectal Temperature (101.0 ± 0.04 °F) was significantly lower in cows under shade+fans+sprinklers than that on shade and fans and shade, and similarly, mean Respiration Rate was also lower (21.2 breaths/min) in cows under shade+fans+sprinklers followed by shade+fans and shade. It was concluded that milk production and physiological performance in Sahiwal cows can be improved by fan-assisted ventilation during hot dry summer in subtropical regions (Ahmad et al., 2018). Similarly, milk production decreased with an increase in ambient temperature. Average dry matter intake in group S (shade), SF (shade and fan), and SFS (shade, fan, and sprinkler) were 75%, 80%, and 90% of the total feed offered to the experimental animals, respectively. The mean rectal temperatures (°F) were 101.69, 101.19, and 100.85 in group S, SF, and SFS, respectively. Heat stress had a pronounced effect on blood glucose level as indicated by the mean glucose concentration in group S and SFS being recorded at 78.04 mg/dl and 90.47 mg/dl, respectively. It is concluded that the buffaloes should be provided with sprinklers and fans to minimize heat load and maximize production during the hot dry season (Ahmad et al., 2019).

The highest average body weight was observed in the group of calves provided ceiling fans alone as thermal relief whereas, in trials with adult buffaloes, the combined use of fans with showers proved better. Blood glucose, total lipids, and phospholipids content were found to consistently increase, whereas cholesterol levels decreased, except for a few where a slight increase was observed (Younas et al., 1982).

Lactating and cyclic Holsteins, in each of two consecutive summers, were assigned randomly to pens, in a free-stall barn, either with or without overhead fans to review the effect of overhead fan cooling on some endocrine and behavioral responses, during the normal estrous cycle (Younas et al., 1993). Rectal temperatures were lower within the group cooled by fans than those animals within the control group, each summer. Luteal progesterone secretion was greater within the fan group in each summer and the area under the secretory phase curve was significantly increased during the second summer. There was a tendency for more pre-ovulatory surges of LH, and higher estrous response rates in the animals which were provided with the overhead fans during the second summer. Consequently, during the summers, the provision of comforts through overhead fan cooling to lactating dairy cows for several weeks before anticipated breeding proved to be enormously efficient in reproductive performance.

9 Discussion

Due to the ongoing climate change, whether arrays grow more, unpredictable and intense climate-related adversities may become more common. Livestock production is likely to be adversely affected by climate change and competition over land and water, thus undermining food security in a very sensitive context. On the other hand, the combined dynamics of population growth—which is expected to increase from 7.2 to 9.6 billion by 2050 (UN, 2013)—and rising living standards, mean that the demand for agricultural products will increase by about 70% in the same timescale (FAO., 2009). Meanwhile, the total global cultivated land area has not changed since

1991 (O'Mara, 2012), and this reflects increased productivity and intensification efforts.

The demand for livestock is rapidly growing in terms of livestock products, and this evolution is directly proportionate to the influx of rural population into urban centers and peripheries. Rates of food insecurity in Pakistan will upsurge even greater in the coming years, especially in a context marked by the impacts of climate change on all determinants of food security. It also will have a strong impact on the agriculture sector, which is extremely vital and on which the livelihoods of the majority of Pakistanis depend. Climate change is also worsening the risks of hunger and malnutrition through extreme weather events, with long-term and gradual climatic hazards. Food security and climate resilience bear a significant priority for the country, which has also been reflected in the National Development Plan, Vision 2025. Moreover, during and post-COVID-19 pandemic, the importance of food security and climate change issues for policy-making processes will be increased. Accordingly, countries like Pakistan require a change in policies regarding livestock production. Keeping in view the effects of climate change, the production systems in the country need to be re-visited and revolutionized. There is no best way than managing the dairy animals wisely and economically during the hot summers enabling them to dissipate their body heat and facilitating their comfort as much as possible. These measures will enormously help minimize the impacts of heat stress/anxiety and other environmental effects on dairy animals.

However, it is hoped that those farmers who would implement adaptation measures become more resilient to the impacts of climate change, especially those on food security. It is pertinent to mention that due to the preemptive and effective control of Simple Measure of Gobbledygook (SMOG) by public institutions in the past few years in Pakistan, the efficiency of the livestock has been improved. In this current and post-COVID-19 pandemic, the significance of climate change and food security has increased. In the years to come, this will also have implications for livestock productivity and food security in Pakistan.

10   Recommendations

Based on the above analysis, the following recommendations intend to address the issues emerging in the livestock sector due to climate change. These could help livestock farmers survive the negatively impacted situation and sustainably undertake their farming activities, thus contributing to fostering the climate resilience of the livestock sector and the food security in Pakistan. Climate-smart agriculture is an approach used in the contemporary world, aiming at increasing agriculture productivity and earnings, while building resilience to climate vulnerability and changes, concurrently. This framework is relevant to Pakistan and can be implemented through many measures, including:

- Integrate measures like the prediction of monsoon, protection from rain and extreme heat, early warning systems for floods, and effective flood relief measures.
- Adopt defined management choices, which will enable the livestock to grow faster, like feeding in a better and balanced way (as it constitutes 70–80% of the cost of milk or meat production), breeding wisely, and weeding the uneconomical.
- Exercise vigilance on pronounced estrous behavior, employ cogent husbandry measures for improved conception rates, and reproduce on time and produce at their maximum, while they are milking.
- Improved feeding management like grazing during the cool hours, offering succulent fodder varieties, and adopting the feed bunk management system, we can maximize dry matter intake (DMI) and productivity.
- Educate farmers on the use of silage and hay for their livestock, and the provision of high-quality nutrient density.
- Provision of improved/value-added housing for the livestock, for a better output.
- Risk analysis and devising some innovative methods to provide the ideal comfort for better livestock productivity.
- Use of ecological modeling and modernizations with appropriate strategic management decisions for the livestock farmers to combat the climatic changes and climatic stress factors.
- Strict genetic selection.
- There should be measures and actions by the government in enhancing food obtainability, refining food access, facilitating food consumption, and guaranteeing food permanence at all echelons.
- Educate farmers on avoiding excess feeding, for normal rumen function, with the provision of fresh and clean water for their livestock.
- Provide extensive extension services by the government and NGOs to train and educate farmers on the use of modern tools like tunnel ventilation, provisions of sprinklers, showering, misting to provide comfort to their livestock in harsh summers.
- Crop and livestock productivity, market access, and the effects of climate change are all extremely location-specific. International development agencies and national governments should work to ensure that technical, financial, and capacity-building support reaches local communities.

It is quite heartening to mention that Pakistan is already taking cogent initiatives to safeguard its environment from the ill effects of Climate Change. In 2005, Pakistan adopted its National Policy on Climate Change. This policy stresses the need to encourage the use of ozone-friendly technologies and phase out the ozone-depleting substances, in line with the provisions of Montreal Protocol, (1987) (an international treaty designed to protect the ozone layer by phasing out the production of numerous substances that are responsible for ozone depletion).

11 Conclusion

The above analysis leads to the conclusion that Pakistan is experiencing the impacts of climate change, mainly on its agricultural and livestock (body-weight, meat, and milk production, etc.) sectors, and this is already a reality to be managed. This in turn threatens the future food security of the country. The best possible option would be to educate and train crops-producing farmers and farmers who own or sustain at least 2–3 animals, through far-reaching extension activities. Similarly, separate educational, awareness, and training programs are needed for dairy farmers. For the farmers managing livestock at their respective farms, there is a dire need to design and implement valid, workable, and sustainable strategies to help them meet national needs and strengthen the economy of Pakistan from a climate change perspective. Such strategies must be developed with the active participation of concerned stakeholders. The priority should be given to such domestic strategies which could remove physical and economic barriers to access food by providing adequate infrastructure, provisioning of special and sufficient food transport and storage facilities, ensuring stable cold chain management, enhancing political/social stability, and curbing poverty. Such measures, if adopted and appropriately implemented, would certainly help farmers overcome climate change impacts, including those on livestock, and ensure national food security.

Acknowledgements The authors are extremely thankful to Prof. Dr. Jan W Hopmans, Distinguished Professor Emeritus, Soil Science and Irrigation Water Management, University of California Davis, USA, and Dr. Michael R. Reed, Emeritus Professor and former Director of the International Programs for Agriculture at the University of Kentucky, USA, for reviewing and proof-reading the text, making helpful comments, and offering valuable suggestions for improvement.

References

Abid, H., & Jayant, K. R. (2012). Status and factors of food security in Pakistan. *International Journal of Development Issues, 11*(2), 164–185. https://doi.org/10.1108/14468951211241146

Ahmad, M., Bhatti, J. A., Abdullah, M., Javed, K., Ali, M., Rashid, G., Uddin, R., Badini, A. H., & Jehan, M. (2018). Tropical Anim. *Health and Products, 50*, 1249–1254. https://doi.org/10.1007/s11250-018-1551-5

Ahmad, M., Bhatti, J. A., Abdullah, M., Ullah, R., ul Ain, Q., Hasni, M. S., Ali, M., Rashid, A., Qaisar, I., Rashid, G., & Uddin, R. (2019). Different ambient management intervention techniques and their effect on milk production and physiological parameters of lactating Nili-Ravi buffaloes during hot dry summer of subtropical region. *Tropical Animal Health and Products, 51*, 911–918. https://doi.org/10.1007/s11250-018-1774-5

Ali, A., & Erenstein, O. (2017). Assessing farmer use of climate change adaptation practices and impacts on food security and poverty in Pakistan. *Climate Risk Management, 16*, 183–194. https://doi.org/10.1016/j.crm.2016.12.001

Ali, S., Liu, Y., Ishaq, M., Shah, T., Alyas, A., & Din, I. U. (2017a). Climate change and its impact on the yield of major food crops: Evidence from Pakistan. *Foods, 6*(6), 39. https://doi.org/10.3390/foods6060039

Ali, S., Liu, Y., Ishaq, M., Shah, T., Ilyas, A. A., Din, I. U. (2017b). "Climate change and its impact on the yield of major food crops: Evidence from Pakistan". *Foods, 6*, 39. https://doi.org/10.3390/foods6060039

Asif, M. (2013). *Climatic change, irrigation water crisis, and food security in Pakistan.* Independent thesis advanced level (degree of master two years). Uppsala University. https://pdfs.semanticscholar.org/8c58/9f922896ba3a84122273a5adf88ebc254d13.pdf

Aydinalp, C. & Cresser, M. S. (2008). The effects of global climate change on agriculture. *American-Eurasian Journal Agricultural and Environment Science, 3*(5), 672–676. ISSN 1818-6769 © IDOSI Publications.

Baig, M. B., Burgess, P. J. & Fike, J. H. (2020). Agroforestry for healthy ecosystems: Constraints, improvement strategies, and extension in Pakistan. *Agroforest System* https://doi.org/10.1007/s10457-019-00467-4

Baig, M. B., Shabbir, A. S., & Gary, S. S. (2013). Making rain-fed agriculture sustainable through environmentally friendly technologies in Pakistan: A review. *International Soil and Water Conservation Research, 1*(2), 36–52. https://doi.org/10.1016/S2095-6339(15)30038-1

Baumgard, L. H., & Rhoads, R. P. (2007). *The effect of heat stress on nutritional and management decisions.* Proceedings southwest nutrition conference, (pp. 191–199). https://www.researchgate.net/publication/236133490_Effect_of_Climate_Change_on_Livestock_Production_in_Pakistan

Baumgard, L. H., Wheelock, J. B., Sander, S. R., Moore, C. E., Green, H. B., Waldron, M. R., & Rhoads, R. P. (2011). Post absorptive carbohydrate adaptations to heat stress and monensin supplementation in lactating Holstein cows. *Journal of Dairy Science, 94*(11), 5620–5633. https://doi.org/10.3168/jds.2011-4462

Briefing Paper. (2019). Global climate risk index. http://www.germanwatch.org/sites/germanwatch.org/files/20-2-01e%20Global%20Climate%20Risk%20Index%202020_14.pdf

CGIAR. (2017). *Facing climate change by securing water for food, livelihoods, and ecosystems.* Accessed May 21, 2017. Available online: http://www.iwmi.cgiar.org/wp-content/uploads/2013/02/sp11.pdf

Charlotte, S. (2011). Review of climate change adaptation practices in South Asia. Oxfam research. Reports. http://www.oxfam.org/sites/www.oxfam.org/files/rr-climate-change-adaptation-south-Asia-161111-en.pdf

Chase, L. E. (2006). Climate change impacts on dairy cattle. Department of Animal Science Cornell Univ, 14853. ClimateChangeImpactsDairyCattle.pdf.

Chapman, S. C., Chakraborty, S., Dreccer, M. F., & Howden, S. M. (2012). Plant adaptation to climate change: Opportunities and priorities in breeding. *Crop and Pasture Science, 63*, 251–268.

Chaudhry, Q. Z. (2017). *Climate change profile of Pakistan.* Asian development bank 6 ADB Avenue, Mandaluyong city, 1550 Metro Manila, Philippines. Viewed in April 2020 at https://www.adb.org/sites/default/files/publication/357876/climate-change-profile-pakistan.pdf

Climate Risks and Food Security Analysis. (2018). *A special report for Pakistan.* WFP, Islamabad, Pakistan. https://reliefweb.int/sites/reliefweb.int/files/resources/Climate_Risks_and_Food_Security_Analysis_December_2018.pdf

Cumming, R. Jr., Rashid, S., & Gulati, A. (2006), "Grain price stabilization experiences in Asia: What have we learnt?". *Food Policy, 31*, 302–12. https://www.academia.edu/36233406/International_Journal_of_Development_Issues_Status_and_factors_of_food_security_in_Pakistan_Article_information

Dawn, I. (2020, April 21). Pakistan social and living standards measurement (PSLM).

De Rensis, F., & Scaramuzzi, R. J. (2003). Heat stress and seasonal effects on reproduction in the dairy cow—A review. *Theriogenology, 60*(6), 1139–1151. https://doi.org/10.1016/s0093-691x(03)00126-2

Delgado, C. (2005). Rising demand for meat and milk in developing countries: Implications for grasslands-based livestock production. In *Grassland: A global resource. ILRI-IFPRI, joint program on livestock market opportunities c/o IFPRI*, 2033 K St., 20006. https://www.ifpri.org/publication/productivity-and-efficiency-grassland-based-livestock-production-latin-america-cases

Dikmen, S., & Hansen, P. J. (2009). Is the temperature-humidity index the best indicator of heat stress in lactating dairy cows in a subtropical environment? *Journal Dairy Science, 92*, 109–116. https://doi.org/10.3168/jds.2008-1370. ©American Dairy Science Association. https://pubmed.ncbi.nlm.nih.gov/19109269/

Eckstein, D., Hutfils, M.-L., & Winges, M. (2018). Briefing Paper. Global climate risk index 2019. Who suffers most from extreme weather events? Weather-related loss events in 2017 and 1998 to 2017. Germanwatch e.V. Available at: www.germanwatch.org/en/crchromeextension://ohfgljdgelakfkefopgklcohadegdpjf/https://germanwatch.org/files/Global%20Climate%20Risk%20Index%202019_2.pdf

FAO. (2008, June 3–5). Soaring food prices: Facts, perspectives, impacts, and actions required, paper presented at high-level conference on world food security: The challenges of climate change and bio-energy, rome.

FAO. (2009). Global agriculture towards 2050. High level expert forum issues paper. FAO. http://www.fao.org/fileadmin/templates/wsfs/docs/Issues_papers/HLEF2050_Global_Agriculture.pdf

FAO. (2009). The state of food and agriculture: Livestock in the balance, FAO. http://www.fao.org/3/a-i0680e.pdf

FAO. (2012). WHO, World food program, and the international fund for agricultural development. The state of food insecurity in the world 2012. Economic growth is necessary but not sufficient to accelerate reduction of hunger and malnutrition. Rome, FAO. *Advances in Nutrition, 4*(1), 126–127. https://doi.org/10.3945/an.112.003343, Published: January 4, 2013.

FAO. (2014). *State of food insecurity in the world: Strengthening the enabling environment for food security and nutrition.* FAO. http://www.fao.org/3/a-i4030e.pdf

FAO. (2015). Women in agriculture in Pakistan. In S. Durre et al., http://www.fao.org/3/a-i4330e.pdf

FAO. (2020, April 17). GIEWS country brief Pakistan. Food security snapshot. http://www.fao.org/giews/countrybrief/country.jsp?code=PAK

Farooqi, A. B., Khan, A. H. & Hazrat, M. (2005). Climate change perspective in Pakistan. *Pakistan Journal Meteorite, 2*(3), 11–21. http://www.pmd.gov.pk/rnd/rnd_files/vol2_Issue3/2.%20CLIMATE%20CHANGE%20PERSPECTIVE%20IN%20PAKISTAN.pdf

Frank, S., Havlík, P., Soussana, J. F., Levesque, A., Valin, H. & Wollenberg, E., et al. (2017). Reducing greenhouse gas emissions in agriculture without compromising food security? *Environment Research Letters.* https://ccafs.cgiar.org/publications/reducing-greenhouse-gas-emissions-agriculture-without-compromising-food-security-0

Fratkin, E., Roth, E. A. & Nathan, M. A. (2004). Pastoral sedentarization and its effects on children's diet, health, and growth among Rendille of northern Kenya. *Human Ecology, 32*(5), 531–559. https://www.academia.edu/8179439/Pastoral_Sedentarization_and_Its_Effects_on_Children_s_Diet_Health_and_Growth_Among_Rendille_of_Northern_Kenya

Fuquay, J. W. (1981). Heat stress as it affects animal production. *Journal Animal Science 52*(1), 164–74. https://doi.org/10.2527/jas1981.521164x. https://pubmed.ncbi.nlm.nih.gov/7195394/

Garnett, (2009). Livestock-related greenhouse gas emissions: impacts and options for policymakers. *Environment Science Policy, 12*, 491–503, 52–164.

Gwazdauskas, F. C. (1985). Effects of climate on reproduction in cattle. *Journal Dairy Science, 68*(6), 1568–78. https://doi.org/10.1016/j.envsci.2009.01.006. https://awfw.org/wp-content/uploads/pdf/FCRN-Livestock2009.pdf

Government of Pakistan, Ministry of National Food Security and Research. (2018). *National food security policy Islamabad.* Pakistan. http://www.mnfsr.gov.pk/frmDetails.aspx

Hansen, P. J. (2007, September 1). Exploitation of genetic and physiological determinants of embryonic resistance to elevated temperature to improve embryonic survival in dairy cattle during heat stress. *Theriogenology, 68*(Suppl 1), S242–9. https://doi.org/10.1016/j.theriogenology.2007.04.008. Epub May 7, 2007.

Hasegawa, T., Fujimori, S., Havlík, P., Valin, H., Bodirsky, B. L. & Doelman, J. C., et al. 2018. Risk of increased food insecurity under stringent global climate change mitigation policy. *Nature*

Climate Chang [Internet], *8*(8), 699–703. Available from: https://doi.org/10.1038/s41558-018-0230-x

Headey, D. & Fan, S. (2008). "Anatomy of a crisis: The causes and consequences of surging food prices". *Agricultural Economics, 39,* 375–91. https://www.annualreviews.org/doi/pdf. https://doi.org/10.1146/annurev-resource-100815-095303

Her, E., Wolfenson, D., Flamenbaum, I., Folman, Y., Kaim, M. & Berman, A. (1988, April). Thermal productive and reproductive responses of high yielding cows exposed to short term cooling in summer. *Journal Dairy Science, 71*(4), 1085–92. https://doi.org/10.3168/jds.S0022-0302(88)79656-3

Henry, B., Charmley, E., Eckard, R., Gaughan, J. B., Hegarty, R., (2012). Livestock production in a changing climate: Adaptation and mitigation research in Australia. *Crop Pasture Science, 63,* 191–202. https://bioone.org/journals/crop-and-pasture-science/volume-63/issue-3/CP11169/Livestock-production-in-a-changing-climate-adaptation-and-mitigation/10.1071/CP11169.short

Humera, A., Tanvir, A., Munir, A. & Muhammad, I. Z. (2010). Gender and development: Roles of rural women in livestock production in Pakistan. *Pakistan Journal Agriculture of Science, 47*(1), 32–36. ISSN (Print) 0552-9034, ISSN (Online) 2076-0906. http://pakjas.com.pk/papers%5C9.pdf http://www.pakjas.com.pk

Hussain, S., & Lee, S. (2009). A classification of rainfall regions in Pakistan. *Journal of the Korean Geographical Society, 44*(5), 605–623.

Hussain, S. M. I., Fuquay, J. W., & Younas, M. (1992). Estrous cyclicity in non-lactating and lactating Holsteins and Jerseys during a Pakistani summer. *Journal of Dairy Science, 75*(11), 2968–2975. https://doi.org/10.3168/jds.S0022-0302(92)78060-6

International Fund for Agricultural Development (IFAD). (2010). *Livestock and climate change.* http://www.ifad.org/lrkm/events/cops/papers/climate.pdf

Iglesias, A., Avis, K., Benzie, M., Fisher, P., Harley, M., Hodgson, N., Horrocks, L., Moneo, M. & Webb, J., (2007). *Adaptation to climate change in the agricultural sector.* AEA Energy & Environment and Universidad de Politécnica de Madrid. https://ec.europa.eu/info/sites/info/files/food-farming-fisheries/key_policies/documents/ext-study-adapt-climate-change-full-text_2007_en.pdf

Igono, M. O., Bjotvedt, G., & Sanford-Crane, H. T. (1992). Environmental profile and critical temperature effects on milk production of Holstein cows in the desert climate. *International Journal Biometeorol, 36*(2), 77–87. https://doi.org/10.1007/BF01208917

ILRI (International Livestock Research Institute). (1995). *Livestock policy analysis.* ILRI training manual 2. ILRI. (p. 264). http://dlc.dlib.indiana.edu/dlc/bitstream/10535/34/1/Livestock_policy_analysis.pdf

IPCC. (2007). Climate change 2007, Impacts, adaptation and vulnerability. IPCC Fourth assessment report. https://www.ipcc.ch/site/assets/uploads/2018/03/ar4_wg2_full_report.pdf

IPCC. (2013). The physical science basis, working group I contribution to the fifth assessment report of the intergovernmental panel on climate change. www.cambridge.org/9781107661820

IPCC. (2018). Summary for policymakers. In: Masson-Delmotte, V., Zhai, P., Portner, H. O., Roberts, D., Skea, J. & Shukla, P. R., et al. *Global warming of 15 °C. An IPCC Special Report on the impacts of global warming of 15 °C above pre-industrial levels and related global greenhouse gas emission pathways, in the context of strengthening the global response to the threat of climate change* (p. 32). World Meteorological Organization. https://archive.ipcc.ch/pdf/special-reports/sr15/sr15_spm_final.pdf

IPCC. (2020). Special report on climate change, desertification, land degradation, sustainable land management, food security, and greenhouse gas fluxes in terrestrial ecosystems. https://www.ipcc.ch/site/assets/uploads/sites/4/2020/02/SPM_Updated-Jan20.pdf

Iqbal, M. & Ahmad, M. (1999). An assessment of livestock production potential in Pakistan: Implications for livestock sector policy [with comments]. *Pakistan Development Review, 1,* 615–628. https://www.pide.org.pk/pdr/pdf/PDR/1999/Volume4/615-628.pdf

Jamil, M., Mansoor, M., Anwar, F., Muhammad, S. & Awan, A. A. (2018). A review on rangeland management in Pakistan: Bottlenecks and recommendations. *Pakistan Journal Science Industrial*

Research Series B: Biology Science, 61B(2), 115–120. https://v2.pjsir.org/index.php/biological-sciences/article/view/357

Jeswani, H. K. & Mulugetta, Y. (2008). How warm is the corporate response to climate change?. *Evidence from Pakistan and the UK. Business Strategy and Environment, 17*(1), 46–60. https://www.researchgate.net/publication/229538705_How_Warm_Is_the_Corporate_Response_to_Climate_Change_Evidence_from_Pakistan_and_the_UK

Kazdere, C. T., Murphy, M. R., Silanikove, N. & Maltz, E. (2002). Heat stress in lactating dairy cows: A review. *Livest Products Science, 77*, 59–91. https://www.ncbi.nlm.nih.gov/pmc/articles/PMC4823286/

Khan, M. A. (2000). Food security in Pakistan, Asian productivity organization, Tokyo. https://www.researchgate.net/publication/235293188_Status_and_factors_of_food_security_in_Pakistan

Kumar, A. G., Roy, D. & Gulati, A. (2010). Liberalizing food-grains markets: Experiences, impact, and lessons from South Asia, Oxford University Press, Oxford, IFPRI Issue Brief 64, August. https://www.ifpri.org/publication/liberalizing-foodgrains-markets-0

Lobell, D., Schlenker, W. & Costa-Roberts, J. (2011). Climate trends and global crop production since 1980. *Science, 333*, 616–620. https://pdfs.semanticscholar.org/9668/ad7b86da465515bfac8ed3330526f0f20e28.pdf

Maplecroft. (2010). Big economies of the future, Bangladesh, India, Phillipine, Veitnam, and Pakistan-most at risk from climate change. https://maplecroft/989.com/about/news/ccvi.html. https://www.preventionweb.net/news/view/16004

Ministry of Health and UNICEF's National Nutritional Survey. (2018). Household food insecurity in Pakistan by province/region.

Mitloehner, F. M., Morrow, J. L., Dailey, J. W., Wilson, S. C., Galyean, M. L., Miller, M. F. & McGlone, J. J. (2001). Shade and water misting effects on behaviour, physiology, performance, and carcass traits of heat-stressed feedlot cattle. *Journal Animal Science, 79*, 2327–2335. https://pubmed.ncbi.nlm.nih.gov/11583419/

Montreal Protocol. (1987). Agreed on 16th September 1987, and entered into force on 1st January 1989. https://www.unenvironment.org/ozonaction/who-we-are/about-montreal-protocol

Moyo, S. & Swanepoel, F. J. C. (2010). Multifunctionality of livestock in developing communities in the role of livestock in developing communities: Enhancing multifunctionality. In F. Swanepoel, A. Stroebel & S. Moyo (Eds.), *Co-published by the technical centre for agricultural and rural cooperation (CTA) and the University of the Free State*. https://www.researchgate.net/publication/265348731_The_Role_of_Livestock_in_Developing_Communities_Enhancing_Multifunctionality

Mukherjee, J., Pandita, S., Huozha, R. & Ashutosh, M. (2011). In-vitro immune competence of Buffaloes (Bubalus bubalis) of different production potential: Effect of heat stress and cortisol. *Veterinary Medical International, 2011*, 869252. (E-pub). https://doi.org/10.4061/2011/860252

Myers, R. B. (2006). "On the costs of food price fluctuations in low-income countries". *Food Policy, 31*, 288–301. https://www.researchgate.net/publication/235293188_Status_and_factors_of_food_security_in_Pakistan

Nardone, A., Ronchi, B., Lacetera, N., Ranieri, M. S., & Bernabucci, U. (2010). Effects of climate change on animal production and sustainability of livestock systems. *Livestock Science, 130*, 57–69. https://doi.org/10.1016/j.livsci.2010.02.011

Nardone, A. (2000). Weather conditions and genetics of breeding systems in the Mediterranean area. In: XXXX *International sysmposium of societa Italiana per il Progresso della Zootecnia*. (pp. 67–92).

National Environment Policy of Pakistan. (2005). Government of Pakistan, ministry of environment. https://mowr.gov.pk/wp-content/uploads/2018/05/National-Environmental-Policy-2005.pdf, redesignated as ministry of climate change http://www.mocc.gov.pk/

National Food Security Policy Pakistan. (2018). Ministry of national food security and research of Pakistan.

Naqvi, S. M. K. & Sejian, V. (2011). Global climate change: Role of livestock. *Asian Journal of Agricultural Science, 3*(1), 19–25. https://www.researchgate.net/publication/49605076_Global_ Climate_Change_Role_of_Livestock

O'Mara, F. P. (2012). The role of grasslands in food security and climate change. *Annals Bot-London, 110*, 1263–1270. https://academic.oup.com/aob/article/110/6/1263/112127

PARC. (1988). Pakistan agricultural research council (PARC) Establishment of the embryo transfer technology facility, with coordinated research project on improvement of sheep and wool.

Pakistan Economic Survey. (2019–20). Finance division, economic advisor's wing, government of Pakistan, Islamabad. http://www.finance.gov.pk/survey_1819.html0

Pakistan Economic Survey (2019–20). *Pandemic hits livestock value chain.* Dawn News, Islamabad, June 14, 2020. https://www.dawn.com/news/1563363/pandemic-hits-livestock-value-chain

Pakistan Economic Surveys (2011–12, 2012–13, 2013–14, 2014–15, 2015–16, 2016–17, 2017–18, 2018–19, and 2019–20).

Pakistan Social and Living Standards Measurement (PSLM) survey. Dawn News, Islamabad. April 21, 2020. https://www.dawn.com/news/1550962

Polley, H. W., Briske, D. D., Morgan, J. A., Wolter, K., Bailey, D. W., & Brown, J. R. (2013). Climate change and North American rangelands: Trends, projections, and implications. *Rangeland Ecological Management, 66*, 493–511. https://doi.org/10.2111/REM-D-12-00068.1

Rahut, D. B. & Ali, A. (2018). Impact of climate-change risk-coping strategies on livestock productivity and household welfare: empirical evidence from Pakistan. *Heliyon, 4.* https://doi.org/10.1016/j.heliyon.2018.e00797

Reynolds, C., Crompton, L. & Mills, J. (2010). Livestock and climate change impacts in the developing world. *Outlook Agric, 39*, 245–248. https://doi.org/10.5367/oa.2010.0015

Rojas-Downing, M. M., Nejadhashemi, A. P., Sean, T. H., & Woznicki, A. (2017). Climate change and livestock: Impacts, adaptation, and mitigation. *Climate Risk Management, 16*, 145–163. https://doi.org/10.1016/j.crm.2017.02.001

Ronchi, B., Stradaioli, G., Verini Supplizi, A., Bernabucci, U., Lacetera, N., Accorsi, P. A., Nardone, A. & Seren, E. (2001). Influence of heat stress and feed restriction on plasma progesterone, estradiol-17β LH, FSH, prolactin and cortisol in Holstein heifers. *Livestock Products Science, 68*, 231–241. https://www.ajol.info/index.php/ajb/article/view/94340/83720

Rosegrant, M., Cline, S., Li, W. & Sulser, T. (2005). *Looking ahead long-term prospects for Africa's agricultural development and food security.* 2020 Discussion Paper 41. International Food Policy Research Institute. https://ageconsearch.umn.edu/record/42255/files/vp41.pdf

Roth, Z., Meidan, R., Shaham-Albalancy, A., Braw-Tal, R., Wolfenson, D. (2001). Delayed effect of heat stress on steroid production in medium-sized and pre-ovulatory bovine follicles. *Animal Reproducation Science, 121*, 745–751. Schmitt EJP. https://www.journalofdairyscience.org/articicle/S0022-0302(11)00215-3/pdf

Saif, U. (2017). Climate change impact on agriculture of Pakistan—A leading agent to food security. *International Journal Environment Science Nature Research, 6*(3), 076–079. https://doi.org/10.19080/IJESNR.2017.06.555690. https://ideas.repec.org/a/adp/ijesnr/v6y2017i3p76-79.html

SBP. (2003–04). *Annual report.* State bank of Pakistan, Islamabad. https://www.dawn.com/news/374030/sbp-annual-report-for-2003-04-today

Seerapu, S. R., Kancharana, A. R., Chappidi, V. S. & Bandi, E. R. (2015). *Effect of microclimate alteration on milk production and composition in Murrah buffaloes.* Veterinary World, EISSN: 2231-0916 144. Veterinary World, EISSN: 2231-0916. https://www.ncbi.nlm.nih.gov/pmc/articicles/PMC4774824/#

Shafiq, M., & Kakar, M. A. (2007). Review: Effects of drought on livestock sector in Balochistan province of Pakistan. *International Journal Agriculture and Biology, 9*(4), 657–665. 1560–8530/2007/09–4–657–665. http://www.fspublishers.org

Shahzad, U. (2015). *Global warming: Causes, effects and solutions.* Durreesamin journal, (vol. 1.4). https://www.researchgate.net/profile/UmairShahzad/publication/316691239_Global_Warming_Causes_Effects_and_Solutions/links/590ca678aca2722d185bff31/Global-Warming-Causes-Effects-and-Solutions.pdf

Simelton, E. (2011). "Food self-sufficiency and natural hazards in China". *Food Security*, *3*, 35–52. https://www.researchgate.net/publication/226531849_Food_selfsufficiency_and_nat ural_hazards_in_China

Stott, G., Wiersma, H. F., & Woods, J. M. (1972). Reproductive health program for cattle subjected to high environmental temperatures. *Journals America Veterniary Medicine Asso-ciation*, *161*, 1339. https://www.researchgate.net/publication/236133490_Effect_of_Climate_C hange_on_Livestock_Production_in_Pakistan

Sulehri, A. Q. & Ramay, S. A. (2009). Food security where we are (Current status) and where we want to go (Way FORWARD), parliament of Pakistan, strengthening democracy through parliamentary development. UNDP. http://www.pk.undp.org/content/dam/pakistan/docs/Democratic%20Gove rnance/Projectbriefs2015/DGU_PB_SDTPD_201501.docx

Supple, K. R., Saeed, I., Razzaq, A. & Sheikh, A. D. (1985). *Barani farming systems of Punjab. Constraints and opportunities for increasing productivity*. Agricultural Economics Research Unit, NARC. https://repository.cimmyt.org/xmlui/bitstream/handle/10883/3821/12216.pdf

Tao, S., Bubolz, J. W., do Amaral, B. C., Thompson, I. M., Hayen, M. J., Johnson, S. E. & Dahl, G. E. (2011). Effect of heat stress during the dry period on mammary gland development. *Journal Dairy Science*, *94*(12), 5976–86. https://pubmed.ncbi.nlm.nih.gov/22118086/

The Express Tribune. (2020, 15 August). https://tribune.com.pk/story/2259219/livestock-contri butes-60-to-agricultural-gdp

Thornton, P., van de Steeg, J., Notenbaert, A. & Herrero, M. (July, 2009). The impacts of climate change on livestock and livestock systems in developing countries: A review of what we know and what we need to know. *International Livestock Research Institute (ILRI), P.O. Box 30709, Nairobi, Kenya. Agricultural Systems*, *101*(3), 113–12. https://www.sciencedirect.com/science/ article/pii/S0308521X09000584

Thornton, P. (September, 2010). Livestock production: Recent trends, future prospects. *Hil PK Thornton—Transaction R. Society B*, *365*(1554), 2853–2867. https://www.ncbi.nlm.nih.gov/pmc/ articles/PMC2935116/#. https://doi.org/10.1098/rstb.2010.0134

United Nations (UN). (2001). Resident coordinator of the UN systems' operational activities for development in Pakistan. Drought update #13. https://reliefweb.int/report/pakistan/drought-pak istan-update-no-12

United Nations (UN). (2013). *World population projected to reach 9.6 billion by 2050*. United Nations Department of Economic and Social Affairs. http://www.un.org/en/development/desa/ news/population/un-report-world-population-projected-to-reach-9-6-billion-by-2050.html

UNFCCC. Paris Agreement. United Nations Treaty Collection. (2015). Available at: https://tre aties.un.org/pages/ViewDetails.aspx?src=TREATY&mtdsg_no=XXVII-7-d&chapter=27&cla ng=_en

UNFCCC Decision 4/CP.23. Report of the conference of the parties on its twenty-third session, held in Bonn from 6 to 18 November 2017. United Nations Framework Convention on Climate Change (UNFCCC).

USAID, food assistance fact sheet—Pakistan (2 April, 2020). https://www.usaid.gov/pakistan/ food-assistance#:~:text=More%20than%2020%20percent%20of,to%20extreme%20weather% 20and%20disasters

USAID. (2017). Climate-resilient development: A framework for understanding and addressing climate change. https://doi.org/10.1080/17565529.2015.1124037

USAID, (2018). USAID country profile. Property rights and resource governance-Pakistan. https:// www.land-links.org/wp-content/uploads/2010/09/USAID_Land_Tenure_Pakistan_Profile_R evised_April-2018.pdf

West, J. W. (1999). Nutritional strategies for managing the heat-stressed dairy cow. *Journal Animal Science*, *77*(Suppl 2), 21–35. https://doi.org/10.2527/1997.77suppl_221x. https://pubmed.ncbi. nlm.nih.gov/15526778/

West, J. W. (2003). Effects of heat-stress on production in dairy cattle. *Journal Dairy Science*, *86*, 2131–2144. https://www.sciencedirect.com/science/article/pii/S002203020373803X

WFP (World Food Program). (2 December, 2015). Food insecurity and climate change map. https://www.wfp.org/publications/2015-food-insecurity-and-climate-change-map

Wolfenson, D., Roth, Z. & Meidan, R., (2000). Impaired reproduction in heat-stressed cattle: Basic and applied aspects. *Animal Report Science, 60–61,* 535–547. https://doi.org/10.1016/s0378-432 0(00)00102-0. https://pubmed.ncbi.nlm.nih.gov/10844222

Younas, M., Chaudhry, N. A. & Khan, B. B. (1979). Effect of heat stress on respiration rate, pulse rate, body temperature, and the bodyweight of buffalo calves. *Pakistan Journal of Science and Research, 31*(3–4), 181–186. http://agris.fao.org/agris-search/search.do?recordID=PK1980055 4424

Younas, M., Chaudhry, N. A. & Khan, B. B. (1982). Blood picture as affected by thermal stress in buffalo calves. *Journal of Animal Science Pakistan, 4*(1-2), 32–36. https://www.researchgate.net/publication/236133490_Effect_of_Climate_Change_on_LivestockProduction_in_Pakistan

Younas, M., Fuquay, J. W., Smith, A. E., & Moore, A. B. (1993). Estrous and endocrine responses of lactating Holsteins to forced ventilation during summer. *Journal of Dairy Science, 76*(2), 430–436. https://doi.org/10.3168/jds.S0022-0302(93)77363-4

Zia, M. S., & Baig, M. B. (1997). Fertilizer management and use efficiency under rain-fed agriculture. *Science, Technology and Development, 16*(2), 24–28. https://doi.org/10.1016/S2095-633 9(15)30038-1

Hammad Ahmed Hashmi DVM, M.Sc (Hons), MBA, CMILT (UK), is Chief Consultant at Paws and Claws Animal Consultancy Pakistan (R&D). In his present assignment, he is working as Additional Director Security and OIC DHA Lahore Kennel Club. He is a Veteran of the Pakistan Army's Remount Veterinary and Farms Corps, where he served for 26 years meritoriously, and got retired at the rank of Lieutenant Colonel. His expertise has allowed him to be an External Examiner of the University of Veterinary and Animal Sciences, Lahore for MPhil. He is Honorary Technical Editor of Agro Veterinary News, Pakistan. During his time in the field, he managed to author two technical books related to Poultry and Dog Management. His articles have been published in local and international journals earning him the respect of his peers and colleagues. He also bears multiple memberships and certifications on his credit of high international repute in a variety of fields.

Azaiez Ouled Belgacem holds a Ph.D. in Biological Sciences and M.Sc. in Natural Resources Management. His experience extends to more than 25 years in the area of arid and desert ecosystems in the MENA region. He joined the Food and Agriculture Organization of the United Nations (FAO) in Saudi Arabia as Technical Adviser in November 2020. During 2011–2020, he was Rangeland Senior Scientist cum Regional Coordinator of the ICARDA program in the Arabian Peninsula (Dubai Office). He also served as Head of the Rangeland Program and Director of Plant resources at The Arab Center for the Studies in Arid Zones and Dry Lands (ACSAD), Damascus and Head of the Rangeland and Mapping Unit in the Institute of Arid Areas (IRA), Tunisia. He coordinated/participated in several research projects with the German Technical Cooperation (GIZ), European Union, IFAD, AFESD, ACSAD, FAO, OSS, and GCC on natural resources management, biodiversity conservation, combating desertification and adaptation to climate change, and rehabilitation and management of rangelands. He published more than 120 scientific research papers.

Mohamed Behnassi is Professor at the Faculty of Law, Economics, and Social Sciences, Ibn Zohr University of Agadir, Morocco. He is as well a Senior Researcher of international law and politics of environment and human security focusing on some specific regions such as the MENA and the Mediterranean. He has a Ph.D. in International Environmental Law and Governance (Hassan II University of Casablanca, 2003) and a Diploma in International Environmental Law and Diplomacy (University of Eastern Finland and UNEP, 2015). He is currently the Founding Director of the Center for Environment, Human Security and Governance (CERES)—Former North-South Center for Social Sciences (NRCS)—which is a member of MedThink 5+5 aiming at shaping relevant research and decision agendas. From 2015 to 2018, he was the Director of the Research Laboratory for Territorial Governance, Human Security and Sustainability (LAGOS) in the same university. Recently, he was appointed as Expert for the Intergovernmental Science-Policy Platform on Biodiversity and Ecosystem Services (IPBES), the National Center for Scientific and Technical Research (CNRST/Morocco), and Mediterranean Experts on Climate and Environmental Change (MEDECC). He is among the Lead Authors who elaborate the 1st Assessment Report (MAR1): *Climate and Environmental Change in the Mediterranean Basin—Current Situation and Risks for the Future* (MEDECC, 2021). Dr. Behnassi has published 15 books, including *Human and Environmental Security in the Era of Global Risks* (Springer, 2019); *Climate Change, Food Security and Natural Resource Management: Perspectives from Africa, Asia and the Pacific Islands* (Springer, 2019); *Environmental Change and Human Security in Africa and the Middle East* (Springer, 2017); *Sustainable Food Security in the Era of Local and Global Environmental Change* (Springer, 2013). In addition, Dr. Behnassi has organized many international conferences covering the above research areas and managed many research and expertise projects on behalf of various national and international organizations. Behnassi is regularly requested to provide scientific expertise nationally and internationally. Other professional activities include social compliance auditing and consultancy by monitoring human rights at work and the sustainability of the global supply chain.

Khalid Javed is a renowned Animal Husbandry Expert with a special area of expertise in animal breeding and genetics. He graduated in Animal Husbandry in 1982, did his Master's and Doctorate degrees in Animal Breeding and Genetics from the University of Agriculture, Faisalabad. He is Ex-Chairman, Department of Livestock Production, University of Veterinary and Animal Sciences, Lahore. He also worked as Convener and member of various committees constituted by the Government of Punjab regarding animal breeding and genetics and was also appointed as Technical Member in Livestock Breeding Tribunal. He is Senior Editor of the Journal of Animal and Plant Sciences (The JAPS) (ISI Thomson Impact Factor 0.481; JCR 2019). He is also on the Editorial Board of different internationals journals. He authored a book on Animal Breeding and Genetics for LAD students and edited another on Livestock Production. He actively contributed to the establishment of UVAS Ravi Campus Pattoki as Resident Officer and as Project Director of the

Small Ruminant Research Centre. After 35 years dedicated service, Dr. Javed retired in January 2019 from the University of Veterinary and Animal Sciences, Lahore.

Mirza Barjees Baig is a Professor at the Prince Sultan Institute for Environmental, Water and Desert Research, King Saud University, Saudi Arabia. He earned his MS degree in International Agricultural Extension in 1992 from the Utah State University, Logan, Utah, USA and was placed on the 'Roll of Honor'. He completed his Ph.D. in Extension for Natural Resource Management from the University of Idaho, USA and was honored with the '1995 outstanding graduate student award'. Dr. Baig has published extensively on the issues associated with natural resources in the national and international journals. He has also made oral presentations about agriculture and natural resources and role of extension education at various international conferences. Food waste, water management, degradation of natural resources, deteriorating environment and their relationship with society/community are his areas of interest. He has attempted to develop strategies to conserve natural resources, promote environment and develop sustainable communities. Dr. Baig started his scientific career in 1983 as a researcher at the Pakistan Agricultural Research Council, Islamabad, Pakistan. He served at the University of Guelph, Ontario, Canada as the Special Graduate Faculty from 2000–2005. He served as a Foreign Professor at the Allama Iqbal Open University (AIOU), Pakistan through the Higher Education Commission from 2005–2009. He served as a Professor of Agricultural Extension and Rural Society at the King Saud University, Saudi Arabia from 2009–2020. He serves as well on the Editorial Boards of many international journals and the member of many international professional organizations.

Chapter 9
Water Scarcity Threats to National Food Security of Pakistan—Issues, Implications, and Way Forward

Muhammad Umar Munir, Anwar Ahmad, Jan W. Hopmans, Azaiez Ouled Belgacem, and Mirza Barjees Baig

Abstract Pakistan is primarily an agrarian country with an agriculture sector that is a major source of economic activities, foreign exchange earnings, and the livelihood of the majority of population, caretaker of food and nutritional security, a means to combat rural poverty, and a supplier of raw material for the industries. Out of the total area of 79.6 million hectares, 22.1 million hectares are cultivated of which almost 80% is irrigated and supported with world's largest contiguous canal irrigation system called Indus Basin Irrigation System (IBIS). However, dependency of this system on transboundary waters is more than 77%. The country has a huge rural population of 132.2 million (more than 64% of total population) which is engaged in some way in on-farm or off-farm activities related to agriculture. Population growth and urbanization are exerting more pressure on the already looming water crisis. This situation is catalyzed by the ever-changing climate. It is estimated that about 70% percent of the total average flows in the Indus system are fed by snow and glacier melt in the Hindu-Kush Karakoram (HKK) part of the Greater Himalayas. Variation in the trends and timing of snowfall and changes in snow and ice melt are erratically occurring due to the climate change, which would have grave implications for managing the basin's water resources. This disturbance in the balance of primary source of irrigation—i.e., the IBIS—would have serious implications on the agriculture sector of Pakistan which, in turn, would be a threat for the national food

M. U. Munir (✉) · A. Ahmad
Pakistan Council for Research on Water Resources, Islamabad, Pakistan
e-mail: munir.m@live.com

J. W. Hopmans
Soil Science and Irrigation Water Management, University of California Davis, Davis, USA
e-mail: jwhopmans@ucdavis.edu

A. O. Belgacem
Arabian Peninsula Regional Program, International Center for Agricultural Research in the Dry Areas, Dubai, UAE
e-mail: a.belgacem@cgiar.org

M. B. Baig
Prince Sultan Institute for Environmental, Water & Desert Research, King Saud University, Riyadh, Saudi Arabia
e-mail: mbbaig@ksu.edu.sa

© The Author(s), under exclusive license to Springer Nature Switzerland AG 2021   241
M. Behnassi et al. (eds.), *Emerging Challenges to Food Production and Security in Asia, Middle East, and Africa*, https://doi.org/10.1007/978-3-030-72987-5_9

security of the country. The water scarcity is attributed to many factors, including global warming and climate change, and leads to visualize future trends of water and food stock availability. This chapter is a review and aims at establishing links among water scarcity, climate change, and food security. The discussion has led to proposing some policy guidelines which may help different stakeholders better understand these challenges within the perspective of overcoming water scarcity and food insecurity.

Keywords Water scarcity · Climate-smart agriculture · Food security · Climate change · Sustaining productivity · Virtual water · Water demand

1 Introduction

Water is a fundamental necessity for sustaining life, social well-being, and driving economy. In the wake of climate change, it is a major concern for many countries, especially those suffering from water scarcity. Pakistan is the sixth largest populous country, but its renewable water resources are only 247 BCM (200 MAF), thereby ranking it at 35th in the world. It maintains the largest contiguous irrigation system of the world called 'Indus Basin Irrigation System (IBIS)'. The system comprises three reservoirs (Terbella, Mangla, and Chashma), 23 barrages/headworks, 12 link canals, 45 canal commands, canals length of 60,000 km, and 110,000 water courses having length of 1.6 million km making supplies to the 4th largest irrigated chunk of land. The estimated cost of IBIS is US\$ 300 billion. The Climate of Pakistan is continental characterized by extreme summers and winters. Annual rainfall is about 250 mm and major part (70%) of it falls in monsoon season (July–September). Annual surface water flows are about 170 BCM (138 MAF), but the transboundary and glacial melt dependency is high. The economy of the country is heavily dependent on agriculture that contributes about 20% of GDP, employs about 40% and supports about 70% of the population. However, the country has the least productive agriculture and one of the inefficient irrigation systems. For instance, the water productivity of rice is 0.45 kg/m^3, which is 55% below average value in Asia (1 kg/m^3), and that of wheat is 0.76 kg/m^3 that is 24% less than world average (1.0 kg/m^3). The overall efficiency of the irrigation system is only 40%. These miseries are likely to be further aggravated by water scarcity and climate change. Pakistan has witnessed 0.6 °C rise in temperature over the last century and is amongst the top vulnerable countries over the current century (IPCC 2007).

The World Resources Institute (WRI) has scored and ranked future water stress—a measure of competition and depletion of surface water in 167 countries by 2020, 2030, and 2040. The WRI identified 33 countries likely to face extremely high-water stress by 2040, half of which are in the Middle East. Pakistan is ranked at 23rd place that is likely to suffer from a water crisis with a score of 4.48 out of 5.00 (WRI 2015). The Asian Development Bank (ADB) reported that the total renewable water resources in Pakistan has already reached to the limit of 1000 cubic meters

Fig. 1 Water situation in Pakistan (2020). *Source* www.worldwater.io

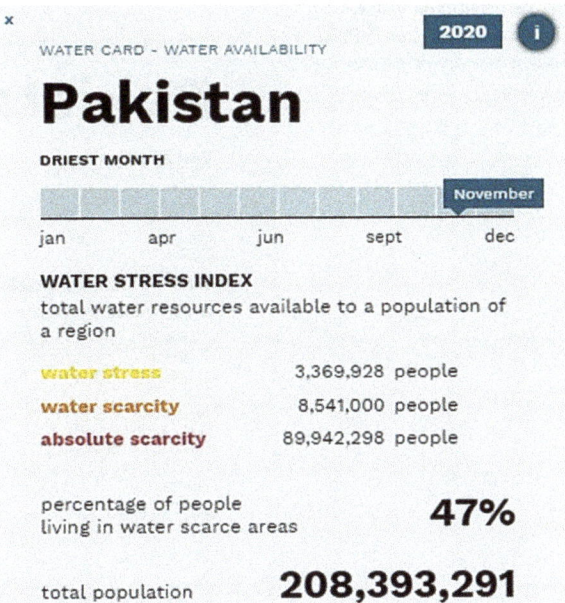

per person per annum, below which serious economic and social consequences are likely to occur (ADB 2016). Almost 50% of the total population is predicted to be living in water scarce areas by 2030 as illustrated in Figs. 1 and 2.

Pakistan and India agreed on the Indus Water Basin Treaty in 1960. Under this Treaty, Pakistan owns the stream flow from the three western rivers while India owns three Eastern rivers. However, later on India started constructing dams on Pakistani allocated rivers resulting in reduced water discharge of reservoirs on the Pakistan side of the Indus basin. Pakistan, being an agrarian country, depends highly on its agricultural products, but the mode of irrigation still practiced results in major wastage of water. The crops like rice and sugarcane are mostly given flood irrigation, for which massive pumping of groundwater is carried out causing depletion of the natural ground water aquifer and having very less water-use efficiency. This situation is more severe in the desert areas of Pakistan which results in migrations as well as deaths of number of people and livestock. Excess land along with natural ground water aquifer is available in the desert area; however, the socio-economic conditions in the area limit the farmers to use the land to its fullest potential.

Fig. 2 Predicted water
situation in Pakistan (2030).
Source www.worldwater.io

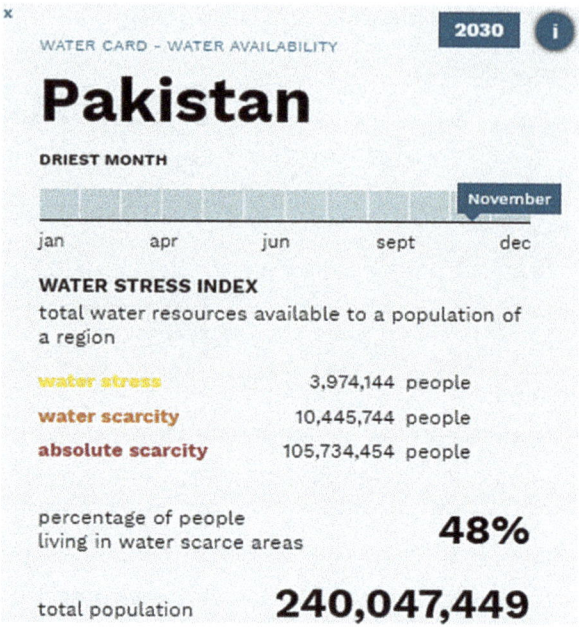

2 Agriculture Sector of Pakistan—An Overview of Production and Water Availability Status in Pakistan

Pakistan is primarily an agrarian country and agriculture is the backbone of its economy with about 20% contribution to national GDP. The contribution from different sectors of agriculture to this GDP includes: major crops 4.7%; other crops 2.2%; Livestock 11.6%; Fishery: 0.4%; and Forestry: 0.5%. In Pakistan, the agriculture sector till now is producing enough commodities to ensure the national food security, but feeding a fast-growing population will be a challenge in the future. In addition, the country has a huge rural population of 132.2 million (more than 64% of total population), which is engaged in some way in on-farm or off-farm activities related to agriculture. Besides, the sector is also a major source of economic activity; source of livelihood of majority of population (employs more than 40% of the workforce); caretaker of food and nutritional security; a means of poverty alleviation especially rural poverty; a supplier of raw material to industries; and a major source of foreign exchange earnings (the agricultural sector along with the agro-based industry contribute up to 80% of the country's total export earnings).

The total geographical area of the country is 79.6 million ha with a sizeable agriculture economy of US$ 67.7 billion. The cultivated area is 22.01 million ha of which 80% is supported with the world's largest canal irrigation system.

Pakistan is blessed with a highly diversified climate suitable for the cultivation of a number of field and horticultural crops, but unfortunately it failed to harness its potential. The agriculture sector has been growing at an average rate of 3% in last

Table 1 Agriculture growth (in %, relative to previous year GDP)

Sector	2013–14	2014–15	2015–16	2016–17	2017–18	2018–19	2019–20P
National	4.1	4.1	4.6	5.2	5.5	1.9	−0.4
Agriculture	2.5	2.13	0.15	2.18	4.00	0.58	2.67
Crops	2.64	0.16	−5.27	1.22	4.69	−4.69	2.98
Livestock	2.48	3.99	3.36	2.99	3.70	3.82	2.58
Forestry	1.88	−12.45	14.31	−2.33	2.58	7.87	2.29
Fishing	0.98	5.75	3.25	1.23	1.62	0.80	0.60

P: provisional (July–March)
Source Pakistan Economic 2019–20

Fig. 3 Sectoral growth in 2019–20 (provisional = July–March). *Source* Pakistan Economic 2019–20

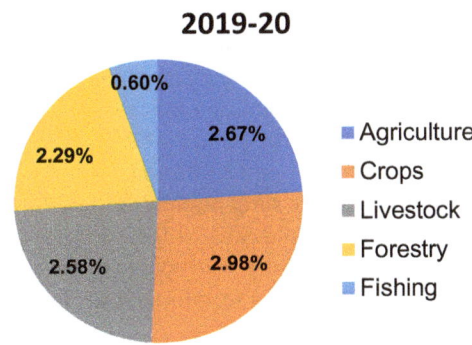

decade and succeeded to attain food self-sufficiency in all major crops, however the trade statistics depict that the sector's growth is being utilized to satisfy the needs of local consumption only because of similar growth rate in population. Though the country has achieved food security through a self-sufficiency in staple food crops, yet it is still facing several challenges that need to be addressed to boost the productivity, profitability, competitiveness, and sustainability of the agricultural sector. Currently, the main objective of agriculture sector in the country is diversification of farm productivity into high-value crops for poverty alleviation in rural areas. Punjab is the most populous province of the country having about 29% of its reported, 57% of the cultivated, and 69% of the cropped area of Pakistan and, therefore, serves as the country's food basket. The agriculture sector showed a growth of 2.67% during 2019–20 as compared to 0.58% in 2018–19 (Pak Economic Survey 2020) (Table 1 and Fig. 3).

3 National Food Security in Pakistan

Despite its ever-growing agricultural production, the country is still facing high levels of food insecurity. According to a global report published jointly by the FAO, WFP, UNICEF, WHO and IFAD in 2019, 20.3% of Pakistan's population (40.0 million people) is undernourished and/or food insecure. The prevalence of malnutrition amongst children is also very high, with an estimated 40% children stunted, 28% underweight, 18% wasted, and 10% overweight. Further, around one-fourth (24%) of the country's population is living below the national poverty line and 39.0% is poor based on multidimensional poverty index (MPI). Despite the growth in production of staple crops, Pakistan has experienced a sharp decline in food security in recent years. In 2010, the country experienced its worst natural disaster in decades when flooding submerged almost one-fifth of the country's area. The floods destroyed infrastructure, left almost 20 million people without access to food, clean water, and health services, and severely damaged the country's agricultural sector (WFP, 2010). The food-insecure population rose from 38% of the total population to 50% (83 million people) between 2003 and 2009; it is estimated that this number has risen further to 90 million people in the aftermath of the 2010 floods (WFP, 2010).

The Pakistan's rural population faces multiple challenges, with two-thirds of the total population and 80% of the poor population living in rural areas (IFAD, 2012). Poverty is particularly widespread in the country's mountainous areas where isolated communities, rugged terrain, and ecological fragility make agricultural production difficult and where a lack of access to markets and services contribute to widespread chronic poverty. Chronic poverty, recurring disasters, and political and economic volatility drive under-nutrition and food insecurity.

Pakistan is also prone to extreme weather and disasters. Drought has become a frequent phenomenon in the country, affecting livelihoods and household food security in parts of Balochistan and Sindh provinces, as reported by the Food Security Cluster (FSC, 2019). The FSC estimated that drought affected approximately 5 million people in 26 districts of Balochistan and Sindh provinces in 2019. According to the Integrated Food Security Phase Classification as of July 2019, approximately 1.3 million people were experiencing Crisis (IPC 3) and Emergency (IPC 4) levels of acute food insecurity in seven drought-affected districts in Sindh Province. The current food security situation in Pakistan, according to the World Food Programme, is illustrated in Table 2.

Table 2 Food security indicators

Food security indicator	Value
Calorie supply per capita (2013)	2440
Population undernourished (2014–16)	19.90%
Children undernourished (2011)	32.00%
Under 5 mortality rate (per 1000) (2016)	78.80

Source WFP Reports 2006–2008

Pakistan has one of the world's largest groundwater aquifers (4th after China, India, and the USA). It provides more than 60% of the country's irrigation water supplies and over 90% of its drinking water. Almost 100% water used in industry comes from groundwater and the number of tube wells has increased from 0.2 million to over 1.2 million over the last two and a half decades. The groundwater has played a major role in increasing the overall cropping intensity in the country from about 63% in 1947 to over 120% in 2018 (Khan et al., 2016). It is the only reliable resource that provides resilience against droughts and climate change impacts. However, this resource is freely accessible without any regulatory framework. This indiscriminate drilling and operation of tube wells has resulted in groundwater depletion and secondary salinization (Qureshi & Barrett-Lennard, 1998).

Though the country is blessed with a huge network of water management/use, including the world's largest irrigation system, yet the need of water to produce large amounts of agricultural commodities for a continuously increasing population, places high demand for developing new water resources/reservoirs in Pakistan. At present, the annual per capita availability of water in Pakistan is estimated at about 1100 m³, which is very close to the limits of chronic water scarcity; because below 1000 m³/person, countries begin experiencing chronic water stress (Qureshi & Ashraf, 2019).

Pakistan is blessed with highly diversified climate that is suitable for the cultivation of a number of field and horticultural crops, but unfortunately failed to harness its production potential. With every passing day, new challenges for agriculture sector emerge like the shortage of water for irrigation to meet the requirements of present cropping intensity.

4   Water Scarcity and Its Impact on Agricultural Food Production

Water is vital for the life and soul of a country's economy, especially for an agrarian one like Pakistan. It acts as a catalyst of economic growth and development. It is the life blood for the survival of human beings, agriculture, household economy, and industry. Pakistan is among those vulnerable states which fall in the bracket of water crisis. Many recent reports show that the future of this country appears rough and tough due to the water enigma. Ironically, after the construction of two major dams (Tarbela and Mangla in 1970), Pakistan is unable to construct any other dam despite having substantial number of feasible sites for dam construction and hydropower generation.

A vast majority of the country's available water resources (almost 90%) are used in agriculture, while the remaining share is split equally between industry and domestic use. The water availability is already below the scarcity level of 1000 m³/person and climatic changes in the region may further worsen the situation. Thus, Pakistan is facing a serious water crisis. Water scarcity is defined as an imbalance between demand and availability (FAO, 2010) and exists when the demand for water exceeds the supply (Molle & Molinga, 2003). It can be defined either in terms of existing and potential supply of water or in terms of present and future demands for water or both (IWMI, 1998). It can also be defined as a relative concept and, therefore, be regarded partly as a 'Social construct' because determining water scarcity varies from a country and a region to another and within the social construct; the scarcity is determined both by the availability and consumption patterns (IWMI, 1998).

Pakistan is rapidly moving from being classified as water 'stressed' to water 'scarce' and with its annual water availability fall below 1000 m^3/person. IIt may in fact have already crossed this threshold. Food security is directly related to the water security as 50–70 times more water is required to grow food than the water used for domestic purposes. About 90% of the food production in Pakistan comes from irrigated agriculture, whereas dry-land (rain-fed) agriculture contributes only 10% due to low rainfall. Furthermore, the average yield of crops per unit water use is much lower than international levels, and there is a significant gap between actual and potential yield.

With water availability per person increasingly declining, and demand for food production continuously increasing, Pakistan faces not only a water crisis but also serious concerns regarding its future food security. This situation also has clear implications for the government's efforts to become an upper middle-income country by 2025 and achieve long-term peace and security. The United Nations estimate that water demand in Pakistan is growing at an annual rate of 10%. That is, the demand is projected to rise to 274 MAF by 2025 but total water availability by 2025 is not likely to change from the current 191 MAF. This gap of about 81 MAF is almost two-thirds of the entire Indus River system's current annual average flow (Qureshi & Ashraf, 2019).

Pakistan is highly vulnerable to climate change with impacts becoming increasingly tangible. According to the Global Climate Risk Index, Pakistan is the world's 7th most vulnerable country that is negatively affected by climate change during the period 1996–2015 (Jan et al., 2017) and it faces an average annual loss of 3.8 Billion US $. According to Pakistan's Meteorological Department, there are 7259 glaciers in northern Pakistan, covering 11,780 km^2 with ice volume of 2066 km^3. Water from melted glaciers contributes to more than 60% of the flows in the rivers, yet climate change is forecasted to impact these flows and/or affect seasonal availability in the long-term (Rasul, 2016). Jan et al. (2017) stated that various challenges of climate change for agriculture included uncertainty in availability of irrigation water, increased variability of monsoon, severe water stress conditions in arid and semi-arid areas which will negatively affect agriculture sector and human health.

Pakistan has one of the world's largest groundwater aquifers (4th after China, India and the USA) (Qureshi & Ashraf, 2019). The groundwater has played a major role in increasing the overall cropping intensity in Pakistan from about 63% in 1947 to over 120% in 2018 (Khan et al., 2016). It is the only reliable resource that provides resilience against droughts and climate change impacts. However, this resource is freely accessible without any regulatory framework. This indiscriminate drilling and operation of tube wells has resulted in groundwater depletion and secondary salinization (Qureshi & Barrett-Lennard, 1998). The rates of groundwater exploitation have increased rapidly in the past several decades, and some areas, mainly Balochistan, are mining aquifers beyond their capacity of natural replenishment. Aquifer mining is supplementing the surface water that reaches Pakistan's farmland, but this indiscriminate pumping and heavy use of pesticides are contaminating the aquifer, whereas tube well water salinity is also increasing.

The present irrigation system is almost one hundred-year-old, hence it is becoming inefficient with every passing day and its management and maintenance are a big challenge. The problems with the management of this irrigation system include social as well as technical ones, as management issues have results in enormous losses in agricultural productivity. Currently, Pakistan is experiencing large water wastage. The old methods of irrigation with flood water are still being used by the farmer with water application efficiencies of about 50 to 60%. The recent floods bestowed Pakistan a big resource of water, but the lack of water reservoirs precluded its use and availability subsequently. The mismanagement of water is having its biggest impact on Pakistan's agricultural sector. According to the World Bank, 43% of Pakistan's employment is in the agricultural sector (WDI, 2015). This prosperous industry relies on the single largest contiguous irrigation system in the world. While this is an impressive feat, Pakistan also has one of the lowest crop yields per unit of water in the world (water-use efficiency). This is alarming because Pakistan uses a whopping 92% of its water resources on its agriculture industry (Kamal, 2009).

Around a quarter of the water delivered for irrigation is lost because of poor farming practices (Kamal, 2009), mostly blamed on the Warabandi (periodic distribution of water at fixed intervals) system of water management in rural areas. According to this system, each farmer has a specific day to irrigate his or her field. The quantity of water used is irrelevant –each farmer pays a flat fee. Although this system was intended to be equitable in the face of water shortages, in reality, farmers who have first access can take a lion's share of the water. Since water is not priced based on usage, there is nothing to discourage the excess and overuse. As such, large and powerful farmers have greater access to water from the Indus, which forces small farmers to rely on tube wells to extract groundwater (Khan et al., 2016). In turn, the over-extraction of groundwater negatively affects the salt content of soil, leading to further environmental degradation. This inequality in water distribution also negatively affects crop yields, since small farmers do not have access to adequate water supplies.

There is a significant food security threat to Pakistan in upcoming years due to inadequate or non-existing water management measures. Therefore, there is a dire need to recognize the challenge of water management keeping in view the threats of food security expected to be faced by the country. It can be done in multiple ways, however, at a time efficient organization and planning to protect the national food security is one of the most critical goals to achieve through the widely accepted approach of integrated water resources management (IWRM, 2000). In this regard, the Government and other stakeholder's agencies should take immediate steps. The Government is showing a high concern in addressing the impacts of water scarcity given the that this issue is extremely sensitive and extensive in dimension. In this context, the issues to be immediately addressed include: increasing the water storage capacity; minimizing water wastages at various levels; rehabilitating/improving the available irrigation network; increasing water productivity; developing an appropriate regulatory framework for surface and groundwater management; devising and implementing appropriate crop zoning; and rationalizing pricing structure for water usage in all sectors.

4.1 Water Scarcity Indicators

Water scarcity can be classified on the basis of five contexts: physical water scarcity; economic water scarcity; institutional water scarcity; managerial water scarcity; and political water scarcity. These types of water scarcity can occur concurrently enhancing both the severity and impacts of water scarcity (IWMI, 1998).

The *physical water scarcity* occurs when water availability is limited by natural availability (Molle & Molinga, 2003). It can be referred to as a situation where there is not enough water to meet all the demands of the population. FAO maps refer physical water scarcity to a situation where 75% of rivers' flows is diverted to agriculture, industry, and domestic uses (Qureshi & Ashraf, 2019). This definition does not include the demand for water in dry areas. However, the FAO maps also define approaching physical water scarcity on the basis of 60% rivers' flows diverted. This definition will imply physical water scarcity in the near future. Physical water scarcity can happen in a particular region or country where there is an excessive development of hydraulic infrastructure (IWMI, 2000).

The *economic water scarcity* refers to a situation where human, institutional, and financial capital limit access to water even though the water is naturally available locally to meet human demands (IWMI, 2000). It is largely caused by lack of investment in water or insufficient human capacity to satisfy water demands.

The political water scarcity is caused by political disputes and negligence. It exists when political forces bar the people from accessing to available water resources.

There are number of factors that lead to this phenomenon which are explained by the as: population growth; (lack of institutional capacity, storage competition, and water demand); climate change and vulnerability; poverty and economic policy; political realities; pollution and poor water quality; cultural and sociological issues; international disputes; sectoral competition and water demand; inappropriate land use; and legislation and water resource management.

The water availability status of a country can be assessed by the following indicators:

- Falkenmark Indicator provides relationship between available water and the human population. According to this indicator, Pakistan crossed the water scarcity line during 2005 and it will achieve the absolute water scarcity line by 2025 (Ashraf, 2016).

- Water Resources Vulnerability Index (WRVI) compares annual water availability with the total annual withdrawals (%). Pakistan, with a WRVI Index of 77%, is among the top ten countries with the largest water withdrawal for agriculture.
- IWMI's Physical and Economic Water Scarcity Indicators. In Pakistan, the shortfall, which was 11% in 2004, is estimated to reach 31% by 2025 (GoP, 2001). Pakistan has sufficient water resources but unfortunately it is lacking storage structures.

About 92% of Pakistan is classified as semi-arid to arid, and the vast majority of Pakistanis are dependent on surface and groundwater resources from a single source: the Indus River basin. Since independence in 1947, Pakistan's population has more than quadrupled; by 2100 it is expected to increase by tenfold. About 90% of the country's agricultural production comes from land irrigated by the Indus Basin Irrigation System (Qureshi, 2011), firmly linking national food security to water levels in the Indus River basin.

4.2   The Water Storage Situation

One of the major issues contributing to water scarcity is the lack of water storage facilities in Pakistan. The per capita water storage of Pakistan is far less than other countries. The current available resources are 138 million-acre feet (MAF) with a storage capacity of 13.7 MAF which is only 10% of available water resources against the world average of 40%. The present water storage capacity of Pakistan in term of days is 30 while that of India is around 120 days and Egypt about 1000 days. The per capita storage of Australia and USA is over 5000 m^3, China 2200 m^3, Egypt 2362 m^3, Turkey 1402 m^3, Iran 492 m^3 while in Pakistan it is only 159 m^3. The per capita water storage of Pakistan is indeed far less than most of the other countries. Furthermore, the increased silting of dams is among the factors which is aggravating water shortfall. Existing dams are rapidly becoming redundant. Reports suggest that Pakistan's existing dams will exhaust their reservoir capacity by the next decade (Table 3) Ashraf (2016). There is a huge need to enhance reservoir capacity

Table 3   Storage capacity and loss of major reservoirs

Reservoir	Live storage capacity		Storage loss	
	Original	Year 2013	Year 2013	Year 2025
Tarbela	9.69 (1974)	6.58 (68%)	3.11 (32%)	4.16 (43%)
Mangla (post raising)	8.24 (2012)	7.39 (90%)	0.85 (10%)	1.16 (20%)
Chashma	0.72 (1971)	0.26 (36%)	0.46 (64%)	0.64 (78%)
Total	18.65	14.23 (76%)	4.42 (24%)	5.96 (37%)

Source Ashraf (2016)

by building smaller and large dams in order to avoid severe water crisis. Long-term policies must be formulated to harness the benefits of naturally available water.

4.3 Virtual Water

Virtual water is the volume of water used to produce consumer products and the total volume of water refers to all of the water used in the production of a finished product. Virtual water is essentially the cost of a commodity in terms of water. Every product we consume contains virtual water. Wheat is responsible for 22% of groundwater depletion with an average water footprint of 1800 L per kilogram. Rice uses 40% of all global irrigation and 17% of groundwater and has an average water footprint of 2500 L per kilogram. Cotton has a heavy use of irrigated water, which can turn arid environments in Southern Punjab and Northern Sindh into deserts. In Pakistan, the cotton crop uses an average of 9800 L per kilogram, which is much higher than other cotton producing countries like USA (Water Aid, 2019). The production of cloth from cotton crop is also highly water intensive. Pakistan was the largest virtual water exporter with 7.3 billion cubic meters back in 2010. Nearly 10% of the UK's virtual water imports come from Pakistan with negative implications for the later. Most of these imports are based on crops grown in lower Indus aquifer where the amount of abstracted groundwater is more than 18 times the amount of natural water recharge through rainfall and glacial melting (Water Aid, 2019) In the long run, the country has to shift its focus from water-intensive crops and manufacturing to reduce the export of virtual water.

5 Causes of Water Shortage

(A)   *Lack of Population Control*

The increase in population is one of the most important factors in the reduction of per capita water availability (Table 4). At the same time, population control has been a neglected subject both at the public and private levels. There is an urgent need to

Table 4 Per capita water availability in Pakistan

Year	Population (million)	Per capita water availability (m^3)
1951	34	5650
2003	146	1200
2010	168	1000
2025	221	800

Source GoP MTDF (2005–10)

Fig. 4 Population versus per capita water availability. *Source* GoP MTDF (2005–10)

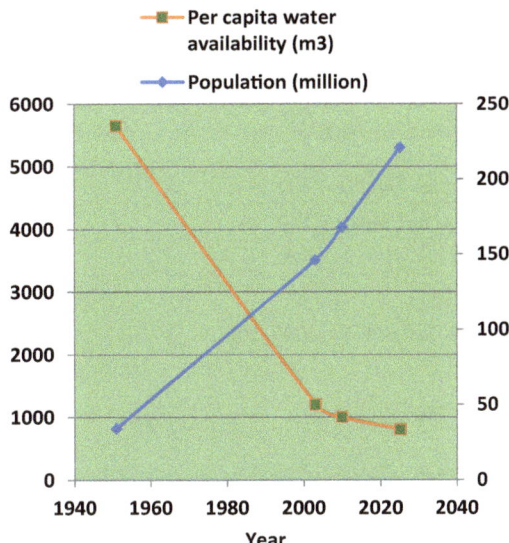

take some stern steps to control the ever-increasing population by involving all the stakeholders. Figure 4 shows a trend in water availability and population growth in Pakistan.

Figure 4 describes the trends of future water availability on the basis of cubic meter per capita per person and clearly depicts that food security of Pakistan in future will be deteriorated more due to increasing population pressure leading to scarce water situation. The said figure shows that water availability is in constant decline and in year 2025, the water availability will be around the levels of absolute scarcity. On the contrary, there is a constant increase in the future demands of water for irrigation, agriculture and livestock, and human population and supply will be low in comparison with the demand. The gaps of supply and demand cannot be filled because water is a natural resource which can neither be produced nor be generated. As amid rapid growth in population (Fig. 4 and Table 4) the supply of food will not be able to feed a big chunk of population, consequently, the food security will deteriorate further for the people of Pakistan. This situation may lead to hunger, famines, conflict, civil wars, extreme poverty, misery, diseases, and an utter helplessness for people and state as well.

(B)   *Low Water Productivity*

Water productivity is defined as the physical or economic output per unit of water application (Cai & Rosegrant, 2003). The average yield of crops per unit water use is much lower in Pakistan than the international levels, and there is a significant gap between actual and potential yields (Table 5 and Fig. 5).

The water productivity can be improved by increasing the yield per unit of used water or by reducing the amount of water used for the same yield. However, in

Table 5 Yield gap for major crops

Crop	Progressive farmers' yield (t/ha)	National average yield (Avg. of last 3 years) (t/ha)	Yield gap (%)
Wheat	4.6	2.6	43.5
Cotton	2.6	1.8	30.8
Sugarcane Sindh	200	54.5	72.8
Sugarcane Punjab	130	49.9	61.6
Maize	6.9	2.9	58.5
Rice	3.8	2.1	45.6

Source GoP MTDF (2005–10)

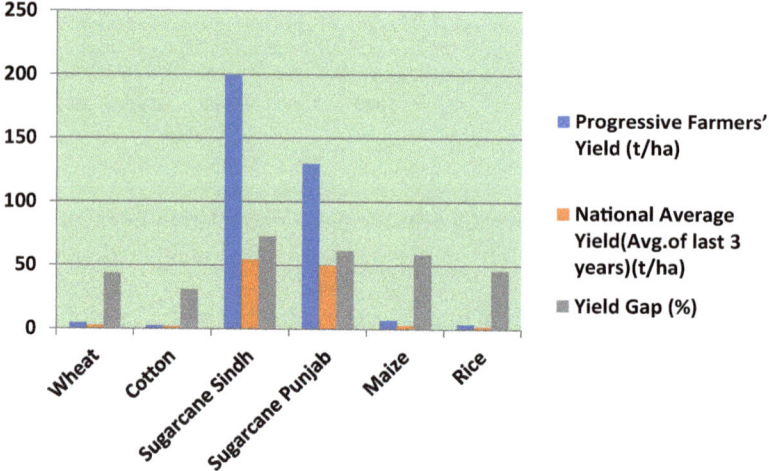

Fig. 5 Yield gap of major crops. *Source* GoP MTDF (2005–10)

Pakistan, there is apotential for both increasing crop yield and reducing the depth of water applied (delta of water) using appropriate methods and techniques like precision land levelling, proper layout of the field, using high-efficiency irrigation technologies, appropriate irrigation methods such as bed planting, and by adopting proper irrigation scheduling (Ashraf, 2016).

(C)   *Water Losses at the Field Level*

Pakistan has one of the largest contiguous irrigation systems in the world, covering about 17 Mha of land. However, at the same time, it is one of the most inefficient irrigation systems where more than 60% water is lost in the channels and during application in the field (Table 6 and Fig. 6).

The following measures can help reduce the water losses at the tertiary level:

Table 6 Water losses in the irrigation system

Location	Delivery at head (MAF)	Losses (MAF)	Losses (%)
Canals	106	16	**15**
Distributary and minor	90	6	**7**
Watercourses	84	26	**31**
Fields	58	17	**29**
Crop use	41	–	–
Total	**379**	**65**	**17**

Source GoP (2001)

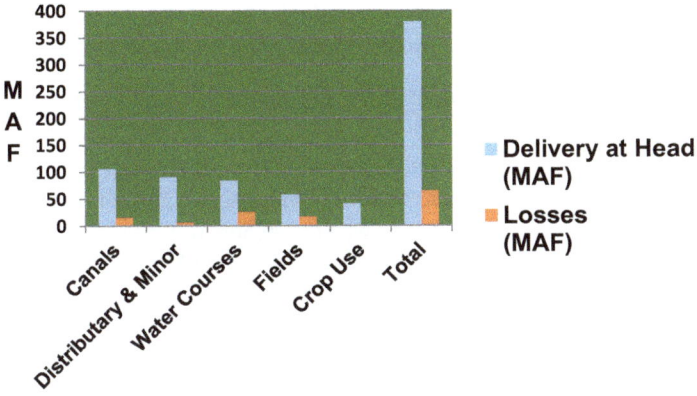

Fig. 6 Delivery at head versus water losses. *Source* GoP (2001)

- Laser levelling of fields and ridge/bed planting at field level.
- Improving water courses, farm outlets, to reduce losses from water channels.
- Where feasible, use of rain gun, drip irrigation and sprinkle irrigation may be encouraged, especially in the hilly areas, sandy soils, and for high value crops.
- Direct seeding of rice should be encouraged which can reduce water use by 15%.
- Growing rice on ridges/beds can save up to 50% water and 20% increase in yield besides having a number of environmental benefits (Farooq, 2009). It would also reduce methane emission from the rice fields.
- Applying irrigation according to the critical stages of the crop.

(D) *Wastage as Un-attended Water*

In the plains, sometime farmers do not need water, particularly during winter and monsoon rains. Farmers generally stop the canal water entry into their cropped fields. The excess water flows downstream through the watercourses, but instead could be stored in farm-scale ponds on small reservoirs for the subsequent growing season.

(E)   *Subsidized Water Prices*

The low water price is an important reason for wastage and overuse of water by the farmer. Canal irrigation water is being supplied to the farmers almost free (i.e., Rs. 135 per acre). This rate is equal to the price of about 4 kg of wheat grain in the market. Therefore, a rational system of water pricing needs to be introduced.

(F)   *Inappropriate Crop Zoning*

High delta crops such as rice and sugarcane are grown in areas that are classified as arid zones, where surface water is insufficient and groundwater is too deep and saline. Cultivation of these crops in such areas creates huge pressures on groundwater, resulting into its depletion and secondary salinization. Logically, rice and sugarcane should be restricted to be grown in those areas that are sub-humid or semi-arid where sufficient water is available with limited dependency on stored water reserves.

(G)   *Mismatch between Irrigation Water Supplies and Crop Needs*

Warabandi is a fixed system of canal water supplies while each crop has a critical physiological stage of development at which water shortage will drastically reduce the yield. This system results in water wastage when water is applied at the inappropriate growth stage. There is an urgent need to develop system that will allow enough flexibility to provide water according to the crop needs at critical stages.

(H)   *Water Pollution by Domestic and Industrial Sectors*

No doubt, water use in domestic and industrial sectors is insignificant as compared to the agricultural sector, but it has important implications for the society and the ecosystem. As more than 90% domestic and almost 100% industrial water comes from groundwater, the water overdraft results in lowering of the ground water table beyond its sustainable availability. It has been estimated that about 2.5 MAF of municipal and 1.5 MAF of industrial wastewater is generated annually and only 3% is being treated, whereas 1.5 MAF is disposed of directly into fresh water bodies (Khalil & Kakar, 2011). About 30,000 ha of land is irrigated with wastewater and 25% vegetables consumed in Pakistan are produced through wastewater irrigation (Khalil & Kakar, 2011). The disposal of untreated wastewater into the surface water bodies has great impact on the whole ecosystem. Ultimately, the contaminants of these surface water bodies become part of drinking water and food supplies, thereby resulting in serious health hazards by causing multiple diseases (Khalil & Kakar, 2011). The sectors responsible for causing water pollution should be charged for the treatment of the polluted water.

(I)   *Lack of Coordination*

Over the time, many promising land and water management technologies have been developed. However, these technologies could not reach the stakeholders, mainly due to a lack of coordination between research and development agencies. Therefore, there is a need to increase coordination at the highest level among water storage, water distribution, and water management agencies.

6 Climate Change and Future Water Scenario of Pakistan

Climate change has become a reality and Pakistan is highly vulnerable to it. According to the Global Climate Risk Index, Pakistan is the world's 7th most vulnerable country negatively affected by climate change during the period 1996–2015 and it faces an average annual loss of 3.8 Billion US $ (Jan et al., 2017). In addition, Pakistan is among the top five countries that have the least clean air (Krepon, 2015). Water from melted glaciers contributes to more than 60% of the flows. Jan et al. (2017) stated that various challenges of climate change for agriculture included increase in temperature, uncertainty in availability of irrigation water, increased variability of monsoon, severe water stress conditions in arid and semi-arid areas, and increase in extreme events such as floods, drought, heat and cold waves, and cyclones which will negatively affect agriculture sector and human health. Due to an increase in frequency and intensity of floods and droughts, new reservoirs need to be developed to make the country climate-resilient.

Climate change is likely to enhance the country's water crisis. When climate change and its implications for Pakistan's water resources are discussed, the conversation normally revolves around the expected decline in water flow in the Indus River Basin as the glaciers of the Hindu Kush-Karakorum-Himalaya mountains are retreating. This concern is understandable given that snow and ice melt runoff, currently generates between 50 and 80% of average water flows in the Indus River Basin (Yu et al., 2013). Inevitably, climate change will lead to significant changes in hydrologic patterns in the Indus River Basin. But the scientific evidence suggests that the volume of water flowing in the Indus River and its tributaries is likely to remain relatively stable or will even increase at least till 2050 (Amir & Habib, 2015). The most significant change could be a shift in the timing of peak flow to slightly earlier in the year, along with a potential increase in variability from one year to the next. Such changes could in fact help to somewhat alleviate Pakistan's growing water stress for the time-being.

A largely overlooked factor in the discussions around water and climate change in Pakistan is the possible impacts of climate change on the country's steadily growing water demand. Rising temperatures will increase the agriculture sector's already substantial demand for water as evapotranspiration rates increase and soil moisture levels decline. Higher temperatures will also affect the country's growing thermal power production sector, which provides approximately 65% of the country's energy. The thermal sector is highly dependent on water for steam production and, subsequently, for cooling the steam. As higher air temperatures decrease, the efficiency of the thermal conversion process will decrease and greater volumes of water will be required by this sector to maintain the production at sustainable levels (Makky & Kalash, 2013). The key priority areas of research in this scenario are as follows:

- Use of remote sensing tools for collecting data.
- Understanding glacier behaviour using a combination of *in-situ* and remote sensing observations, pale climatic analysis, and modelling.
- Understanding other factors influencing hydrology, including the role of avalanches, debris and dirt cover, including black carbon deposit.
- Understanding water balance using state-of-the-art tools, including improved hydro-meteorological observation networks.
- Understanding the roles of socio-economic, institutional, and policy-related factors Involving stakeholders to develop adaptive water management strategies.
- Crop varieties development that can tolerate high temperatures and drought at the critical growth stages.
- Due to inadequate storage, Pakistan has lost more than 90 MAF of water during the floods of 2010, 2012 and 2014. This shows the amount of water available in the system. Therefore, there is a dire need to develop small dams as well. Besides providing irrigation water, small dams have several advantages. They recharge the groundwater, provide water for domestic and municipal purposes, control erosion, and help develop aquaculture. However, the only way to inject water into the system is the construction of large storage reservoirs.
- The improvement of watercourses is very important as more than 60% of water is lost within the system—*i.e.,* from canal head to the fields. These losses further aggravate the problem of water scarcity, particularly towards the tail end. The watercourse improvement increases the conveyance efficiency with equitable distribution of water among the head, middle and tail end farmers.

The potential impacts of climate change on water demand have been highlighted in many recent research studies (such as Amir & Habib, 2015); Makkay & Kalash, 2013). These studies suggest that higher temperatures will lead to a significant increase in water demand compared to a business-as-usual scenario. The immediate threat posed by climate change to Pakistan's water sector, therefore, is on the demand side. The findings reinforce the need for the country to focus on improving the efficiency of infrastructure with which it uses its water—to make sure that every drop counts.

The recently completed studies (Qureshi & Ashraf, 2019), also highlight the potential benefits of investing in efforts to improve the efficiency of water use—particularly in the irrigated agriculture sector, where the opportunities for improvement are significant. The Indus River Irrigation System is characterized by large inefficiencies at the canal, watercourse, and field levels. Only about 30% of water flowing through the system is delivered to farms, and farmers at the tail end of the system rarely get water. Water management is weak because: water prices and recovery rates don't generate the revenue needed to cover operation and maintenance costs; there is an absence of regulatory enforcement; and farmers continue to follow traditional flood irrigation practices that overwater crops and have led to waterlogging of soils in parts of the Indus Basin.

Greater efforts to promote the uptake of high-efficiency irrigation systems by smallholder farmers, along with infrastructure investments such as canal upgrades and precision land levelling, would be important steps to improve the situation. At the same time, much more effort is needed to understand the water demand challenges facing Pakistan. There is a general absence of water demand data and its analysis, particularly for different provinces and sectors. More research is also needed in areas such as water pricing to develop and implement systems that promote more efficient water use.

7 Dryland Agriculture

Dry-land (rainfed) area constitutes about 20% of the total cultivable area of Pakistan. However, it contributes to only 10% to the total crop production. The maximum investment in agriculture sector has been in irrigated areas, whereas the rainfed areas have almost been neglected. It has been estimated that there is a potential of about 18 MAF of water from seasonal upland high flows. This water is available during a short period of two to three months, and innovative water management practices could help boost the socio-economic conditions of the local communities in these rainfed areas (Qureshi & Ashraf, 2019).

There are two dry land farming systems—*Sailaba,* hill-torrents irrigation (spate irrigation) and *Khushkaba* (which is exclusively rainfed):

- The *Sailaba* irrigation system (spate irrigation) constitutes the major portion of the country's dry-land farming system. The run-off from the hill-torrents is directed through a network of indigenously managed infrastructural system. The water availability in the spate system depends on the occurrence and distribution of rainfall in the catchment areas along with the hydrological aspects of watersheds responsible for run-off process.
- The *Khushkaba* system is exclusively dependent on the incident of rainfall and localized runoff. This is the second largest water harvesting system in the dry lands. Nevertheless, risk of crop failure under this system is relatively high due to insufficient soil-moisture conditions. This system is more vulnerable to drought in comparison with *Sailaba*. Productivity of this farming is low, resulting in food insecurity of the poor community. However, livestock is an integral part of these systems, income from which is a major source for many of the poor farming communities.

A number of technologies have been developed for dry land agriculture. By adopting these technologies, the land and water productivities of these areas can be increased manifold. Data gathered from a number of regions (Wyn Jones et al., 2006) show that wheat yields of 4 to 5 tons per hectare can be achieved with the use of 300–400 mm water in rainfed conditions.

8 Implementation of the 2018 National Water Policy

The Council of Common Interest (CCI) approved long awaited National Water Policy (NWP) in 2018 along with a Water Charter signed by the Prime Minister and the Chief Ministers of the four provinces. In the Water Charter, the Federal and Provincial leaders have shown their commitment in the following words: "The Charter is a Call to Action and the declaration of a water emergency. We must look beyond our differences and come together as a nation to rise to the challenge that is before us. We have done so before, and we can do it again. We will seize the day and secure our collective future. This is our promise to the coming generations".

The NWP is a national consensus document which can be used as a guiding principle. There are 33 objectives covering almost all aspects of water including: water resources development and management (both surface and groundwater); development of regulatory framework; urban water management; hydropower development; flood and drought management; rainwater harvesting; capacity building; and institutional arrangements. It has set targets and timelines for some of the important tasks such as: the development of new water reservoirs (up to 10 MAF); reduction in conveyance losses by 33%; enhancing water use efficiency by 30%; real-time river/canal flow monitoring to develop transparent water accounting system by 2021; and so on. A mechanism has been proposed in the NWP for its implementation through the National Water Council (NWC) to be chaired by the Prime Minister and a Steering Committee (SC) with Secretariat at the Ministry of Water Resources (MoWR) to be chaired by the Federal Minister. However, the current NWC and SC, as given in the NWP, are skewed towards engineering profession. It is proposed that the NWC and SC should include members from the Ministry of National Food Security and Research Division (dealing with the agriculture sector which is the largest water user), the Ministry of Climate Change, relevant research organizations such as PCRWR, and IWMI who can assist the SC on emerging issues.

A two-pronged strategy is required to address the issue of misuse of water in domestic and industrial sectors that includes:

- Reduce the groundwater extraction by providing water on volumetric basis and imposing the water tariff accordingly.
- Groundwater recharge should be an integral component of any water development schemes.
- Rainwater collected from rooftops, public parks and play grounds may be diverted to aquifer through recharge wells.
- Infrastructure may be developed for wastewater treatment and recycling as also required to meet the target of 50% wastewater recycling under SDGs.

9   Conclusions and Way Forward

Water scarcity is posing great challenges to national food security in Pakistan. A country which is already food insecure will face dire consequences in the future. The impact of water scarcity is acute because of two major issues: climate change and population growth. Indiscriminate use of water is causing the water tables to recede fast in the aquifers and climate change is negatively impacting the agricultural productivity. In this scenario, one of the most difficult tasks for Pakistan is the tackling of water scarcity to ensure its national food security. In such a context, the following measures will help address this challenge:

- Sustainable development and management of water resources should be given top priority in the national policy. Building multipurpose and environmental-friendly water reservoirs, including large dams, is extremely important to store excess water from melting glaciers and runoff from Monsoon rainfall. These dams will also provide relatively cheap hydropower besides controlling floods and droughts caused by climate change.
- Building small and medium size dams and farm-scale ponds is equally important to recharge the ground water, to store runoff and any excess water especially in the dry land areas.
- Immediate measures should be taken to minimize water losses from water courses and at the farms. There is a dire need to repair and to maintain the existing water conveyance infrastructure. This can be done by properly improving the water courses to minimize seepage and leakage, while field losses should be minimized by adopting laser levelling and appropriate sowing techniques such as bed planting, etc.
- Appropriate high-efficiency irrigation technologies, like drip irrigation and sprinkler irrigation, may be promoted for efficient use of available water. Conservation of water must be prioritized and farmers should be provided with water conserving agriculture technologies.
- Farmer may be encouraged to use tunnel farming, where possible, especially for vegetables.
- A pricing system of water use for agriculture, industrial, and domestic purposes needs to be developed and effectively implemented.
- Pollution of the surface water bodies and groundwater aquifers affects the whole ecosystem in general, and human health in specific. Therefore, the industry and other sectors polluting the water bodies should be charged by imposing a pollution tax, using the 'polluter pays principle'.
- Research efforts in crops and related sectors should be enhanced to produce varieties having flexibility and tolerance to drought, heat, and salinity in line with the current climate change scenario. Shifts in food diet habits may be employed from staple food to other foods which require lesser water in its production.
- Productivity enhancement of major crops is an urgent need by developing varieties compatible with a changing environmental scenario. An alternate cropping system needs to be developed to reduce irrigation requirements during the dry period.

Cropping system with optimal need of water may be promoted and high delta crops may be discouraged.

- Long-term studies should be initiated to assess causes of sea water intrusion and extent of damage caused to the coastal ecosystem, especially to the mangroves, aquatic life and land in the coastal area.
- Studies should be undertaken to develop innovative technologies to create flexibility in the existing rigid water availability system for farmers so as to provide water at the critical stages of crop growth.
- A regulatory framework should be devised and strictly implemented for the installation and operation of tube wells to reduce and control the over-extraction of groundwater.
- Appropriate crop zoning and cropping pattern should be adopted and implemented.
- Further development of cultivable area should be restricted only to less water intensive sectors such as forestry, wildlife, and pastures for livestock instead of crop sector.
- Water-use efficiency needs to be increased by developing appropriate varieties and crop production technologies. Physical inputs such as seeds, fertilizers, credit and pesticides should be made available at farmer's doorsteps to enhance the agricultural productivity.
- A comprehensive climate change strategy should be developed with water-demand solutions at the forefront of its efforts. This focus will help overcome the country's immediate and growing water crisis. It will also help reduce Pakistan's vulnerability to more variable water flows and the inevitable longer-term impacts of climate change on the essential water resources of the Indus River basin.
- Appropriate balance between centralized and decentralized water management must be achieved. The promotion of private sector's involvement in water resources management will be effective to alleviate the crisis.
- There is a lack of coordination between the water-related research and development departments. All such departments should be placed under the umbrella of Ministry of Water Resources to implement the NWP in its true letter and spirit.
- There is a need to improve the coordination at the policy and planning levels between the agencies responsible for water storage, water distribution, water management, crop productivity, and food security, on one hand, and water users, on the other hand. Currently, the representation on the NWC and SC is skewed in favour of engineering professionals. Members from Ministry of National Food Security and Research Division, Ministry of Climate Change, and the relevant research organizations like PCRWR and IWMI should be included in the NWC and the Steering Committee.
- An effective public awareness campaign needs to be undertaken through the media and the extension wing of the Provincial Departments to promote the judicious use of water and the sustainable productivity in agriculture. Media can also play an important role in creating awareness about saving water during domestic and industrial uses.

Since absolute water scarcity and food insecurity in Pakistan are approaching fast, a way forward is imminent. Climate change may be taken as a challenge and a way can be paved for harnessing its positive impacts by developing appropriate technologies.

References

Amir, P., & Habib, Z. (2015). Estimating the impacts of climate change on sectoral water demand in Pakistan. *Action on Climate Today.*

Ashraf, M. (2016). Managing water scarcity in Pakistan: Moving beyond rhetoric. In *Proceedings of AASSA-PAS Regional Workshop on Challenges in Water Security to Meet the Growing Food Requirement* (pp. 3–14). Pakistan Academy of Sciences.

Asian Development Bank (ADB). (2016). *Asian water development outlook 2016: Strengthening WATER SECUrity in Asia and the Pacific.* Asian Development Bank.

Cai, X., & Rosegrant, M. (2003). World water productivity: current situation and future options. In J. W. Kijne, R. Barker, & D. Molden (Eds.), *Water productivity in agriculture: Limits and opportunities for improvement.* (pp. 163–178). CABI.

FAO. (2010). *Enduring farms: Climate change, smallholders and traditional farming communities.* FAO.

Farooq, M., Kobayashi, N., Wahid, A., Ito, O., & Basra, S. M. (2009). 6 Strategies for producing more rice with less water. *Advances in Agronomy, 101*(4), 351–388.

Food Security Cluster. (2019). *Drought situation of Pakistan.*

Government of Pakistan (GoP). (2001). Ten years perspective development plan 2001–2011 and three year development program 2001–2004. *Planning Commission, 53*(1), 1–15.

IFAD. (2012). *Report on Islamic Republic of Pakistan Country strategic opportunities programme.*

Inter-Governmental Panel of Climate Change (IPCC). (2007). *Climate change.* University Cambridge Press.

International Water Management Institute (IWMI). (1998). World water demand and supply, 1990 to 2025: Scenarios and issues. *Volume 19 of IWMI Research Report*, Colombo, Sri Lanka.

International Water Management Institute (IWMI). (2000). *Water Issues for 2025. A research perspective.*

Jan, M. Q., Kakakhel, S., Batool, S., Muazim, K., & Chata, I. A. (2017). Global warming: Evidence, causes and consequences and mitigation. *Journal of Development Policy, Research and Practice, 1*, 61–81.

Kamal. (2009). Pakistan's water crisis, and Pakistan's water challenges: Entitlement, access, efficiency, and equity. *Running on Empty.* Wood Words Wilson Centre for Scholars, Asia Program.

Khalil, S., Kakar, M. K. (2011). Agricultural use of untreated urban wastewater in Pakistan. *Asian Journal of Agriculture and Rural Development*, 21–26.

Khan, A. D., Iqbal, N., Ashraf, M., & Sheikh, A. A. (2016). *Groundwater investigation and mapping in the Upper Indus Plain* (p. 72). Pakistan Council of Research in Water Resources (PCRWR).

Kijne, J. W., Barker, R., & Molden, D. (Eds.) (2003). *Water productivity in agriculture: Limits and opportunities for improvement* (pp. 163–178). CABI, Oxford.

Krepon, M. (2015). *Global warming up; Arms control down.* The Stimson Centre's South Asia Programme.

Makky, M., & Kalash, H. (2013). *Potential risks of climate change on thermal power plants.*

Maxwell, S. (1995). *Measuring food insecurity: The frequency and severity of coping strategies.* IFPRI FCND Discussion Paper No. 8.

Molle, F., & Mollinga, P. (2003). Water poverty indicators: Conceptual problems and policy issues. *Water Policy, 5*, 529.

MTDF. (2005–10). *Mid-term development framework*. Planning Commission, Government of Pakistan.

National Water Policy. (2018). *Ministry of water resources* (p. 41). Government of Pakistan.

Pakistan Economic Survey. (2019–20). Finance Division, Government of Pakistan.

Qureshi, A. S. (2011). Water management in the Indus Basin in Pakistan: Challenges and opportunities. *Mountain Research and Development, 31*(3), 252–260.

Qureshi, R. H., & Ashraf, M. (2019). *Water security issues of agriculture in Pakistan* (p. 41). Pakistan Academy of Sciences (PAS).

Qureshi, R. H., & Barrett-Lennard, E.G. (1998). *Saline agriculture for irrigated land in Pakistan: A hand book*. Australian Centre for International Agriculture Research.

Rasul, G. (2016). Implications of climate change for Pakistan. In *Proceedings of AASSA—PAS Regional Workshop on Challenges in Water Security to Meet the Growing Food Requirement* (pp. 133–139). Pakistan Academy of Sciences.

USAID. (2009). *Food and agriculture system in Pakistan*. USAID.

Water Aid. (2019). *Beneath the surface: The State of the World's Water 2019*.

World Bank. (2005, November). *Pakistan Country water resources assistance strategy water economy running dry*.

World Bank. (2006). *Reengaging in agricultural water management: Challenges and options* (p. 218). The World Bank.

World Bank. (2008). *World development report 2008: Agriculture for development*. The World Bank.

World Food Programme. (2010, September). *Pakistan flood impact assessment report*.

World Resources Institute (WRI). (2015). *Aqueduct projected water stress country rankings, 2015*. World Resources Institute (WRI), Washington DC.

Wyn Jones, R. G., Gorham, J., & Hollington P. A. (2006). *Proceedings of International Conference on Sustainable Crop Production on Salt-affected Soils*. University of Agriculture, Faisalabad. pp 9–16

Yu, W., Yang, Y. C., Savitsky, A., Alford, D., & Brown, C. (2013). *The Indus Basin of Pakistan, the impacts of climate risks on water and agriculture*.

Muhammad Umar Munir graduated in Civil Engineering from the National University of Science and Technology, Islamabad, Pakistan with specialization in construction management, water resources management and Hydrology. In terms of professional career, he is working as a research scholar at Pakistan Council of Research in Water Resources (PCRWR) in Islamabad, emphasizing his research focus on issues of Integrated Water Resource Management (IWRM) of groundwater in arid and semi-arid areas of Pakistan. Umar Munir has also worked extensively on issues of High Efficiency Irrigation Systems (HEIS) such as drip and sprinklers, land and Water Productivity, Rainwater Harvesting, Supplemental Irrigation and Water Use Efficiency, etc. He has also undertaken work on soakaway pits and inverted wells coupled with check structures for flash flood mitigation and groundwater recharge with a focus on rooftop rainwater harvesting techniques. He is member of Pakistan Engineering Council and is professional trained from reputed national and international institutions in engineering related to his research work.

Anwar Ahmad received his Master of Philosophy in Anthropology from Quaid-i-Azam University, Islamabad, Pakistan and is currently working with the Pakistan Council of Research in Water Resources (PCRWR) as Anthropologist. He is integral part of research teams/programmes working on different issues of water in different establishments of PCRWR and related institutions from public and private sectors of Pakistan. His research interests include: impact of climate change on availability of water resources; flood disaster management; unequal distribution of water; water access and management in different geographical and cultural settings of Pakistan; lifestyle dimension of water insecurity; phenomenology of water insecurity; measures of water access and adequacy; Integrated Water Resource Management (IWRM); utility of locally developed water insecurity scale; conflicting knowledge systems for water at multiple levels; and water politics. He is adequately trained in designing and conducting qualitative and quantitative research work coupled with comprehensive socio-cultural and economic understanding on water issues.

Jan W. Hopmans is Professor Emeritus of Vadose Zone Hydrology at the University of California, Davis. His research and teaching focused on soil hydrology, irrigation water and nutrient management, and climate change impacts on California hydrology. He has about 200 peer-reviewed publications in soil science and water resources journals. He is Fellow of the Soil Science Society of America, the American Geophysical Union, and the American Association of the Advancement of Science (AAAS). He was Chair of the Department of Land, Air and Water Resources and Chief Editor of Vadose Zone Journal. Since 2009, he served as Associate Dean for the College of Agricultural and Environmental Sciences (CAES), being Director of the CAES International Programs from 2015 to 18. He was the 2014 President of the Soil Science Society of America. He served as Interim Associate Vice Provost of Global Affairs in 2015 and was the Interim Director of the UC Davis World Food Center in 2016. He retired from the University of California as Distinguished Professor in 2018.

Azaiez Ouled Belgacem holds a Ph.D in Biological Sciences and M.Sc. in Natural Resources Management. His experience extends to more than 25 years in the area of arid and desert ecosystems in the MENA region. He joined the Food and Agriculture Organization of the United Nations (FAO) in Saudi Arabia as Technical Adviser in November 2020. During 2011–2020, he was Rangeland Senior Scientist cum Regional Coordinator of the ICARDA program in the Arabian Peninsula (Dubai Office). He also served as Head of the Rangeland Program and Director of Plant resources at The Arab Center for the Studies in Arid Zones and Dry Lands (ACSAD), Damascus and Head of the Rangeland and Mapping Unit in the Institute of Arid Areas (IRA), Tunisia. He coordinated/participated in several research projects with the German Technical Cooperation (GIZ), European Union, IFAD, AFESD, ACSAD, FAO, OSS, and GCC on natural resources management, biodiversity conservation, combating desertification and adaptation to climate change, and rehabilitation and management of rangelands. He published more than 120 scientific research papers.

Mirza Barjees Baig is a Professor at the Prince Sultan Institute for Environmental, Water and Desert Research, King Saud University, Saudi Arabia. He earned his MS degree in International Agricultural Extension in 1992 from the Utah State University, Logan, Utah, USA and was placed on the 'Roll of Honor'. He completed his Ph.D. in Extension for Natural Resource Management from the University of Idaho, USA and was honored with the '1995 outstanding graduate student award'. Dr. Baig has published extensively on the issues associated with natural resources in the national and international journals. He has also made oral presentations about agriculture and natural resources and role of extension education at various international conferences. Food waste, water management, degradation of natural resources, deteriorating environment and their relationship with society/community are his areas of interest. He has attempted to develop strategies to conserve natural resources, promote environment and develop sustainable communities. Dr. Baig started his scientific career in 1983 as a researcher at the Pakistan Agricultural Research Council, Islamabad, Pakistan. He served at the University of Guelph, Ontario, Canada as the

Special Graduate Faculty from 2000 to 2005. He served as a Foreign Professor at the Allama Iqbal Open University (AIOU), Pakistan through the Higher Education Commission from 2005 to 2009. He served as a Professor of Agricultural Extension and Rural Society at the King Saud University, Saudi Arabia from 2009 to 2020. He serves as well on the Editorial Boards of many international journals and the member of many international professional organizations.

Chapter 10
Climate Change Impact on Water Resources and Food Security in Egypt and Possible Adaptive Measures

Mahmoud A. Abdelfattah

Abstract Climate change, food/nutritional, and water insecurities are among the major challenges humanity is currently facing at global and national levels. In this context, the Middle East and North Africa (MENA), of which Egypt is part of, is particularly vulnerable to climate-induced impacts which have increased its profile as a water-stressed region. Yet addressing adaptive strategies to deal with increased hydrological risk remains a low priority in national policies. In Egypt, climate change impacts and adaptation strategies are increasingly becoming major concerns in different areas, including water resources and food security. Indeed, climate change impacts on water availability for agriculture in arid regions are increasingly perceived as one of the greatest challenges to supply sufficient food for an ever-increasing population while sustaining the already stressed marginal environment. Egypt has reached a unique juncture; its traditional water resources, mainly surface and groundwater, are fully utilized, whilst its water demands grow due to increasing population and rising living standards. The total population has increased from 71 million in 2006 to 91 million in 2016 and currently (2020) exceeds 100 million. Population is expected to reach 150 million by 2050. The high population growth rate and Nile upstream storage projects will certainly aggravate the problems associated with the water sector allocation. Future cuts in supply are expected to be directed to the agriculture sector. On the other hand, climate change will affect agriculture and food production in complex ways, directly through changes in agroecological conditions and indirectly by affecting growth and income distribution, and thus demand for agricultural produce. It has thus become imperative to improve the efficiency of the country's limited water resources. In other words, Egypt has to produce more food with less water. The present chapter provides a comprehensive review on the global impact of climate change on water resources, agriculture, and food security, with particular emphasis on Egypt. It also highlights climate change and number of water resources

M. A. Abdelfattah (✉)
Soils and Water Department, Faculty of Agriculture, Fayoum University, Fayoum, Egypt
e-mail: maa06@fayoum.edu.eg; mahmoud.abdelfattah@fao.org

Institute of Strategic Research and Studies for Nile Basin Countries, Fayoum University, Fayoum, Egypt

Food and Agriculture Organization of the United Nations (FAO), Fayoum, Egypt related issues in Egypt including water supply (conventional and non-conventional), water demands, water quality, challenges facing water resources sector, construction of the Grand Ethiopian Renaissance Dam (GERD), and climate change adaption and mitigation measures.

Keywords Climate change · Water · Food security · Adaption · Mitigation · GERD · Egypt

1 Introduction

Global food security, threatened by climate change, is one of the most important twenty-first century challenges to the supply of sufficient food for the increasing population while sustaining the already stressed marginal environment (Lal, 2005). Climate change affects social-ecological systems in a variety of ways, and water is the primary resource through which these impacts are felt in various sectors of life such as agriculture and associated food security and eco-system services. These impacts are clearly obvious through the increasing frequency and intensity of storms, floods, and droughts and higher average temperatures (at least $+2\ °C$) since pre-industrial time. Increasing variability in the global water cycle implies greater water stress at different times and in different areas. Water-related impacts of climate change also include negative effects on food security, human health, energy production, environment, and biodiversity. These, in turn, have already led and will further lead to rising societal inequities, social unrest, mass migration and conflict (Miletto & Connor, 2020).

Global water resources affected and is affected by climate change can reduce the predictability of water availability and impair water quality. It also increases the occurrence of extreme weather events, threatening sustainable socio-economic development and biodiversity worldwide. This has profound implications for sustainable water resources at national and global levels. Climate change exacerbates the ever-growing challenges associated with water scarcity and the sustainable management of water resources. Conversely, the way water is managed influences the drivers of climate change (Houngbo, 2020). Water, therefore, is the ultimate connector in the global commitments towards a sustainable future (UNESCO UN-Water, 2020).

North Africa, being considered a climate change hotspot, has received great attention in the recent years (Diffenbaugh & Giorgi, 2012). Climatologists have emphasized the high wider temporal variability of rainfall amounts/frequencies, the associated drought periods and heat waves (Cook et al., 2016; Lelieveld et al. 2016). In this region, the combination of climate change and strong population growth is very likely to further aggravate the already scarce water situation (Abdelfattah, 2013; Schilling et al., 2020). Alboghdady and El-Hendawy (2016) stated that a 1% increase in temperature in the winter results in a 1.12% decrease in agricultural production in the Middle East and North Africa (MENA).

Egypt, as an arid country, is suffering severe water stress due to limited supplies, growing population and increased competition on water from the upper Nile basin countries (Shendi et al., 2013). The uncertain climate change impacts on the Nile flow add another challenge for water management. Besides, the projected high temperature would increase the local water demands, especially on the agricultural sector (Nour El-Din, 2013). Das Gupta et al. (2007) found that 10% of Egypt's population would be affected by sea-level rise and 12–15% of the agricultural land of the Nile Delta could be lost. Even if direct inundation is prevented, rising sea levels will change the freshwater-saline interface, rendering some of the fertile coastal agricultural areas increasingly difficult to cultivate. Egypt's Mediterranean coastal cities are particularly vulnerable, as climate change impacts will interact with the ongoing coastal erosion (EEAA 1999; Kreimer et al. 2003). The present chapter aims at summarizing the global impacts of climate change on water resources, food production, agriculture, with particular emphasis on Egypt, and possible adaption and mitigation measures.

2   Global Impacts of Climate Change

Global climate change coupled with demographic growth will profoundly affect the availability and quality of water resources, particularly in the MENA region (Conway & Hulme, 1996; Suppan et al., 2008; Alpert et al., 2008; Sánchez et al., 2004; Milly et al., 2005; Gao & Giorgi, 2008; Evans 2008a, b). Climate change is likely to intensify and accelerate the hydrological cycle resulting into more water being available in some parts of the world and less in other parts. The latter situation mostly concerns the Global South, where weather patterns in these parts of the world, are predicted to be more extreme. Those regions adversely affected will experience more intense droughts and/or possible flooding (Abdelfattah et al., 2010; Elsaeed, 2012; Pain & Abdelfattah, 2015; Shahid et al., 2010). As a largely arid region, the MENA is particularly vulnerable to climate-induced impacts on water resources, yet promoting adaptation strategies to deal with the increased hydrological risk remains a low priority in national policies (Abdelfattah & Shahid, 2007; Shahid et al., Shahid, Abdelfattah, et al., 2013, b). It is increasingly clear that climate change will interact with other social, economic and political variables to exacerbate social and political vulnerabilities (Sowers et al., 2011). Acceleration in the hydrological cycle will likely make droughts longer and rainfall events more variable and intense, raising probabilities of flooding and desertification (Abdelfattah, 2009; Abdelfattah et al., 2009). In Africa, climate change is arguably one of the most important challenges, largely due to geographic exposure, low income, greater reliance on climate-sensitive sectors such as agriculture, and weak capacity to adapt to a changing climate (Al-Muaini et al., 2019; Belloumi, 2014; Ochieng et al., 2016). The Impacts of climate change on global water resources, agriculture and food security and the adaption and mitigation measures are outlined below.

2.1 Impacts on Global Water Resources

Climate change affects global water resources in multiple ways, with complex spatiotemporal patterns, feedback effects, and interactions between physical and human processes (Bates et al., 2008). It has already caused significant impacts on water resources, food security, hydropower, and global human health (Magadza, 2000). These effects will add challenges to the sustainable management of water resources, which are already under severe pressure in many regions of the world (WWAP, 2012) and subject to high climate variability and extreme weather events. Notably, they affect the availability, quality and quantity of water for basic human needs, threatening the effective enjoyment of the human rights to water and sanitation for potentially billions of people. Although the effects of climate change can be highly distinctive at the local scale (IPCC, 2019), current trends and future projections indicate major shifts in climate, and more extreme weather events in many parts of the world (IPCC, 2014). It is therefore paramount to consider the potential impacts of a changing climate when managing water which is a resource fundamental to sustainable development. The hydrological changes induced by climate change imply major risks for society, not only directly through alterations in the hydrometeorological processes that govern the water cycle, but also indirectly through risks for food security, economic development, energy production, and social inequalities, among others (Stewart et al., 2020).

It is projected that water availability will increase in some parts of the world, which will have its own effects on increased crop production and improved food security through increased water allocation to the farmers. Crop production may increase if irrigated areas are expanded or irrigation is intensified, at the cost of increasing rate of environmental degradation. Since climate change impacts on soil–water balance will lead to changes of soil evaporation and plant transpiration, consequently, the crop growth period may shorten in the future impacting on water productivity (Kang et al., 2009; El-Keblawy & Abdelfattah, 2014; El-Keblawy, 2015). Global water use has increased by a factor of six over the past 100 years and continues to grow steadily at a rate of about 1% per year, with increasing population, economic growth and shifting consumption patterns (Stewart et al., 2020). In 2012, the Organization for Economic Cooperation and Development (OECD) projected that water demand would increase by 55% globally between 2000 and 2050, mainly as a function of growing demands from manufacturing, thermal power generation, and domestic use (OECD, 2012). A different study concluded that the world could face a 40% global water deficit by 2030 under a business-as-usual scenario (WRG 2030, 2009).

Available data indicate that about four billion people live under conditions of severe physical water scarcity for at least one month per year (Mekonnen & Hoekstra, 2016), around 1.6 billion people, or almost a quarter of the world's population, face economic water shortage, which means they lack the necessary infrastructure to access water (UN-Water, 2014), the climate change will further increase their problems for water availability. On the other hand, climate change will aggravate the situation of currently water-stressed regions, and generate water stress in regions where water resources are still abundant today (Stewart et al., 2020). Physical water scarcity is often a seasonal phenomenon, rather than a chronic one; however, climate change is likely to cause shifts in seasonal water availability throughout the year in several places (IPCC, 2014). Water quality will be adversely affected as a result of higher water temperatures and reduced dissolved oxygen; consequently, the self-purifying capacity of freshwater bodies is being increasingly reduced.

2.2 Impacts on Global Agriculture

The specific challenges for agricultural water management are twofold. The first is the need to adapt existing modes of production to deal with higher incidences of both water scarcity and excess. The second is to 'decarbonize' agriculture through climate mitigation measures that reduce greenhouse gas (GHG) emissions and enhance water availability (De Souza et al., 2020). The scope for adaptation in rainfed agriculture is largely determined by the ability of crop varieties to cope with shifts in temperature and to manage soil–water deficits. Supplementary irrigation allows cropping calendars to be rescheduled and intensified, thus providing a key adaptation mechanism for land that previously relied solely on precipitation. The largest contribution to agricultural GHG emissions, in terms of equivalent tons of CO_2, is made by the release of livestock methane through enteric fermentation and manure deposited on pasture. Agriculture has two main avenues for mitigation of GHGs: carbon sequestration through organic matter accumulation above and below the ground; and emission reduction through land and water management, including adoption of renewable energy inputs such as solar pumping (De Souza et al., 2020).

Climate-Smart Agriculture (CSA) is a recognized suite of well-informed approaches to land and water management, soil conservation and agronomic practice that sequester carbon and reduce GHG emissions. CSA practices help retain soil structure, organic matter and moisture under drier conditions, and include agronomic techniques (including irrigation and drainage) to adjust or extend cropping calendars to adapt to seasonal and interannual climate shifts (Abdelfattah & Kumar, 2015; Dawoud & Abdelfattah, 2009; De Souza et al., 2020). Changes in temperature and precipitation associated with continued GHG emissions will bring changes in land suitability and crop yields. In temperate latitudes, higher temperatures are expected to bring predominantly benefits to agriculture and the areas potentially suitable for cropping will expand, the length of the growing period will increase, and crop yields may rise. In drier areas, climate models predict increased evapotranspiration and lower soil moisture levels (IPCC,). As a result, some cultivated areas may become unsuitable for cropping and some tropical grassland may become increasingly arid.

2.3 Impacts on Global Food Security

Climate change affects agriculture and food production in complex ways. It affects food production directly through changes in agroecological conditions and indirectly by affecting growth and income distribution, and thus demand for agricultural produce. Impacts have been quantified in numerous studies and under various sets of assumptions (IPCC, 2007b). Global and regional weather conditions are also expected to become more variable than at present, with increases in the frequency and severity of extreme events such as cyclones, floods, hailstorms, and droughts. By bringing greater fluctuations in crop yields and local food supplies and higher risks of landslides and erosion damage, they can adversely affect the stability of food supplies and thus food security. If climate fluctuations become more pronounced and more widespread, droughts and floods, the dominant causes of short-term fluctuations in food production in semiarid and sub-humid areas, will become more severe and more frequent. In semiarid areas, droughts can dramatically reduce crop yields and livestock numbers and productivity (IPCC, 2001). How strongly these impacts will be felt will crucially depend on whether such fluctuations can be countered by investments in irrigation, better storage facilities, or higher food imports. In addition, a policy environment that fosters freer trade and promotes investments in transportation, communications, and irrigation infrastructure can help address these challenges early on. The main concern about climate change and food security is that changing climatic conditions can initiate a vicious circle where infectious disease causes or compounds hunger, which, in turn, makes the affected populations more susceptible to infectious disease. The result can be a substantial decline in labor productivity and an increase in poverty and even mortality. Essentially all manifestations of climate change, be they drought, higher temperatures, or heavy rainfalls have an impact on the disease pressure, and there is growing evidence that these changes affect food safety and food security (IPCC, 2007b).

3 Impacts of Climate Change on Egypt

Egypt has reached a unique juncture; its traditional water resources are fully utilized, whilst demands for water continue to grow in response to the increasing population and rising standards of living. The rapidly growing population in Egypt will require more water to live, where current water resources are fully utilized. The climate change will add further pressure on water resources in Egypt. The renewed freshwater annual per capita consumption decreased to about 650 m^3, which is significantly below the 1000 m^3 water scarcity threshold (Dawoud, 2019). Population predictions for 2025 will bring Egypt's per capita share down to 500 m^3/capita/year (Attia, 2018). Egypt is an arid country where rainfall is scarce, and the desert covers most of the land. In addition to its fixed Nile quota, its deep groundwater reservoir is not renewable and the higher the exploitation rate, the shorter the period of use will be (El Bedawy,

2014). Water shortage is the main constraint and a major limiting factor facing the implementation of the country's future economic development plans (Allam & Allam, 2007). Consequently, water management in Egypt is facing unprecedented population pressures and alarming levels of water pollution. Meanwhile, water is closely linked with numerous aspects of the national economy and social stability, and at the same time has very direct effects on the health and livelihoods of many citizens (Luzi, 2010).

In order to understand the vulnerability of the water sector in Egypt to climate change, it is important to know the water balance, supply (conventional and nonconventional) and demands, water quality, challenges facing water sector, and how this sector is managed. Egypt's water demand for irrigation, industry, and domestic consumption already exceeds the supply of the Nile. This is substituted by recycling fresh water more than once, which raises the over water use efficiency to 85% (MWRI, 2010, 2015 and 2017). The water balance implies that there is a shortage in fresh water resources, and also reflects the high efficiency of the system as well as its sensitivity to the deterioration of water quality, a problem that may arise due to the expansion in nonconventional water use. The sections below summarize Egypt's physical setting (including geography and climatic conditions), characteristics of the water sector in Egypt in terms of resources, demands, challenges, quality, and management.

3.1 Physical Setting of Egypt

3.1.1 Geography

Egypt covers an area of about one million square kilometers, and is located between 22° to 32° North and 24° to 37° East (Fig. 1). Most of the country lies within the wide band of desert that stretches eastwards from Africa's Atlantic Coast across the continent and into southwest Asia. The Nile Valley and Delta; the most extensive oasis on earth, was created by the sediments and deposits of the Nile along thousands of years until the construction of the High Aswan Dam in 1968. Only 35,000 km^2 of the total land area is cultivated and permanently settled. Egypt's geological history has produced four major physical regions: the Nile Valley and the Nile Delta, the Western Desert, the Eastern Desert, and Sinai Peninsula. The Nile Valley and Nile Delta are the most important regions, being the country's only cultivable regions supporting about 95% of the population. The Nile valley extends approximately 900 km from Aswan to the outskirts of Cairo. The delta has been formed through annual supply of nutrients and sediment deposits for thousands of years by the Nile, forming a topsoil of about 20 m in depth over the original shallow sea bed. Intensive farming has been going on in the delta for 5000–6000 years. With the construction of the High Aswan Dam, the delta no longer receives nutrients and sediments, and heavy fertilization is used instead. The Western Desert covers an area of some 700,000 km^2, thereby accounting for around two-thirds of Egypt's total land area. This immense desert to the west of the Nile occupies the area from the Mediterranean Sea southwards to the Sudanese border. The Eastern Desert covers an area of approximately 220,000 km², and is relatively mountainous and uninhabited. The Sinai Peninsula is a triangular-shaped peninsula, about 61,100 km². Similar to the desert, the peninsula contains mountains in its southern sector that are a geological extension of the Red Sea hills, the low range along the Red Sea coast that includes Catherine mountain, the country's highest point; at 2642 m above sea-level.

3.1.2   Climate

Egypt's climate is hot, dry, deserted and is getting warmer. During the winter season (December-February), Lower Egypt's climate is mild with some rain, primarily over the coastal areas, while Upper Egypt's climate is practically rainless with warm sunny days and cool nights. During the summer season (June- August), the climate is hot and dry all over Egypt. Summer temperatures are extremely high, reaching 38–43 °C with extremes of 49 °C in the southern and western deserts. The northern areas on the Mediterranean coast are much cooler, with a maximum of about 32 °C. The average daily temperature ranges from 17 to 20 °C along the Mediterranean to more than 25 °C in Upper Egypt along the Nile (SNC, 2010). Egypt. From 1961 to 2000, the mean maximum air temperature increased 0.34 °C/decade, while the mean minimum air temperature increased 0.31 °C/decade (SNC, 2010). Rainfall in Egypt is very low, irregular and unpredictable. Annual rainfall ranges between a maximum of about 200 mm in the northern coastal region to a minimum of nearly zero in the south, with an annual average of 51 mm.

3.2   Water Resources in Egypt

Conventional water resources in Egypt include the Nile water, rainfall, deep ground-water, and desalinated water. About 98% of Egypt's freshwater resources originate outside of its borders, such as the Nile River and groundwater aquifers. The Nile River provides the country with about 93% of its water requirements (Dawoud, 2019). This is considered to be one of the main challenges for the domestic water policy and decision-makers. The total annual renewable fresh water resources are about 59.25 billion m³, as per 2017 records. The estimated virtual water, the water embedded in the production of food and fiber and non-food commodities including energy, was about 34 billion m³ and the total projected water demand was about 114 billion m³ with a deficit in water resources balance of about 20 billion m³. Nonconventional water resources include agricultural drainage water, desalinated brackish groundwater and/or seawater, non-renewable groundwater aquifers, and treated municipal wastewater. These resources represent about 22% of the total avail-able water resources, and are generally used for agriculture, landscaping and industry through specialized processes (Dawoud, 2019). The sections below summarize water

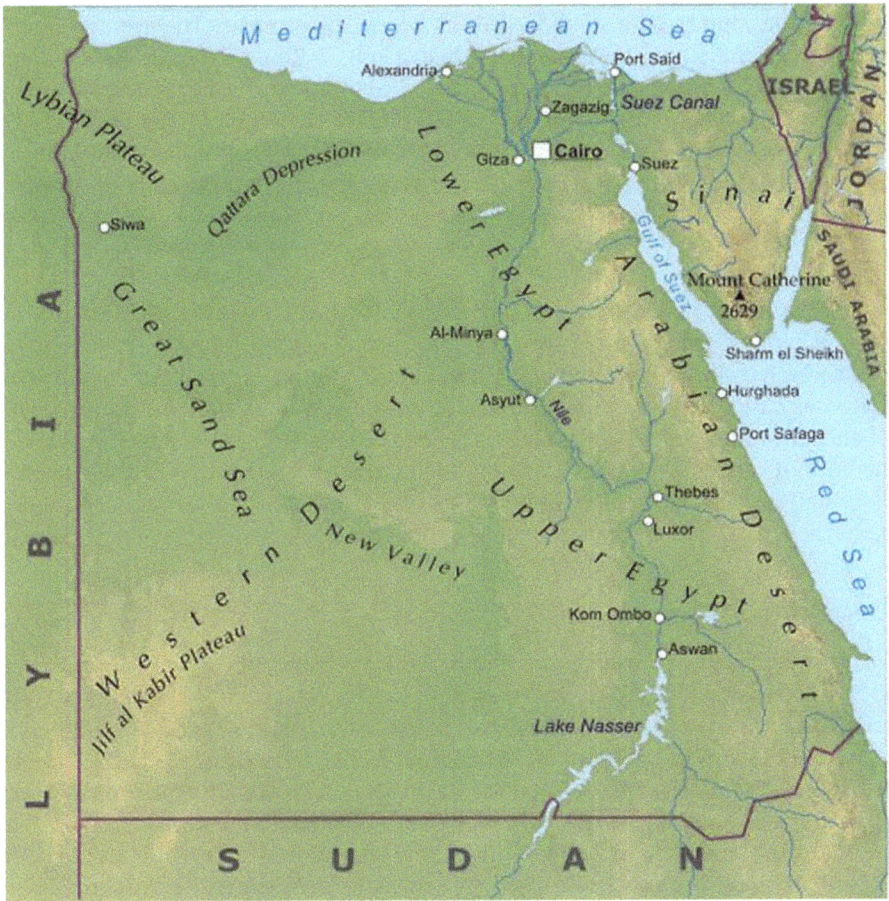

Fig. 1 Location map of Egypt

supply (conventional and non-conventional resources), water demand, water quality and challenges facing water resources sector in Egypt, followed by the way it is impacted by climate change.

3.2.1 Water Supply

a. *Conventional Water Resources*

The River Nile

Egypt almost entirely depends on the River Nile water source for its water supply. Egypt receives more than 98% of its freshwater resources from outside its international borders, such as the Nile River and groundwater aquifers. Indeed, the Nile River provides the country with some 93% of its water requirements. The average annual natural flow of the river at Aswan is estimated at 84 billion m^3. This flow is subject to wide seasonal variations. Egypt's annual share of the river water is determined by the 1955 agreement with Sudan at 55.5 billion m^3. The High Aswan Dam (HAD) is the major storage facility on the river. It started its operation in 1968 ensuring Egypt's control over its share of water and guiding its full utilization. Downstream HAD, Nile water is diverted into an intensive network of canals.

Rainfall

Rainfall in Egypt is very scarce except in a narrow band along the northern coastal areas, where insignificant rain-fed agriculture is practiced (Attia, 2018). Rainfall on Egypt varies from region to another. The annual rainfall on northern coastal strip decreases eastward from about 200 mm at Alexandria to 75 mm at Port Said with an annual average of about 150 mm (Dawoud, 2019). Rainfall occurs in winter in the form of scattered showers along the Mediterranean shoreline. The total amount of rainfall does not exceed 1.5 billion m^3/year. Flash floods, occurring due to short-period heavy storms, are considered a source of environmental damage especially in the Red Sea area and Southern Sinai.

Groundwater Resources

Groundwater in Egypt is also an important (dependent) source of freshwater with about 7.5 billion m^3/year of rechargeable live storage. The country's renewable fresh groundwater resources can be considered as a reservoir in the Nile River system. It is recharged by the seepage from the River Nile, its two branches, the canals network and deep percolation of excess irrigation water. Thus, it is part of Egypt's annual share of Nile water. Pumping of this water for domestic water use and irrigation further increase the water-use efficiency of the Nile water to over 90%. Groundwater also exists in the nonrenewable deep Nubian Sandstone aquifers in the Western Desert region and shared with Sudan, Libya and Chad. There is a total abstraction of about 0.9 billion m^3/year, which is associated with a high development cost due to the huge lifting power needed (Attia, 2018).

b.   ***Nonconventional Water Resources***

Reuse of Drainage Water

Drainage water in Upper Egypt returns to the river and is reused downstream. Reuse of drainage water was adopted in the Nile Delta by mixing it with freshwater in the main and branch canals. Farmers used to pump drainage water for direct irrigation specially for fields at the tail end of the canals. Recently, there has been a decreasing trend in the amounts of water pumped into the sea with a significant increase in the amounts of drainage water reused. The reuse of agricultural drainage water and of treated sewage water cannot be considered new resources. However, they help augment the freshwater supply in certain regions. This recycling process of previously used Nile freshwater improves the overall efficiency of the water distribution system in Egypt (Attia, 2018). The total amount of official drainage reuse increased from 6.0 billion m³/year in 2008 to 7.0 billion m³/year in 2017. This amount is expected to increase up to 9.0 billion m³/year by the year 2020 (Dawoud, 2019).

Treated Domestic Sewage

The national produced wastewater amounted to about 8.2 billion m³/year, about 4.7 billion m³/year of which were untreated, 2.4 billion m³/year were secondary treated, 0.9 billion m³/year were primary treated, and only 0.068 billion m³/year were tertiary treated. The total number of the wastewater treatment plants in Egypt is 358. Out of the 3.368 billion m³/year of treated wastewater, only 0.30 billion m³/year were reused directly for agriculture and irrigation all over Egypt, while the remaining amount was disposed to the national drainage network (Dawoud, 2019).

Desalination

As the need for water is rapidly increasing, and current freshwater resources will not be able to meet all water needs in the future requirements, desalination either for seawater or brackish groundwater has been practiced regularly for over the past 50 years and is a well-established tool for high quality freshwater supply. Recently, there is a considerable effort to increase Egyptian desalination capacity and experience. The total capacity of the desalination plants in Egypt was 144,060 m³/day and there is a plan to increase the daily capacity to about 1,559,330 m³/day by 2030. Desalinated water contributes with about 4.9% of the water resources in Egypt (Dawoud, 2019).

3.2.2 Water Demand

a. *Agriculture*

The agricultural sector is the largest consumer of water in Egypt and exceeds 85% of the total demand. The future expansion programs in the cultivated area depend very much on the availability of additional water resources. Therefore, most land and water policies are concerned with agriculture water-use efficiency. The agriculture land base consists of old land in the Nile Valley and Delta, rain-fed areas, several oases, and new lands reclaimed from the desert. The total irrigated area in 1977 was about 7 million feddans (1 feddan equals 0.42 hectares), and the rain-fed areas along the Mediterranean coast cover about 0.12 million feddans. The NWRP (2005) stated that the total area of irrigated land in the year 2000 was approximately 7.7 million feddans (3.25 million hectares), and expected to be 11 million feddans (4.6 million hectares) by 2017 due to expansion in area, and the implementation of the two mega projects of El-Salam Canal at North Sinai and Toshka at South Valley (Attia, 2018). The estimated agriculture water use was about 61.0 billion m³ in 2017 (Dawoud, 2019).

Table 1 Average virtual water trade in Egypt 1996–2005 (million m³)

Amount/products	Agricultural products		Animal products		Industrial products		Total	
Amount (million m³)	Exports	Imports	Exports	Imports	Exports	Imports	Exports	Imports
	7086.1	32,125.4	2833.5	1562.3	755.5	536	19,675.1	34,223.7
Percentage	66.4	93.8	26.6	4.6	7.0	1.6	100	100

Source Mekonnen and Hoekstra (2011)

b. *Domestic Demand*

The total municipal water use was estimated to be about 9.72 billion m³/year in 2016/2017. A portion of that water is consumed and the rest returns to the system, either through sewage collection networks or by the seepage to the groundwater aquifer systems. The domestic water is delivered to the users through municipal distribution network in urban areas and few villages. The major factor affecting the amount of diverted water for municipal use is the efficiency of the delivery networks, which generally ranges between 80 and 85% (Dawoud, 2019).

c. *Industrial Demand*

The annual estimated water requirement for the industrial sector during the year 2016/2017 was in the order of 4.88 billion m³ (Dawoud, 2019; El-Sadek, 2011).

d. *Virtual Water Resources*

The Egyptian virtual water balance consists of three main items: gross virtual water import; gross virtual water export; and net virtual water import. At present, the estimated annual volume of virtual water is about 34 billion m³/year (MWRI, 2017; Dawoud, 2019), Table 1.

In conclusion, the annual projected water demand is about 114 billion m³ (Table 2). The present available annual renewable water resources are about 59.25 billion m³ and the estimated virtual water quantity is about 34.0 billion m³ which means that there is a gap of 20 billion m³. This gap is covered by the reuse of non-conventional water resources as presented in Table 3.

3.2.3   Water Quality

Water quality surveys of the Nile River showed a general uniform distribution of parameters from Aswan to Cairo, and although the Nile receives enormous loads of different materials, it still maintains its self-purification capacity (NWRP, 2005). However, water quality deteriorates in the Nile branches due to disposal of agricultural drainage as well as decreased flow. Water quality in the canals is supposed to be similar to those of the branches, and they comprise the main source for downstream drinking water treatment plants. However, most canals suffer from industrial and domestic wastes (liquid and solid). Groundwater quality in the Nile system is

Table 2 Egypt's water balance (2017)

Item	Water source	Annual quantity (billion m^3)
Available renewable water resources	Nile river	55.5
	Renewable groundwater aquifers	1.05
	Rain-fed harvesting	2.7
Total available renewable water resources		*59.25*
Virtual water	Estimated virtual water	34.0
Total water resources		*93.25*
Water demand	Projected water demand	114.0
Gap in water resources in Egypt		*20.75*

Sources Ministry of Water Resources and Irrigation (2017) andDawoud (2019)

Table 3 Reuse of non-conventional water resources in Egypt

Non-conventional water resources	Annual quantity billion m^3/year
Non-renewable groundwater aquifers in desert	2.65
Groundwater aquifers in Nile delta and valley	3.40
Agriculture drainage reuse	9.70
Treated wastewater reuse	4.90
Desalination	0.10
Total non-conventional water resources	20.75

Sources Ministry of Water Resources and Irrigation (2017) andDawoud (2019)

reasonable, but pollution had affected some shallow groundwater bodies. Almost 20% of groundwater in the Nile aquifer does not meet drinking water standards, especially at the fringes where there is little or no protective clay cap. In the Western deserts water quality is generally very good, and in the Eastern desert and Sinai it is high saline. The carbonate aquifer contains brackish water in general, but has some fresh water in recharge areas.

Fertilizers are a large source of pollution for soil and water resources. Egyptian farmers consume more than 1.8 million tons of fertilizers annually, mainly using nitrogen, phosphorus and potassium in different forms. Industrial wastewater constitutes 39% of the environmental problems of the industrial sector (SNC, 2010). It contains dissolved industrial organic and inorganic wastes, solids and metals, all having negative and hazardous impacts with direct impact on human health (Nour El-Din, 2013). The Holding Company of Water and Wastewater estimated that in year 2009, 100% of the population in Egyptian cities and 11% of village population would be served by sanitary networks (SNC, 2010).

3.2.4   Challenges Facing Water Resources in Egypt

There are six main challenges facing water management in Egypt, as follows (Nour El Din, 2013):

- The growing population and the related increased water demand for public water supplies and economic activities, in particular agriculture;
- The expected Nile flow reduction which has emerged recently due to the rapid implementation plans of the GERD that represent serious direct and immediate threat to the Egyptians basic water needs;
- The expected impacts of climate change on the Nile flows and the different demands of the water sector;
- Water quality degradation in the canals' network due to the disposal of the domestic, industrial and agricultural wastewater, and the increased population in particular in the Nile Delta;
- The institutional setting of water management which is a governmentally controlled and central by nature. The management of the water sector should be effective and able to deal with the recent rapid expected changes; and
- The sea-level rise that is threatening the coastal zones and the Nile delta in particular, where a significant area is subject to inundation as well as impacting the quality of the coastal fresh water aquifers due to sea water intrusion.

Recently, Dawoud (2019) summarized the most important factors affecting water security in Egypt as follows:

- Increasing demand due to increasing population;
- Decreasing water use efficiency specially in agriculture;
- Relations between countries of the Nile basin towards either cooperation or struggle and political stability of the Nile basin countries;
- Shifts to irrigated agriculture and minimizing pressure on the blue water;
- Change in the economic and economic growth rate;
- Water reservoirs or control utilities;
- Impact of climate change on water sector (supply/demand);
- Water resources quality and pollution; and
- Groundwater abstraction from non-renewable aquifer system.

Most of the challenges mentioned above, except that of climate change, are certain and interacting, and there are measures to avoid its negative consequences. On the other hand, the challenges induced by climate change are uncertain in terms of impacts, magnitude, spatial distribution, and the onset.

3.3 Impacts of Climate Change on Egypt

Being the most downstream country on the Nile, Egypt is affected by climate change impacts, not only within its borders, but also due to changes within the whole basin, which is shared with 10 other countries. Economic developments in upstream countries and the measures they might take to adapt to climate change are likely to put more pressure on water resources in Egypt. In addition, the rapid development of the GERD is a clear example of these development effects. Despite the inter-annual storage in Lake Nasser behind the HAD, the vulnerability of Egypt to changes in river flows due to climate change is acute. El-Raey et al. (1995) identified water resources as one of the three most vulnerable sectors to climate change in Egypt, the others being coastal zones and agricultural resources. In addition, the rapid increase in population and urbanization will exacerbate this vulnerability, given the strong linkages of the Nile River, Nile Delta, Coastal resources, and surrounding deserts.

The Nile waters are highly sensitive to climate change, both in amount of rainfall and variations in temperature. And since these two factors are also interrelated, i.e., temperature changes affecting rainfall, it can be expected that climate change will take the form of changes in levels of precipitation as a result of changes in temperature, or other factors, and that the resulting effect on Nile flows will be from moderate to extreme, with the latter scenario most likely in the long term. Nile water flows are also sensitive to temperature changes. Hulme et al. (1995) argued that changes in temperature affect evaporation and evapotranspiration correspondingly. Increase in evaporation and evapotranspiration as a result of increase in temperature could reduce the levels of water flows in some Nile sub-basins by double or triple the percentage of evapotranspiration. For example, an increase in temperature of 4 °C coupled with a 20% decrease in precipitation could decrease the flow in Nile by 98%. A slightly smaller increase in temperature (2 °C) with the same reduction in precipitation could result in an 88% decrease in Nile flows. Thus, climate changes have a potentially dramatic effect on Nile flows and thus on water resources for Egypt, which is heavily dependent on the Nile for its water supply (Elsaeed, 2012).

Agrawala et al. (2004) concluded that Egypt's vulnerability in terms of water resource dependence on the Nile is tied to trends in population growth, land use, and agricultural and economic activity being almost exclusively focused along the Nile Valley and Delta. As demand increases, due to growth in population and any increases in temperature leading to greater evaporative losses of the country's allocation of Nile water, Nile water availability is likely to be increasingly stressed. Any activity upstream that diminishes available water resources in Egypt, whether manmade by upstream riparian countries or otherwise unaccounted for could seriously exacerbate this stress on Egypt. Countries downstream in the Nile basin, of which Egypt is by far the most populous and the most dependent on the Nile for its water needs, are sensitive to the variability of the runoff from the Ethiopian part of the basin, according to the International Water Management Institute (IWMI) together with Utah State University (Kim et al., 2008) and cited in the Egyptian Environmental Affairs Agency report (EEAA, 2010). Notably, the results obtained on the impact of climate change in the Nile Basin are strongly dependent on the climatic conditions and the models used. Asseng et al. (2018) reported that wheat yields are projected to decline in Egypt due to climate change, despite improvements obtained through crop intensification and expansion of irrigated areas. In addition to losses caused by changes in temperature, rainfall, and extreme CO_2 concentrations, a sea-level rise of 1 m is projected to lead to a loss of one-third of the current agricultural land in the Nile Delta (El-Nahry & Doluschitz, 2010).

3.3.1 Impacts of Climate Change on Egypt's Water Resources, Agriculture and Industry

The impacts of climate change on Egypt's water resources, agriculture, and industry are summarized below (Dawoud, 2019):

a. *Water Resources, Water Supply and Sanitation*

- Drought-affected areas are likely to increase and extreme precipitation events, which are very likely to increase in frequency and intensity, will increase flood risk.
- Higher water temperatures, increased precipitation intensity and longer periods of low flows exacerbate many forms of water pollution, with impacts on ecosystems, human health, and water system reliability and operating costs.
- Reducing the predictability of water availability and increasing the likelihood of damage and disruption to drinking water and sanitation infrastructure. Current water management practices are very likely to be inadequate to reduce the negative impacts of climate change on water supply reliability, flood risk, health, energy and aquatic ecosystems. With less runoff and water for sewage treatment, the effectiveness of sewage treatment may be reduced.

b. Agriculture

- An increased frequency of droughts and floods negatively affects crop yields and livestock. Impacts of climate change on irrigation water requirements may be great, with the potential for higher water needs.
- Sea-level rise, reduced recharge rates and higher evaporation rates will extend areas of salinization of groundwater and estuaries, resulting in a decrease in freshwater availability. This will affect crop yields and ultimately the type of crops cultivated (as a shift to more drought-resistant varieties may be necessary). Added to this, water sources used for irrigation are likely to become more saline, and this will increase salt concentrations of groundwater.

c.   Industry

Infrastructure, such as urban drinking water supply and sanitation, is vulnerable to sea-level rise and reduced regional precipitation, especially in coastal areas. Projected increases in extreme precipitation events have important implications for infrastructure: design of storm drainage, road culverts and bridges, and levees and flood control works, including the sizing of flood control detention reservoirs.

3.3.2   Construction of GERD and Climate Change Impacts

Tension throughout the Nile River Basin has escalated during the past ten years over Ethiopia's decision to construct the GERD. Given the GERD's acknowledged benefits of hydroelectric power generation, it is in Ethiopia's interest to fill the dam's reservoir as quickly as possible. Conversely, given stated concerns about the detrimental impact of dramatic water diversion on economic sustainability, Egypt wants the reservoir filled slowly. Arguments cited in favor of a fast fill rate (i.e., maximum water diverted from the Blue Nile into the GERD Reservoir) include: maximizing energy production for export from Ethiopia; climatic change uncertainty; and upstream storage capacity as protection against increases in regional temperature and evaporation at Lake Nasser. Arguments cited against a fast fill rate strategy include: the devastating effects of Egypt's agricultural yields and farming households; a delayed electricity transmission infrastructure; and maximum energy production, even on the assumption that the distribution infrastructure is in place, is only possible for three to four months annually. With diversion of $11-19$ km^3 annually, which amounts to declines of 20–35% and 13–23% of the Blue Nile respectively, Egypt estimates that two million farmers could be wiped out (Keith, 2016).

A recent climate change estimates suggest the Nile Basin region is quite vulnerable to shifts in temperature and precipitation; water diversion strategies must account for climatic change during the interval in which the reservoir is filled. Evidence supports the conjecture that climate change will increase average temperatures in the Basin (Chen et al., 2013; Di Baldasserre, 2011; Nour El-Din, 2013). Elshamy et al. (2009) report that temperature in the Basin may increase 2–5 °C. Other studies corroborate these findings, suggesting that temperatures in Africa will increase, on average, 3–4 °C (Boko 2007) with the North African region possibly realizing increases in temperature of 7–9 °C by the century's end (Boko et al., 2007; Elshamy et al., 2009). The net effect is that temperature could seriously challenge water resources in the Nile Basin (Yates & Strzepek, 1998; Keith, 2016). Climatologists report that the African continent is likely to be severely impacted by volatility in precipitation and monotonic increases in temperature throughout the twenty-first century (Keith, 2016). By 2020, upwards of 250 million people in Africa are projected to experience increased water stress with agricultural yields potentially decreasing by 50%.

The impact of GERD construction coupled with climate change threats on the Nile Basin will certainly cause a devastating impact on Egypt's agriculture and overall food security, and hence an agreement must be reached concerning its fill rate and overall management and operation. Currently, negotiations are going on between the three countries, Egypt, Ethiopia, and Sudan on these and other issues under the umbrella of the African Union, with great hope that this will reach to a mutual beneficial agreement very soon.

3.4 Adaptation and Mitigation Measures

Adaptation and mitigation are complementary strategies for managing and reducing climate change-induced risks. Adaptation encompasses a combination of natural, engineered and technological options, as well as social and institutional measures to moderate harm or exploit beneficial opportunities from climate change. Adaptation options exist in all water-related sectors and should be investigated and applied where possible. Mitigation consists of human interventions to reduce the sources or enhance the sinks of GHGs. While mitigation options are also available across every major water-related sector, they remain largely unrecognized (Timmerman et al., 2020). Adaptation and mitigation face similar barriers (Hamin and Gurran 2009) and to best deal with the situation, there needs to be a balanced approach between them (Becken, 2005; Laukkonen et al., 2009; Hamin and Gurran, 2009). This is proving to be one of mankind's largest modern challenges (Chen et al., 2019).

The INC (1999) distinguished between adaptations of supply and demand as well as water quality. While the SNC (2010) envisaged the major adaptation measures for the water resources sector, and classified them under measures for adaptation to uncertainty, adaptation to possible increase of inflow, measures to adequately confront reductions in the inflow of water resources, development of new water resources, and finally soft intervention adaptation measures. The adaptation to uncertainty, both SNC (2010) and NSACC (2011) favored the idea of maintaining low levels at the Lake Nasser Reservoir of the High Aswan dam in order to reduce evaporation losses and accommodate future high floods. The other alternative regarding uncertainty of floods is to enhance the Toshka spillway and establish efficient early warning systems. Also, the NSACC (2011), when considering adaptation to increased flow of the Nile, pointed to the construction of reservoirs in the upstream countries under the Nile Basin Initiative. We now believe that these actions are not appropriate after the recent developments and the fast implementation of the GERD. The issues of high floods are becoming less critical than those of water shortage, especially during the filling stage of these dams. In the meantime, it may be important to enhance the Toshka spillway to provide sufficient protection in case of emergency regarding the safety of any of the upstream dams or if GERD released unexpected volumes for any unpleasant circumstances. Elsaeed (2012) concluded that the following adaptive measures are suggested:

- Adapt to uncertainty through maintaining storage at Aswan High Dam at lower elevations and allocate other storage, to receive or absorb surplus water in the event of emergencies;

- Adapt to increase of inflow through reviving the plan to store in upstream lakes in light of the present development of the Nile Basin Initiative;
- Adapt to inflow reduction, Egypt's per capita share of water will be reduced by at least half by 2050 even in the absence of climate change and other related conflicts. Some of the measures that need to be taken according to the National Water Resources Plan (NWRP, 2005) developed by the Ministry of Water Resources and Irrigation are the following: physical improvement of the irrigation system, more efficient and reliable water delivery, better control on water usage, augmented farm productivity and raised farmers income, empowerment and participation of stakeholders, quick resolution of conflicts between users, use of new technologies of weed control, redesign of canal cross sections to reduce evaporation losses, cost recovery systems, and improvements to drainage, change of cropping patterns and on farm irrigation systems;
- Adapt through developing a new water resources management plan which includes: re-evaluate previous upper Nile conservation in light of impacts of climate change projects to increase Nile flows, explore deep groundwater reservoirs in the Sinai Peninsula and the Western desert as potential sources of water if needed, promote rain harvesting as a possible solution to destructive Red Sea area flash floods, desalinate brackish groundwater, increase recycling of treated wastewater (both domestic and industrial), and increase reuse of land drainage water;
- Adapt through soft interventions including promotion of public awareness, develop circulation models, increase research in all fields of climate change and its impact on water systems, encourage exchange of data and information between Nile Basin countries, and enhance precipitation measurement networks (meteorological stations) in upstream countries of the Nile Basin.

Most of the adaptation measures mentioned in the literature are general and descriptive without quantification. In addition, demand projections are not given the same attention as supply predictions; e.g., Nile Flow forecasting. The presented water budget trends take into account the projected demands and some adaptation measures; such as developing additional conventional and non-conventional water resources, to balance the growing demands. These measures included abstractions from the non-renewable deep ground water, more water harvesting from rainfall and flashfloods (which is non-reliable), and more reliance on desalination (which is expensive). Also, it assumed some measures to reduce demands in order to cope with decreasing water availability, such as using less water consuming crops and/or modifying the cropping, reducing the irrigated areas, or reducing irrigation water duties, and adjusting consumption of municipal water. These trends will guide planners on specific adaptation measures regarding when and how much water is needed from different potential sources.

4 Conclusions

Climate change affects, and is affected by global water resources. It reduces the predictability of water availability and affects water quality. It also increases the occurrence of extreme weather events, threatening sustainable social-economic development and biodiversity worldwide. This has profound implications for water resources. Egypt, as an arid country, is suffering severe water stress due to limited supplies, growing population and increased competition on water from the upper Nile Basin countries. The uncertain climate change impacts on the Nile flow add another challenge for water management. Being the most downstream country on the Nile, Egypt is affected by climate change impacts, not only within its borders, but also due to changes within the whole Basin, which is shared with 10 other countries. Conventional water resources in Egypt include the Nile water, rainfall, deep groundwater, and desalinated water. About 98% of Egypt's freshwater resources originate outside of its borders, such as the Nile River and groundwater aquifers. The Nile River provides the country with about 93% of its water requirements. The total annual renewable fresh water resources are about 59.25 billion m^3. The agricultural sector is the largest consumer of water in Egypt and exceeds 85% of the total demand. Nonconventional water resources include agricultural drainage water, desalinated brackish groundwater and/or seawater, non-renewable groundwater aquifers, and treated municipal wastewater. These resources represent about 22% of the total available water resources, and are generally used for agriculture, landscaping and industry through specialized processes. The estimated virtual water was about 34 billion m^3 and the total projected water demand was about 114 billion m^3 with a deficit in water resources balance of about 20 billion m^3. This gap is covered by the reuse of non-conventional water resources. The impact of the GERD construction coupled with the climate change threats will certainly have a devastating impact on Egypt's agriculture and overall food security, and hence a beneficial mutual agreement must be reached concerning GERD's fill rate and overall dam's management. Numerous adaption and mitigation measures are suggested to alleviate the climate change impacts on Egypt's water resources and agriculture.

References

Abdelfattah, M. A., Dawoud, M. A., Shahid, S. A. (2009). Soil and water management for combating desertification—Towards implementation of the United Nations convention to combat desertification from the UAE perspectives. In *Proceedings of the international conference on soil degradation* (pp. 35–45). Riga, Latvia, 17–19 February 2009.

Abdelfattah, M. A., Shahid, S. A., Othman, Y. R., & Kumar, A. T. (2010, January 11–14). Soil salinity mapping through extensive soil survey and using GIS, case study from Abu Dhabi, UAE. In *Proceedings of the international conference on "management of soil and groundwater salinization in arid regions*, Sultan Qaboos University, Muscat, Oman.

Abdelfattah, M. A., & Shahid, S. A. (2007). A comparative characterization and classification of soils in Abu Dhabi coastal area in relation to arid and semi-arid conditions using USDA and FAO soil classification systems. *Arid Land Research and Management, 21*(3), 245–271. https://doi.org/10.1080/15324980701426314

Abdelfattah, M. A. (2009). Land degradation indicators and management options in the desert environment of Abu Dhabi, United Arab Emirates. *Soil Survey Horizons J Soil Science Society of America, 50*, 3–10

Abdelfattah, M. A. (2013). Pedogenesis, land management and soil classification in hyper-arid environments: Results and implications from a case study in the United Arab Emirates. *Soil Use and Management J, 29*(2), 279–294. https://doi.org/10.1111/sum.12031

Abdelfattah, M. A., & Kumar, A. T. (2015). A web-based GIS enabled soil information system for the United Arab Emirates and its applicability in agricultural land use planning. *Arabian Journal of Geosciences, 8*(3), 1813–1827. https://doi.org/10.1007/s12517-014-1289-y

Agrawala, S., El Raey, M. A., My Conway, D., Van Aalst, M., Hagenstad, M., & Smith, J. (2004). Development and climate change in Egypt, focus on coastal resources and the Nile, working party on global and structural policies, Organization for Economic Cooperation and Development, (OECD).

Alboghdady, M., & El-Hendawy, S. E. (2016). Economic impacts of climate change and variability on agricultural production in the Middle East and North Africa region. *International Journal of Climate Change Strategies, 8*, 463–472. https://doi.org/10.1108/ijccsm-07-2015-0100

Allam, M. N., & Allam, G. I. (2007). Water resources in Egypt: Future challenges and opportunities. *Water International, 32*(2), 205–218. https://doi.org/10.1080/02508060708692201

Al-Muaini, A. H., Green, S. R., AbouDahr, W. A., Al-Yamani, W., Abdelfattah, M. A., Pangilinan, R., McCann, I., Dakheel, A., Abdullah, A., Kennedy, L., Dixon, S., Sallam, O., Kemp, P., Dawoud, M. A., & Clothier, B. E. (2019). Sustainable irrigation of date palms in the hyper-arid United Arab Emirates: A review. *ChronicaHorticulturae, 59–04*, 30–36

Alpert, P., Krichak, S. O., Shafir, H., Haim, D., & Osetinsky, I. (2008). Climatic trends to extremes employing regional modeling and statistical interpretation over the Eastern Mediterranean. *Global and Planetary Change, 63*, 163–170

Asseng, S., Kheir, A. M. S., Kassie, B. T., Hoogenboom, G., Abdelaal, A. I. N., Haman, D. Z., & Ruane, A. C. (2018). Can Egypt become self-sufficient in wheat? *Environmental Research Letters, 13*, 1–11. https://doi.org/10.1088/1748-9326/aada50

Attia, B. B. (2018). Securing water resources for Egypt: A major challenge for policy planners. In: A. M. Negm (Ed.), *Unconventional water resources and agriculture in Egypt*; *Hdb Env Chem* (2019) 75, 485–506. Springer International Publishing AG, part of Springer Nature. https://doi.org/10.1007/698_2018_334.

Bates, B. C., Kundzewicz, Z. W., Wu, S. & Palutikof, J. P. (Eds.). (2008). *Climate change and water*. Technical paper of the intergovernmental panel on climate change (IPCC). Geneva, IPCC Secretariat. www.ipcc.ch/publication/climate-change-and-water-2

Becken, S. (2005). Harmonising climate change adaptation and mitigation: The case of tourist resorts in Fiji. *Global Environmental Change Part A, 15*(4), 381–393

Belloumi, M. (2014). *Investigating the Impact of climate change on agricultural production in eastern and southern African countries.* AGRODEP working paper 0003, 2014. Retrieved on October 01, 2014, from https://ebrary.ifpri.org/cdm/ref/collection/p15738coll2/id/128227

Boko, M., Niang, I., Nyong, A., et al. (2007). Climate change 2007: Impacts, adaptation and vulnerability. In: M. L. Parry, O. F. Canziani, J. P. Palutikof, et al. (Eds.), *Contribution of working group II to the fourth assessment report of the intergovernmental* (pp. 433–467). Cambridge University Press, Cambridge.

Chen, H., Shang, S., & Chong-Yu, X. (2013). Prediction of temperature and precipitation in Sudan and South Sudan by using LARSWG in future. *Theoretical and Applied Climatology, 2013*(115), 363–375

Chen, W., Suzuki, T., & Lackner, M. (Eds.). (2019). *Handbook of climate change mitigation and adaptation.* Springer. https://doi.org/10.1007/978-3-319-14409-2.

Conway, D., & Hulme, M. (1996). The impacts of climate variability and climate change in the Nile Basin on future water resources in Egypt. *Water Resource Development, 12*(3), 277–296

Cook, B. I., Anchukaitis, K. J., Touchan, R., Meko, D. M., & Cook, E. R. (2016). Spatiotemporal drought variability in the Mediterranean over the last 900 years. *Journal of Geophysics Research: Atmosphere, 121*, 2060–2074. https://doi.org/10.1002/2015JD023929

Dawoud, M. A., & Abdelfattah, M. A. (2009). Development of an integrated soil and water information system (ISWIS): Abu Dhabi case study. In *Proceedings of the 10th Annual Research Conference* (Vol. I), pp. 693–703. UAE University.

Dawoud, M. A. (2019). Egypt's non-conventional water resources: Desalination sector analysis and proposed development. In *African development bank (AfDB), country report for the Arab Republic of Egypt*.

De Souza, M., Nishimura, Y., Burke, J., Cudennec, C., Schmitter, P., Haileslassie, A., Smith, M., Hülsmann, S., Caucci, S., Zhang, L., & Stewart, B. (2020). Agriculture and food security. In *UNESCO, UN-water, 2020: United Nations world water development report 2020: Water and climate change*, Paris, UNESCO.

Di Baldasserre, G., Elashamy, M., & van Griensven, A. (2011). Future hydrology and climate in the Nile River Basin: A review. *Hydrological Sciences Journal, 2011*(56), 199–211

Diffenbaugh, N. S., & Giorgi, F. (2012). Climate change hotspots in the CMIP5 global climate model ensemble. *Climate Changes, 114*, 813–822. https://doi.org/10.1007/s10584-012-0570-x

Egyptian Environmental Affairs Agency (EEAA). (2010). *Egypt second national communication report*.

El Bedawy, R. (2014). Water resources management: Alarming crisis for Egypt. *Journal of Management and Sustainability, 4*(3), 2014. https://doi.org/10.5539/jms.v4n3p108

El-Keblawy, A. A., Abdelfattah, M. A., & Khedr, A. (2015). Relationships between landforms, soil characteristics and dominant xerophytes in the hyper-arid northern United Arab Emirates. *Journal of Arid Environments, 117*, 28–36. https://doi.org/10.1016/j.jaridenv.2013.10.001

El-Keblawy, A. A., & Abdelfattah, M. A. (2014). Impacts of native and invasive exotic prosopis congeners on soil properties and associated flora in the arid United Arab Emirates. *Journal of Arid Environments, 100–101*, 1–8. https://doi.org/10.1016/j.jaridenv.2013.10.001

El-Nahry, A. H., & Doluschitz, R. (2010). Climate change and its impacts on the coastal zone of the Nile Delta. *Egyptian Environment and Earth Science, 59*, 1497–1506. https://doi.org/10.1007/s12665-009-0135-0

El-Raey, M. (1995). *Inventory and mitigation options, and vulnerability and adaptation assessment, interim report on climate change country studies*.

El-Sadek, A. (2011). Virtual water: An effective mechanism for integrated water resources management. *Agricultural Sciences, 2*(3), 248–261

Elsaeed, G. (2012). Effects of climate change on Egypt's water supply. In H. J. S. Fernando et al. (Eds.), *National security and human health implications of climate change, NATO science for peace and security series C: Environmental security*. Springer Science+Business Media B.V. https://doi.org/10.1007/978-94-007-2430-3_30

Elshamy, M. E., Seierstad, I. A., & Sorteberg, A. (2009). Impacts of climate change on Blue Nile flows using bias-corrected GCM scenarios. *Hydrology and Earth System Sciences, 2009*(13), 551–565

Evans, G. (2008a). Conflict potential in a world of climate change. Address to Bucerius Summer School on Global Governance Berlin, 29 August.

Evans, J. P. (2008). 21st century climate change in the Middle East. *Climate Change, 92*, 417–432

Gao, X., & Giorgi, F. (2008). Increased aridity in the Mediterranean region under greenhouse gas forcing estimated from high resolution simulations with a regional climate model. *Global Planetary Change, 62*, 195–209

GWP. (2000). Integrated water resource management (IWRM): Global water partnerships. TAC background paper No. 4.

Hamin, E. M., & Gurran, N. (2009). Urban form and climate change: Balancing adaptation and mitigation in the U.S. and Australia. *Habitat International, 33*(3), 238–245.

Houngbo, G. F. (2020). Forward of "United Nations world water development report 2020: Water and climate change. In *UNESCO, UN-water, 2020: United Nations world water development report 2020: Water and climate change*, Paris, UNESCO.

Hulme, M., Conway, D., Kelly, P. M., Subak, S., Downing, T. E. (1995). *The impacts of climate change on Africa*, SEI, Stockholm, Sweden.

INC. (1999). Initial National Communication on Climate Change, Egyptian Environmental Affairs Agency (EEAA June, 1999), The Arab Republic of Egypt, Prepared for the United Nations Framework Convention on Climate Change (UNFCCC).

IPCC, 2007a. Intergovernmental Panel on Climate Change. Climate Change. (2007a). *The physical science basis, contribution of working group I to the fourth assessment report of the intergovernmental panel on climate change*. Cambridge Univ Press.

IPCC. (2001). Intergovernmental panel on climate change. In *Climate change: impacts, adaptation and vulnerability, contribution of working group II to the third assessment report of the intergovernmental panel on climate change*, Cambridge Univ Press, Cambridge, UK.

IPCC. (2007b). Intergovernmental panel on climate change. In *Climate change: Impacts, adaptation and vulnerability, contribution of working group II to the fourth assessment report of the intergovernmental panel on climate change*, Cambridge Univ Press, Cambridge, UK.

IPCC. (2014). Climate change 2014: Impacts, adaptation, and vulnerability. Part A: Global and sectoral aspects. In *Contribution of working group II to the fifth assessment report of the intergovernmental panel on climate change*. Cambridge University Press, Cambridge/New York, United Kingdom/USA. www.ipcc.ch/site/assets/uploads/2018/02/WGIIAR5-PartA_FINAL.pd.

IPCC. (2019). *IPCC special report on the ocean and cryosphere in a changing climate*. Geneva, IPCC. www.ipcc.ch/srocc

Kang, Y., Khan, S., & Ma, X. (2009). Climate change impacts on crop yield, crop water productivity and food security—A review. *Progress in Natural Science, 19*, 1665–1674

Keith, B., Ford, D. N., & Horton, R. (2016). Considerations in managing the fill rate of the Grand Ethiopian Renaissance Dam Reservoir using a system dynamics approach. *Journal of Defense Modeling and Simulation: Applications, Methodology, Technology, 1–11*. https://doi.org/10.1177/1548512916680780

Kim, U., Kaluarachchi, J. J., Smakhtin. V. U. (2008). Climate change impacts on hydrology and water resources of the upper blue Nile River Basin, Ethiopia, research report 126, International Water Management Institute.

Kreimer, A., Arnold, M., & Carlin, A. (2003). *Building safer cities: The future of disaster risk*. World Bank.

Lal, R. (2005). Climate change, soil carbon dynamics, and global food security. In R. Lal, B. Stewart, N. Uphoff, et al. (Eds.), *Climate change and global food security* (pp. 113–143). CRC Press, Boca Raton (FL).

Laukkonen, J., Blanco, P. K., Lenhart, J., Keiner, M., Cavric, B., & Kinuthia-Njenga, C. (2009). Combining climate change adaptation and mitigation measures at the local level. *Habitat International, 33*(3), 287–292

Lelieveld, J., Proestos, Y., Hadjinicolaou, P., Tanarhte, M., Tyrlis, E., & Zittis, G. (2016). Strongly increasing heat extremes in the Middle East and North Africa (MENA) in the 21st century. *Climate Change, 137*, 245–260. https://doi.org/10.1007/s10584-016-1665-6

Luzi, S. (2010). Driving forces and patterns of water policy making in Egypt. *Water Policy, 12*, 92–113. https://doi.org/10.2166/wp.2009.052

Magadza, C. (2000). Climate change impacts and human settlements in Africa: Prospects for adaptation. *Environmental Monitoring and Assessment, 2000*(61), 193–205

Mekonnen, M, & Hoekstra, A. (2011). National water footprint accounts: the green, blue and grey water footprint of production and consumption. *Volume 1: Main report, value of water research report series*, No. 50, UNESCO- IHE. Delft, The Netherlands, P11.

Mekonnen, M., & Hoekstra, A. (2016). Four billion people facing severe water scarcity. *Science Advances, 2*(2). https://doi.org/10.1126/sciadv.1500323.

Miletto, M., & Connor, R. (2020). Preface of "United Nations world water development report 2020: Water and climate change". In *UNESCO, UN-water, 2020: United Nations world water development report 2020: Water and climate change*, Paris, UNESCO.

Milly, P. C. D., Dunne, K. A., & Vecchia, A. V. (2005). Global pattern of trends in stream flow and water availability in a changing climate. *Nature, 438*, 347–350.

MWRI. (2010). Water resources development and management strategy in Egypt 2050, in Arabic, Ministry of Water Resources and Irrigation, Cairo, Egypt.

MWRI (2015) Water for the Future, National Water Resources Plan 2017. Available online at: http://extwprlegs1.fao.org/docs/pdf/egy147082.pdf

MWRI (2017) Facts Regarding the Water Situation in Egypt Report

Nour El-Din, M. M. (2013). *Climate change risk management in Egypt and proposed climate change adaptation strategy for the ministry of water resources & irrigation in Egypt*. UNESCO.

NSACC. (2011, December). Egypt's national strategy for adaptation to climate change and disaster risk reduction, Egyptian Cabinet/Information and Decision Support Center/UNDP.

NWRP. (2005). Water for future. National water resources plan 2017. MWRI, Cairo. https://extwprlegs1.fao.org/docs/pdf/egy147082.pdf

Ochieng, J., Kirimi, L., & Mathenge, M. (2016). Effects of climate variability and change on agricultural production: The case of small-scale farmers in Kenya. *NJAS—Wageningen Journal of Life Sciences, 77*(2016), 71–78

OECD. (2012). *OECD environmental outlook to 2050: The consequences of inaction*. OECD Publishing. www.oecd-ilibrary.org/

Pain, C. F., & Abdelfattah, M. A. (2015). Landform evolution in the arid northern United Arab Emirates: Impacts of tectonics, sea level changes and climate. *CATENA, 134*, 14–29. https://doi.org/10.1016/j.catena.2014.09.011

Shahid, S. A., Abdelfattah, M. A., Omar, S. A., Harahsheh, H., Othman, Y. R., & Mahmoudi, H. (2010). Mapping and monitoring of soil salinization using remote sensing, GIS, modelling, electromagnetic induction and conventional methods—Case studies. In *Proceedings of the international conference on soils and groundwater salinization in arid countries*, January 11–14, 2010, pp. 59–97. Sultan Qaboos University, Muscat.

Shahid, S. A., Abdelfattah, M. A., & Taha, F. K. (Eds.). (2013a). Developments in soil salinity assessment and reclamation, innovative thinking and use of marginal soil and water resources in irrigated agriculture, p. 808. Springer Science+Business Media, B.V., Berlin. ISBN 978-94-007-5684-7. https://www.springer.com/978-94-007-5683-0

Shahid, S. A., Taha, F. K., & Abdelfattah, M. A. (Eds.). (2013b). Developments in soil classification, land use planning and policy implications, innovative thinking of soil inventory for land use planning and management of land resources, p. 858. Springer Science+Business Media, B.V., Berlin. ISBN 978-94-007-5331-0. https://www.springer.com/environment/book/978-94-007-5331-0

Shendi, M. M., Abdelfattah, M. A., & Harbi, A. (2013). Spatial monitoring of soil salinity and prospective conservation study for Sinnuris District soils, Fayoum, Egypt. In Shahid, S. A., Abdelfattah, M. A., & Taha, F. K. (Eds.), *Developments in soil salinity assessment and reclamation, innovative thinking and use of marginal soil and water resources in irrigated agriculture*, pp. 199–217, 808. Springer Science+Business Media, B.V., Berlin. ISBN 978-94-007-5684-7. https://www.springer.com/978-94-007-5683-0

SNC. (2010). Egypt's second national communication, Egyptian environmental affairs agency (EEAA-May 2010), under the United Nations framework convention on climate change on climate change.

Sowers, J., Vengosh, A., & Weinthal, E. (2011). Climate change, water resources, and the politics of adaptation in the Middle East and North Africa. *Climatic Change, 104*, 599–627. https://doi.org/10.1007/s10584-010-9835-4

Stewart, B., Buytaert, W., Mishra, A., Zandaryaa, S., Connor, R., Timmerman, J., Uhlenbrook, S., Hada, R. (2020). The state of water resources in the context of climate change. In *UNESCO, UN-water, 2020: United Nations world water development report 2020: Water and climate change*, Paris, UNESCO.

Suppan, P., Kunstmann, H., Heckel, A., & Rimmer, A. (2008). Impact of climate change on water availability in the Near East. In F. Zereini & H. Hotzl (Eds.), *Climate changes and water resources in the Middle East and North Africa.* Springer, Environmental Science and Engineering.

Timmerman, J., Connor, R., Uhlenbrook, S., Koncagül, E., Buytaert, W., Mishra, A., Zandaryaa, S., Webley, N., Amani, A., Stewart, B., Hada, R., & Kjellé, M. (2020). Climate change, water and sustainable development. In *UNESCO, UN-water, 2020: United Nations world water development report 2020: Water and climate change,* UNESCO, Paris.

UNESCO, UN-Water. (2020). *United nations world water development report 2020: Water and climate change.* UNESCO.

UN-Water. (2014). International decade for action, water for life 2005–2015 website. www.un.org/waterforlifedecade/scarcity.shtml

World Resources Group (WRG). (2030). 2009. charting our water future: Economic frameworks to inform decision-making. 2030 WRG. www.mckinsey.com/~/media/mckinsey/dotcom/client_ser vice/sustainability/pdfs/charting%20our%20water%20future/charting_our_water_future_full_r eport_.ashx

WWAP (World Water Assessment Programme). (2012). *The United Nations world water development report 4: Managing water under uncertainty and risk.* UNESCO.

Yates, D. N., & Strzepek, K. M. (1998). Modeling the Nile Basin under climatic change. *Journal of Hydrology Engineering, 1998*(98), 98–108

Dr. Mahmoud A. Abdelfattah brings 28 years of experience in the field of soils and water management in arid and semiarid environments, mostly in the United Arab Emirates, Egypt, Sultanate of Oman, Jordon, and Libya. He worked for the International Center for Biosaline Agriculture (ICBA), Dubai, the UAE University, and the Environment Agency Abu Dhabi (EAD). Currently, he is a professor at the Soils and Water Department at the Faculty of Agriculture, Fayoum University, Egypt and serves as Vice-Dean for the Institute of Strategic Research and Studies for Nile Basin Countries and Director of the Projects Management Unit at Fayoum University. In March 2021, he joined the Food and Agriculture Organization of the United Nations (FAO) as Senior Technical Advisor. His area of expertise/strength focuses on sustainable soil and water management, land degradation and combating desertification, land evaluation and land resources inventories, water resources management, and salinity management. He has published over 78 publications and supervised number of masters and PhD students. He was in charge of resource mobilization to implement 22 research projects that helped mitigate severe environmental problems. He has participated in several capacity building programs and assignments at local, regional, and international levels in number of countries including UAE, The Netherlands, Turkey, Spain, Bahrain, Kuwait, Oman, Jordon, Libya, and Egypt, which were completed in collaboration with number of organizations including FAO, ACSAD, OPEC, OFID, SPUSH, ICBA, INBA, Arab League, and UNEP. He has vast interdisciplinary experience, worked with multi-culture and multi-disciplinary teams, established global network of international partners and consultants, and coordinated with private sector and governments. He is co-leading two EU Erasmus+ funded projects, one with RWTH Aachen University, Germany and another with KU Leuven, Belgium.

Chapter 11
The Combined Impact of Climate Change and the Use of Solar Energy on the Water Consumption in Agriculture: A Case Study from Souss Massa Region

Fouad Elame, Hayat Lionboui, and Rachid Doukkali

Abstract The pressure on energy in general, and electricity in particular, has become very important. This growing demand for energy is the result of a hydro-agricultural development policy launched since the sixties by the Moroccan government. This increasing energy demand is also due to urban and rural electrification. This chapter simulates the impact of climate change and the use of renewable energy on water resources in Souss Massa region, Morocco, using a dynamic management model for irrigation water allocation. To assess the impact of climate and pumping costs changes on water resources use, a simulation was conducted and based on a comparison of water pumping costs. These costs are introduced in a dynamic model and compared to solar energy pumping cost (Scenario B). The results showed that the average shadow price of water remains below MAD $3/m^3$ in the first five years, knowing that it has already exceeded MAD $5/m^3$ compared to scenario A assessing only the climate change impact. The economic price directly affects the amount of water consumed for irrigation. Indeed, there is a 17% decrease of surface water use. This amount was offset by an increase of 23% of groundwater pumping because its operating cost becomes more competitive. This change will adversely affect groundwater resources availability if no constraint limiting water pumping is applied. Add to that the small decrease in the total value added or even a recorded increase in the first five years, despite the impact of climate change. This increasing of the value added is explained by the gain on pumping costs by using solar energy. This chapter, assessing climate change combined to the use of solar energy (Scenario B), shows the potential impact of the use of renewable energy reducing pumping costs and water availability with no constraints limiting water pumping. At the contrary, there will be a massive overexploitation of groundwater. Therefore, an immediate strategy

F. Elame (✉)
National Institute of Agronomic Research, Agadir Regional Center, Rabat, Morocco

H. Lionboui
National Institute of Agronomic Research, Tadla Regional Center, Rabat, Morocco

R. Doukkali
Agronomy and Veterinary Hassan II Institute, Rabat, Morocco
e-mail: m.doukkali@iav.ac.ma

© The Author(s), under exclusive license to Springer Nature Switzerland AG 2021   293
M. Behnassi et al. (eds.), *Emerging Challenges to Food Production and Security in Asia,*
Middle East, and Africa, https://doi.org/10.1007/978-3-030-72987-5_11

has to be implemented for an appropriate combination of the technology and regulation by taxing groundwater at the farm level to deal with the aquifer depletion and environmental scarcities.

Keywords Renewable energy · Water · Climate change · Dynamic management model · Simulation · Economic price

1 Introduction

During the last decades, climate change is presented as the greatest threat that affects the sustainability of our ecosystem. The new debate on climate change considers that the development of new approaches to evaluate agricultural policies is crucial. Climate change models indicate that precipitation could decrease by over 20% by 2050 while evaporation and climate variability will increase (GIEC, 2014), which would result in a more significant water shortage at different basins of Morocco (World Bank, 2010).

Renewable energy and energy efficiency are increasingly present in the scientific and political debate in Morocco. In a context of an increasing demand and a significant variability of oil prices, with a considerable impact on the trade balance, renewable energy represents the most interesting alternative to reduce the economic vulnerability of countries in the energy sector (Teimoori et al., 2019). In order to ensure a sustainable development in Morocco, a new energy strategy was developed with the objective to secure energy supply, to ensure the availability and accessibility of energy at a lowest cost, and to reduce energy dependence by diversifying sources, developing national energy potential, and promoting energy efficiency in all sectors. According to the Ministry of Energy, Mines, Water and Environment, Morocco has decided to increase the share of clean energy in the electricity mix to 52% and to save 15% of the energy consumption by 2030 (ADEREE, 2014).

The use of energy is essential to the agricultural sector, which is considered as an important pillar of the economic growth in Morocco. Indeed, energy is used primarily for water pumping stations and drinking water discharge, industrial and irrigation use, drinking water treatment and sewage treatment. The water sector currently consumes about 1450 Gigawatt (GWh). This consumption would be 6150 GWh in 2030 (ADEREE, 2014), more than four times the current consumption. The irrigation sector uses 900 GWh, of which 230 GWh is used by irrigated agriculture from surface water. The large share of energy needs of around 670 GWh is expressed by irrigated agriculture from groundwater (Badraoui & Berdai, 2011).

However, the energy sector presents an important challenge, especially in terms of air pollution and greenhouse gas emissions, knowing that most of energy sources have an impact on our environment (Babatunde et al., 2019). Regarding this, solar energy is considered as the most competitive source of energy (GIZ, 2011). It is considered as a clean energy which helps save energy resources, without producing wastes or emitting emissions, including CO_2, and preserves the environment without affecting human health.

In this context, this study aims to assess the impact of climate changes and the use of solar energy over the fifteen coming years on water availability and its use in the agricultural sector.

2 Study Area and Data Collection

Located between the Atlantic Ocean and the High Atlas and Anti-Atlas mountains, the Souss-Massa region extends approximately over 53,789 km^2; representing 7.6% of the national territory. The region's climate is generally dry, influenced by the Ocean and the Sahara (HCP, 2014).

Despite its dry climate, the region is experiencing an intensive agriculture. The agricultural area represents 616,500 ha. These lands are mainly in the Souss-Massa plain, where early crops and citrus are the main productions. The high yields of citrus and vegetables are strong points of this region. Their annual production usually exceeds 50% of the national total.

Agriculture is so far the main economic activity of the Souss-Massa with 143,640 ha of irrigated lands of which about 60% is under modern systems dedicated to market gardening 34%, citrus 25%, cereals 10%, and 28% livestock. The production of citrus and early vegetables contributes over 50% of the volume of national exports (Haddouch et al., 2019). The Souss-Massa population stood at 2,676,847 inhabitants as of the 2014 Moroccan census. It represents almost 7.9% of the total population of Morocco.

The Souss-Massa basin is an important basin from an economic point of view. It provides most of the Moroccan agricultural exports and covers more than 10 months of the country's needs from vegetable products (ORMVASM, 2014). Nevertheless, the basin is facing the impact of climate change reflected by recurrent droughts. In addition, the intensification of boreholes and the intensive use of irrigation reduce the groundwater potential and make this resource scarcer, and therefore more expensive to extract. In this context, all development programs will be constrained by water scarcity facing the area.

Given the complexity of the basin and the number of data required for the model development, a considerable effort was made for data collection. Multidisciplinary models, such as the non-linear programming model developed in this work, require several types of data, including parameters estimated empirically, data from surveys and available statistics. The model inputs include hydrological, agronomic, water demand data and economic data directly or indirectly related to water use. A questionnaire was drawn up. Several surveys were carried out among farmers in the study area. Likewise, several institutions were investigated. The model was run over fifteen years as a recursive dynamic model. This model is written and executed using the General Algebraic Modeling System (GAMS) language program.

A structured questionnaire to collect data on the different inputs and outputs of the area's crops was performed. The agricultural holdings were chosen according to different criteria, namely: the perimeter, the farm size, the crop grown, irrigated or non-irrigated, type of irrigation, etc. These criteria were chosen to allow a better representation of all the crops grown as well as the different management methods.

Regarding the use of renewable energy scenario, additional surveys, carried out in the Sous-Massa area, allowed to determine the water pumping costs by pumping system and to set up a comparison between these pumping systems. Then, a statistical analysis was applied to determine the relationship between pumping costs and the wells depths.

3   The Integrated Model Approach

Complexities of water allocation and water use at the basin scale require a holistic approach for water resources management and planning in order to get an optimal and a sustainable water allocation and, at the same time, an efficient water use (McKinney et al., 1999).

The model developed in this study belongs to the integrated river basin models category. It is a more detailed model that includes hydrologic, economic, and agronomic components of the basin. The advantage of this type of model is its ability to reflect the relationship and links between these various components listed above and to simulate the economic consequences due to policy choices (Cai, 1999; Rosegrant et al., 2000; Cai et al., 2003; Ringler & Nguyen, 2004; Lionboui et al., 2014). The model represents an efficient and useful tool for decision support on policy choices on water allocation and setting priorities for institutional reforms that guide water resources allocation.

This integrated economic and hydrologic river basin model is based on real links between different spatial units of the hydrological network. Spatial units are connected by interconnections or nodes. These units represent river flows, reservoirs, aquifers, or water demand sites (agricultural demand area, drinking water, industrial water, etc.) (Elame et al., 2016; Heidecke & Kuhn, 2006; Lionboui et al., 2018).

The water distribution differs between the agricultural, industrial and municipal use. Irrigation water is allocated according to water use efficiency, water requirements and crops productivity, while for the other sectors, such allocation is determined exogenously in the model.

In the case of surface water, basic units are nodes distributed across the basin and represent water supplies, storage entities, and water demand of different sectors. For groundwater, these nodes represent water from aquifers used for agricultural, municipal, and industrial purposes (Fig. 1) (Elame & Farah, 2008).

The model maximizes the agricultural value added at the basin level taking into account a set of hydrological, agronomic, and resources availability constraints (water, land, labor, etc.). These components described above are integrated into a

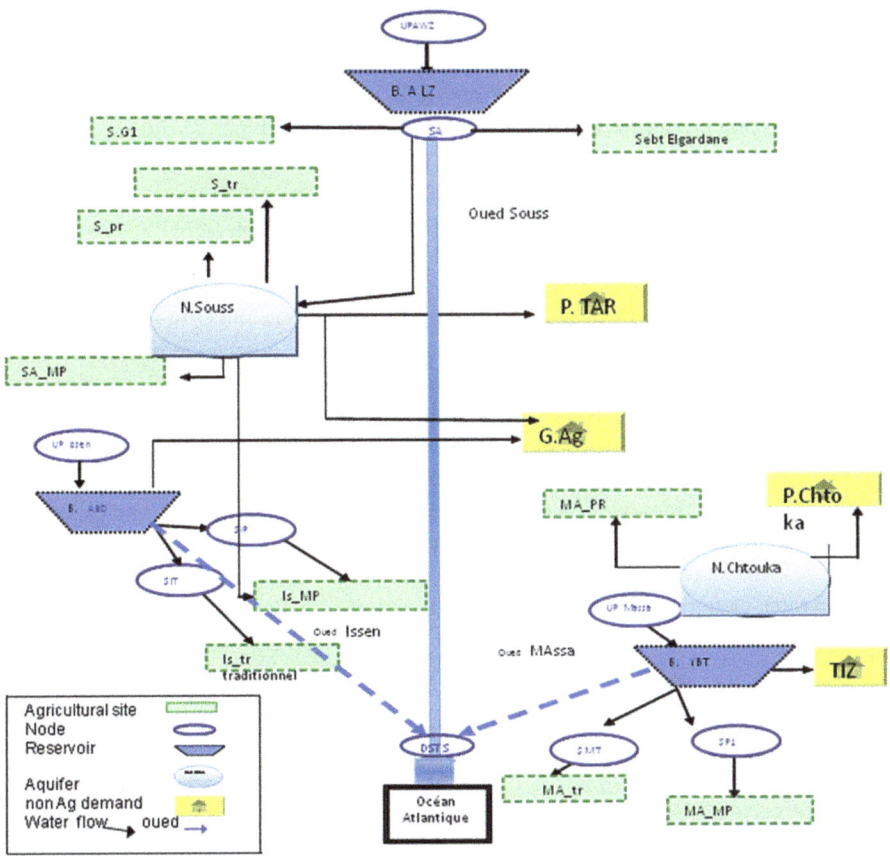

Fig. 1 A schematic representation of the various components and interconnections of the Sous-Massa basin

consistent structure of water allocation, taking into account the functioning of hydro-logical systems, rules for allocating water at demand sites level, and assessment of the environmental consequences and the economic viability of such allocation. Water demand is determined endogenously based on empirical yields and crops production functions (Elame & Doukkali, 2012).

At each agricultural sector, water is allocated to crops according to their growth stages and crops requirements. Water supply is obtained from the water supply–demand balance, result of maximizing the total value added in the basin under phys-ical, technical, and political choices constraints. The positive mathematical program-ming method (PMP) was used for the calibration of this model (Chantreuil & Gohin, 1999; Howitt, 1995). The PMP calibration method can perfectly calibrate the model using a restricted data set. This calibration process allows apprehending missing data and ensures that the model reproduces the allocation of land and water for the basic year (a normal year) (Heckelei, 1997; Rosegrant et al., 2006).

4   Dynamic Development of the Model

For the static analysis of natural resources allocation, the optimization is performed to obtain an objective function at a given time. However, dynamic models are those where we have an explicit consideration of time concept which allows knowing how the current use of a given resource may limit or compromise its future use (Chantreuil & Gohin, 1999). Therefore, time representation in economic models remains very important. The introduction of time can be done in different ways.

In order to take account of climate change and its impacts, the dynamic aspect of the model was developed allowing a good understanding of the system structure and capture dynamics of water resources system behaviour, in particular, in drought years where water is very scarce. This type of modelling is characterized by introducing an equation, which converts the past state of the system on data to be used by the optimization algorithm; a feedback algorithm that determines the actual state of the system according to the optimum solution and the past state.

This modelling framework simulates the impact of climate change and renewable energy use (solar energy) on water allocation and water pumping cost. Moreover, it captures the behaviour of various water users while analysing surface water and groundwater changes and offers a better allocation of water resources.

5   Comparison of Different Pumping Systems
in the Sous-Massa Area

The use of solar energy for water pumping purposes represents a strategic choice for decision-makers and an alternative to reduce pressure on conventional energy. Fuel supply, transportation or higher operating costs are no longer a concern. Indeed, solar water pumping systems require no fuel. Thus, there are no more expenses to bear for energy supply, thus saving time and reducing costs.

For this analysis, we distinguished between two categories of wells; wells with an average depth of less than 100 m (m) and wells with depths greater than 100 m. Farms that have wells with less than 100 m typically use diesel or gas (butane) as an energy source while those with deeper wells use electricity for groundwater pumping. The cost calculation was based on an average energy consumption of 100 Kwh.

The Table 1 gives a comparison between the different systems. We should notice that this analysis is not exhaustive since it does not distinguish between the tradable and non-tradable goods and the economic price of these goods. Noting also that the Moroccan government subsidizes the price of butane, while diesel is no longer subsidized. The results of this analysis are presented in Table 1. The most important costs are related to energy use, maintenance, and materials purchases.

The Table 1 compares the operating cost of four different energy sources used for irrigation. The cost of photovoltaic energy is more interesting than diesel after 5 years of process. The electricity cost is lower because we have not included the

Table 1 Cost comparison of different pumping systems

		Energy sources			
		Gasoil (Liter)	Butane (Bottle)	Photovoltaic	Electricity
Daily consumption		10 L	0.6 bottle	–	
Cost per unit		MAD 8.6/L	MAD 45/bottle	–	0.85
Cost per day		86	27	0	26
Monthly costs		2,580	810	0	46,537.5
Cost per year		31,390	9,855	0	
				0	
Cost after 5 years	Energy costs	156,950	49,275	0	
	Engine maintenance	36,800	36,800	0	
	Materials cost purchases	22,080	22,080	164,560	68,000
	Total costs	215,830	108,155	164,560	114,537.5
Cost per m^3 (5 years)		0.99	0.5	0.7	0.53
Cost per m^3 (10 years)		0.94	0.44	0.32	0.52

Source Author surveys

value added tax and connection costs, which can be very high in the case of a non-connected farm to the electrical grid. Adding to that the electricity taxing system of the cubic meter of water once a certain amount of energy consumed is exceeded.

The cost of butane is also lower because, according to this survey, this energy is a subsidized product in Morocco. For a better competitiveness and a reduction in energy costs, solar pumping system seems more sustainable and competitive because it is eco-friendly and less costly. To confirm these results, a simulation of water pumping cost variation in combination with the impact of climate change was conducted using the dynamic model for water management.

Statistical analysis of surveys of water pumping costs has determined the relationship that exists between water pumping costs and wells depth. The scatter plot shows a linear relationship between the two elements of this analysis. It is a simple regression since we have one explanatory variable. The linear regression gave an estimated linear equation: "$Y = ax + b + r$" in which Y represents the water pumping cost, x the well depth, the variable (a) is equal to 0.0033 and the constant (b) is equal to 0.34. This statistical analysis has also determined the coefficient of determination (R-squared) which measures the amount of information provided by the regression line. It indicates the proportion of the variance in the dependent variable that is predictable from the independent variable. This coefficient is higher than 87.8%, which confirms the significant linear correlation between the two factors. This linear equation will be introduced in the dynamic model and will represent the average pumping cost in areas using groundwater.

The price of groundwater will be computed again by the model then introduced in the objective function. We have to mention that only groundwater pumping cost will be recalculated. However, several variables of the model will be influenced in this simulation including the economic price of water, the amount of water pumped, agricultural land, etc. The equation of the water price pumped from aquifers will be written as follows:

Equation (1):

$$Pwatercost(Dma, farm\_t, E) = (3.3 * 10^{-3} * Profond$$
$$(Dma, farm\_t, E)) + 0.3406 \qquad (1)$$

With: *Pwatercost*: water price; *Profond*: well depth; *E:* index of wells.

6 Impact of Solar Energy Use and Climate Change on Water Resources

In order to assess the impact of climate change and the use of solar energy (Scenario B) on water pumping costs and water resources availability, a simulation was conducted in this context and was based on a comparison of water pumping costs of different wells. These costs are introduced in the dynamic model and compared to the pumping cost by solar energy. Applying the new pumping cost affects the amount of water consumed. The underground water consumed in this simulation becomes greater in comparison with the first scenario without solar energy use (Scenario A). We recorded an increase of 90 million cubic meter of water pumped (around 23% of increase). This change will adversely affect groundwater resources availability if no constraint limiting water pumping is applied.

Regarding surface water, there has been a 58% decrease of surface water use in scenario (B) assessing the impact of solar energy use. A difference of 17% less was recorded compared to the first scenario (A) that evaluates only climate change impact.

This shows that the impact of climate change, influenced mainly by rainfall and water flows, is lower in the case of solar energy use, which remains as an alternative to reduce the impact of climate changes over the next few years (Table 2; Fig. 2).

6.1 The Expected Changes of Economic Indicators and Agricultural Areas

The average economic price (shadow price) of water resources in the Souss-Massa basin in scenario (B) (combination of climate change and solar energy use) remains

Table 2 Results of solar energy use under climate change impact scenario

Year						
	A1	A4	A6	A8	A11	A14
Surface water use (M. m^3)	328.01	243.29	239.05	227.31	216.93	207.75
Groundwater use (M. m^3)	382.05	496.16	496.52	496.84	496.87	496.88
Total agriculture water use (M. m^3)	710.06	739.45	735.57	724.15	713.81	704.63
Consumption water use (M.m^3)	39.29	78.59	117.88	117.88	117.88	117.88
Shadow water price (MAD/m^3)	2.679	2.751	3.310	5.073	8.486	13.716
Total water use (M.m^3)	749.36	818.04	853.45	842.04	831.69	822.51
Agriculture value added (M. MAD)	9210.93	9540.74	9521.21	9382.18	9202.21	8971.96
Use of irrigable crop area (%)	99.54	97.49	97.12	96.18	95.61	95.20
Use of rainfed crop area (%)	100.30	103.44	103.95	105.22	106.01	106.57

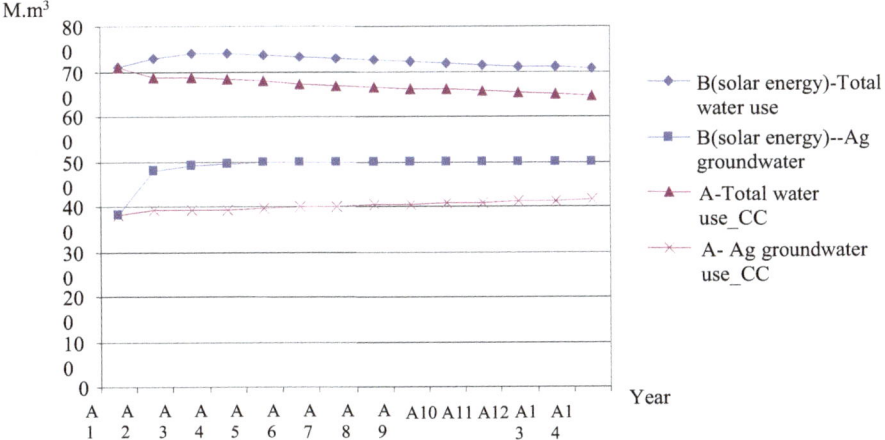

Fig. 2 Water resources use in scenario (**A**) and (**B**)

below MAD 3/m^3 the first five years, while it has already exceeded MAD 5/m^3 in the case of the first scenario (A). The economic price variations, for the two simulations, have the same upward trend with a difference of MAD 2/m^3 less in scenario B (Fig. 3).

In this simulation, the economic price influences the quantities consumed for irrigation purposes. Indeed, there is a reduction in surface water use in the order of 17%. This amount will be offset by an increase in the use of groundwater resources since their operating cost becomes more competitive. Noting that, the increase in underground water resources use was around 20% for the first five years of the simulation.

These changes will have certainly consequences on agricultural areas. It is noted that the evolution of irrigated and non-irrigated areas, in Scenario B, follows the same trend of the evolution of the areas in Scenario A with an average difference of 1% more for irrigated areas. This confirms that the impact of climate change is least perceived in Scenario B (see Fig. 4).

Regarding the value added, a clear difference was recorded between scenarios (A) (B). Agricultural value-added decreases with 2% in scenario B, while it decreases over 8% in scenario A. this result is due to the low cost of water pumping by solar energy.

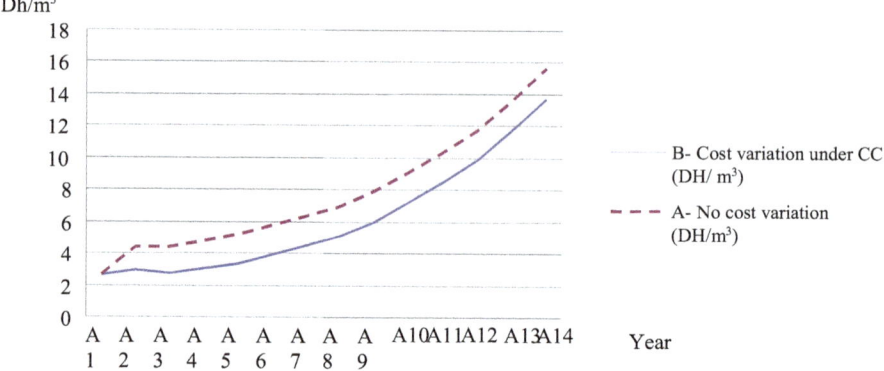

Fig. 3 Water economic price changes (Shadow price)

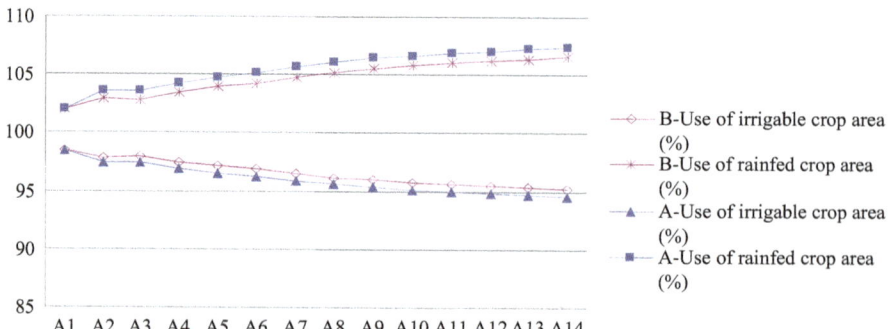

Fig. 4 Evolution of total agricultural land (rainfed and irrigated areas)

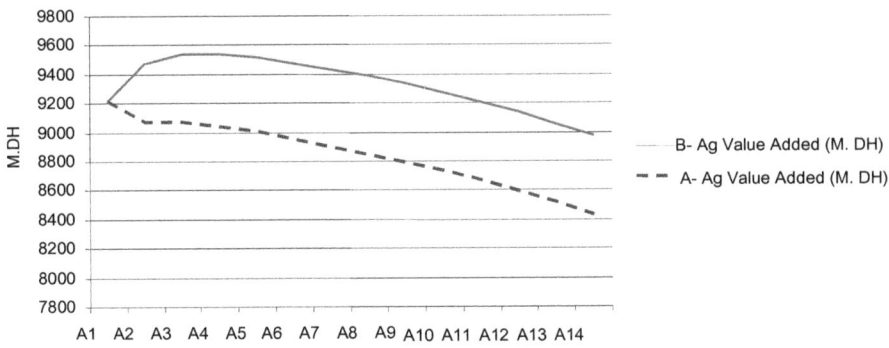

Fig. 5 Changes of the total agricultural value added

A slight decrease in irrigated areas will lead to an even higher value added despite the impact of climate change. The difference of the value added between the first and the last year for both scenarios represents MAD 538 million. For scenario B, there is an increase in the total value added for the first five years despite the impact of climate change. This increase of the value added is explained by the benefit on pumping costs by using solar energy on the one hand and the slight decrease in irrigated areas on the other hand (see Fig. 5).

7   Conclusion

This chapter assessing climate change combined with the use of solar energy, has shown the potential impact of the use of solar energy on reducing water pumping costs, and thus on the use of water resources in general. Furthermore, the use of this energy source, almost unusable until now in Morocco, can reduce energy dependence vis-à-vis the international market, thus reducing the negative impact on the trade balance.

The results show that the average economic price remains below MAD $3/m^3$ the first five years knowing that it has already exceeded MAD $5/m^3$ compared to scenario A. The economic price variations, for the two simulations, have the same upward trend with a difference of 2 dh/m^3 less in scenario B (solar energy use).

The economic price directly influences the amount consumed of irrigation water. Indeed, there is a 17% decrease in the use of surface water. This amount will be offset by an increase in the use of groundwater resources since their operating pumping cost becomes more competitive. Adding to that the small decrease of the total value added, and even a recorded increase in the first five years, despite the impact of climate change. This variation of the value added to the increase is explained by the benefit on pumping costs by using solar energy and the slight decrease in irrigated areas. However, this simulation has shown the adverse impact on the aquifer due to overexploitation, recording an increase of 23% of groundwater use.

The use of renewable energy sources for irrigation purposes will contribute to achieve the objectives assigned by the new Moroccan energy strategy focusing on green energy expansion in order to ensure the country's energy security and to mitigate the impact of climate change. In the meantime, an immediate strategy has to be developed with the aim to accommodate an appropriate combination of the technology and enforcement of regulation and control to deal with the aquifer depletion and environmental scarcities by taxing groundwater at the farm level, thus mitigating the impact of lower energy costs on water consumption.

We, therefore, conclude that the use of integrated water management policies is more effective. Consequently, the Moroccan government can intervene by encouraging the use of solar energy while putting in place a set of restrictions on groundwater pumping in order to reduce the massive use of groundwater and also by restoring the balance between the supply and demand, mainly by increasing water availabilities, whether from underground, surface or non-conventional sources.

References

ADEREE. (2014). Agence national pour le développement des énergies renouvelables et de l'efficacité énergétique, Mars 2014. *Stratégie nationale d'efficacité énergétique à horizon 2030: État des lieux de la consommation d'énergie finale au Maroc.*

Babatunde, O. M., Denwigwe, I. H., Babatunde, D. E., Ayeni, A. O., Adedoja, T. B., & Adedoja, O. S. (2019). Techno-economic assessment of photovoltaic-diesel generator-battery energy system for base transceiver stations loads in Nigeria. *Cogent Engineering, 6*(1), 1684805. https://doi.org/10.1080/23311916.2019.1684805.

Badraoui, M. H, & Berdai, M. (2011). Adaptation du système eau-énergie au changement climatique: Etude nationale—Maroc. Plan Bleu. In *Centre d'Activités Régionales PNUE/PAM.* Sophia Antipolis Janvier 2011.

Cai, X. (1999). A modeling framework for sustainable water resources management. In *Dissertation for the degree of doctor of philosophy.* The University of Texas.

Cai, X., Ringler, C., & Rosegrant, M. W. (2006). Modeling water resources management at the basin level. In *Methodology and application to the maipo river basin.* International Food Policy Research Institute. Research report 149. https://doi.org/10.2499/0896291529RR149.

Cai, X., McKinney, D. C., & Lasdon, L. S. (2003). Integrated hydrologic-agronomic-economic model for river basin management. *Journal of Water Resources Planning and Management, 129*(1), 4–17. https://doi.org/10.1061/(ASCE)0733-9496(2003)129:1(4).

Chantreuil, F., & Gohin, A. (1999). *La programmation mathématique positive dans les modèles d'exploitation agricole.* Mars 1999.

Elame, F., & Farah, A. (2008). Gestion économique de l'eau au niveau des bassins versants: Application d'un modèle intégré de bassin versant (Loukkos et Tadla). In *Agrpo/resear. Impetus, Sous-projet B4.* 03/2008.

Elame, F., & Doukkali, M. R. (2012). Water valuation in agriculture in the Souss-Massa Basin (Morocco). In *Integrated water resources management in the mediterranean region* (pp. 109–122). ISBN 978-94-007-4756-2.

Elame, F., Doukkali, R., & Fadlaoui, A. (2016). Modélisation économique de l'impact des changements climatiques sur les ressources en eau. *NewMedit, 15*(3), (September 2016) 10–18.

GIEC. (2014). Groupe d'experts intergouvernemental sur l'évolution du climat. In *Bilan 2007 des changements climatiques*. Rapport de synthèse.

GIZ. (2011). Étude du potentiel de développement de l'énergie photovoltaïque dans les régions de Meknès-Tafilalet, Oriental et Souss-Massa-Drâa. *GIZ GmbH, Eschborn/Bonn*, Mars et Novembre 2011.

Haddouch, M., Elame, F., Abahou, H., & Choukr-Allah, R. (2017). Socio-economics and governance of water resources in the Souss-Massa river basin. In *The Souss-Massa River Basin, Morocco, Hdb Env Chem*. Springer International Publishing, https://doi.org/10.1007/698_2016_75.

HCP. (2014). Haut-commissariat au plan. *Monographie régionale*. 2014.

Heckelei, T. (1997). Positive mathematical programming: review of the standard approach. *Working paper 97/03*. CAPRI.

Heidecke, C., & Kuhn, A. (2006). Calculating feasible charges for water in the Drâa valley in Southern Morocco. *Journal of Agricultural and Marine Sciences (oman), 11*, 47–54.

Howitt, R. E. (1995). Positive mathematical programming. *American Journal of Agricultural Economics, 77*, 329–342.

Lionboui, H., Benabdelouahab, T., Hasib, A., Elame, F., & Boulli, A. (2018). Dynamic agro-economic modeling for sustainable water resources management in arid and semi-arid areas. In C. M. Hussain (Ed.), *Handbook of environmental materials management*. (pp. 1–26). Springer International Publishing.

Lionboui, H., Fadlaoui, A., Elame, F., & Benabdelouahab, T. (2014). Water pricing impact on the economic valuation of water resources. *International Journal of Education and Reseach, 2*(6), 147–166.

McKinney, D. C., Cai, X., Rosegrant, M. W., Ringler, C., & Scott, C.A. (1999). Integrated basin-scale water resources management modeling: review and future directions. *IWMI SWIM paper no. 6*. International Water Management Institute, Colombo.

ORMVASM. (2014). Office régional de mise en valeur agricole de Souss-Massa. *Annual report 2014*.

Ringler, C., & Nguyen, V. H. (2004). *Water allocation policies for the Dong Nai River Basin in Vietnam: an integrated perspective*. EPTD discussion papers 127, International Food Policy Research Institute (IFPRI).

Teimoori, M., Mirdamadi, S. M., & Farajollah Hosseini, S. J. (2019). Modeling of climate change effects on groundwater resources: the application of dynamic systems approach. *International Journal of Agricultural Management and Development, 9*(2), 107–118.

World Bank. (2010). Rapport sur le développement 2010 dans le monde. *Développement et changement climatique*. Octobre 2010.

Fouad Elame is a researcher at the regional center of agronomic research of Agadir, Morocco. He has several scientific publications in relevant Journals related to economic valuation, natural resources assessment, and market chain analysis. His thesis deals with water management and environmental economics. He is involved in many projects focusing on economic modeling, economic assessment profitability and efficiency. He was in charge of commercial development at the social development agency in 2008 and joined the National Institute of Agronomic Research, the Department of Economy and Rural Sociology, in 2009.

Hayat Lionboui is a researcher at the regional center of agronomic research of Tadla, Morocco. She has several scientific publications in relevant Journals. Her thesis was related to water management and environmental economics. Her research activities are dealing with agro-economic modeling, land and water resources management, and efficiency analysis. She joined the Department of Economy and Rural Sociology at the National Institute of Agronomic Research in 2009.

Rachid Doukkali is a Professor in economics at the Department of Human Sciences, Hassan II Institute of Agronomy and Veterinary Medicine (IAV), Rabat since 1981. He graduated from the IAV as Engineer in Agronomy, and holds a Ph.D. in Applied Economics from the University of Minnesota, USA. His works focus on agricultural policy and management options.

Chapter 12
A Mathematical Model for Control of Drainage in an Irrigation System

Asaf Hajiyev, Yasin Rustamov, and Mohamed Mujithaba Mohamed Najim

Abstract Mathematical models describing the estimation of optimal amounts of water depend on various parameters such as the quality of soil that depends on many random factors. Investigation methods and approaches of the random processes of those parameters play an important role. In irrigation systems, drainage systems are widely used to achieve multiple objectives and it is necessary to pay attention to the choice of mathematical models that are adequately describing the processes of drainage and control of groundwater levels. An algorithm for the regulation and prevention of salinization of arable land and swampy areas in addition to an estimation of the amount of drainage water, have been proposed. Behaviour of the groundwater (temporal variation) is analysed applying the proposed model. Under the given conditions, it was derived that for an isotropic and homogeneous case, the level of the priming relative to axis x is approximately stable. A change in other parameters leads to the reduction of the effectiveness of drainage system and, consequently, to decreasing the level of water. Hence, using some engineering processes, it is necessary to increase the velocity of water through the drainage system.

Keywords Priming · Mathematical model · Drainage system · Non-linear equations · Regression models · Unknown parameters

1 Introduction

Climate change is an important challenge for agricultural development, including the irrigation problems. The deficiency of water in Asia, Africa, and in other parts of the world, implies the necessity to investigate the optimal use of the available water resources, because the condition of the soil and harvest directly depend on the amount

A. Hajiyev (✉) · Y. Rustamov
Azerbaijan National Academy of Sciences, Institute of Control Systems, Baku, Azerbaijan

M. M. M. Najim
Department of Zoology and Environmental Management, Faculty of Science, University of Kelaniya, Kelaniya, Sri Lanka
e-mail: mnajim@kln.ac.lk

© The Author(s), under exclusive license to Springer Nature Switzerland AG 2021   307
M. Behnassi et al. (eds.), *Emerging Challenges to Food Production and Security in Asia, Middle East, and Africa*, https://doi.org/10.1007/978-3-030-72987-5_12

of available water. The modern world is still facing a serious and dangerous health crisis—Covid-19—which has negatively affected both the harvest and the productivity in various fields including the industrial sectors. Under this circumstance, the harvest of various agricultural crops, which strongly depends on existing irrigation systems, plays a crucial role in feeding the world population. There is an abundant literature in this field (El-Hafez et al. 2020; Plaster, 2013; Samiha et al. 2020), but the issues referring to the estimation of the optimal amount of water needed for agriculture and environmental services are not widely researched. Azerbaijan is a fossil fuel-producing country (especially gas and oil), but at the same time, it is an agricultural country which is frequently affected from water deficits. Hence, the irrigation related issues have a high impact on the development of Azerbaijan economy in particular, and on the region in general; therefore, a serious focus on this area of investigation is needed.

Mathematical models describing the estimation of optimal amounts of water depends on various parameters such as quality of soil that depends on many random factors (weather, rain, region), hence it has a random structure. Investigation methods and approaches of the random processes of those parameters play an important role (Knill, 2009). These processes are also related to other research areas such as statistical data analysis of such processes in addition to the conceptualizations and results of regression models with an increasing number of unknown parameters (Hajiyev, 2009, 2011). Similarly, to estimate the reliability of the equipment used in irrigation systems, the methods of the reliability theory (Gnedenko et al., 2002; Hajiyev & Rustamov, 2013) is applied. Other related research areas include environmental issues where differential equations are used to describe the behaviour of such processes. Hence, the approaches of these areas of investigation (Hajiyev, 2014; Hajiyev & Li, 2019) can be used in the research on such processes and some general mathematical models are also presented by Полубаринова et al., (1969) and Аверьянов (1978). However, for specific applications, it is necessary to construct mathematical models, considering the specificity of the regions, climate and climate change, and other related conditions. In irrigation systems, for the control of groundwater, closed drainage systems are widely used. Sometimes, the use of new drainage systems may increase the groundwater levels. Moreover, such drainage systems are characterized by inappropriate planning and the use of non-effective engineering approaches. Hence, it is necessary to pay attention to the choice of the mathematical model, adequately describing these processes and also estimating the reliability of the systems' operation (Rustamov, 2010).

In this chapter, an algorithm for the regulation and prevention of salinization of arable land and swampy areas in addition to estimation of the amount of drainage water, have been proposed.

2 Mathematical Model

The level of underground water plays an important role in ameliorative projects and irrigation systems. Drainage systems are widely used with the ability to control the level of underground water. For the construction of the mathematical model, following parameters are introduced as given in the Fig. 1 also:

- h_1 is the thickness of the priming (measured in m), $h1 = const.$;
- h_2 is the thickness of the filter layer (measured in m), $h2 = const.$;
- $H_0 = h_0 + h1 + 0.5D_0 = const.$—is the distance from the highest point of the soil to the center of the drainage system, where D_0 is the diameter of the drainage tube; $D_0 = const.$ (measured in m);
- $p(x, y, z)$ is the density of the priming, enriched by water (measured in kg/m^3);
- x, y, z—space coordinates (measured in m);
- $h(x, y, z) = y + p(x, y, z)/\rho g$—current condition of the priming water, where g—is the free fall coefficient.

For homogeneous or isotropic cases, it is supposed that ρ is the soil density (measured in kg/m^3) and it is equal to a constant.

- W—volume of rain water (m^3).
- Ω_2—depth of field from the surface of water to the layer of percolation (it is also called as field enriched by water) (m).
- Ω_3—area of field of filter (m^2).

Figure 1, gives the schematic diagram of the process explained above

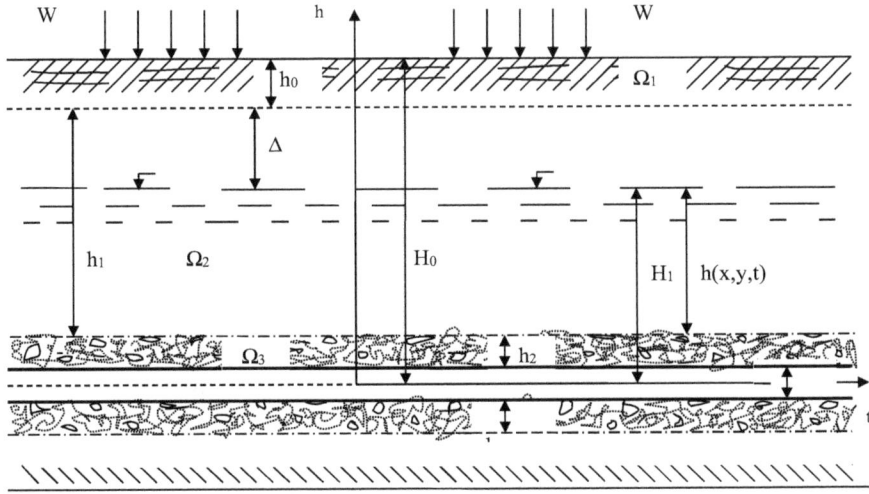

Fig. 1 Schematic diagram showing drainage system to control the level of groundwater

$$H_0 - h_0 \le y \le H_0, -1 \le x \le 1, t > 0, \Delta = const, m;$$

If $H_0-h_0 \le y \le H_0, -l \le x \le l, t > 0$ then, according to (Полубаринова et al., 1969), an equation describing water percolation in the field Ω_1 can be represented in the following form:

$$N_0 \cdot h_t(x, y, t) = (h_x(x, y, t) + h_y(x, y, t))^2 + h_y(x, y, t) \cdot (2h_x(x, y, t) - 1)$$
$$N_0 \cdot f_t(x, y, t) = -h_y(x, y, t) + h_x(x, y, t) \cdot f_x(x, y, t) \tag{1}$$

where, $N_0 = m/k$, m—porosity of the medium, k—is the hydraulic coefficient

$$f(x, y, t) = k \cdot (h(x, y, t) - b(x)),$$

$b(x)$— distance to the impenetrable soil layer and $b(x) = h_0+h_1+D+2 \cdot h_2+1 = const$.

Integrating expression (1) under the following initial and boundary conditions (the initial condition that are taken is $h(x, y, 0) = H_0 - h_0 - \Delta = const$).

$f(x, 0) = f_0(x)$ $x \in [-l; l]$ $f_0(x)$—is the given function and in the capacity of the boundary conditions are taken as $y = H_0$,

$$h(x, H_1, t) = f(x, t), \quad H_1 = H_0 - h_0 - \Delta \tag{2}$$

If $y \in \left[-\frac{D_0}{2}; \frac{D_0}{2};\right]$ then $h_x = v(t), v(t)$— speed of the water in the end of drainage system. If

$$y = H_0 \text{ and } h = -h_2, h_y = 0, \text{g} \tag{3}$$

3 Solution of the Mathematic Model

The expression (1) has a non-linear structure, and finding the exact solution for this non-linear differential equation face some difficulties. In practical problems, scholars and mathematicians are trying to find an approximate solution, but sometimes using analytical approaches, which is a typical situation for applications. Consider:

$$\sigma \cdot \frac{\partial h}{\partial t} = \frac{k}{2}(\frac{\partial^2 h^2}{\partial x^2} + \frac{\partial^2 h^2}{\partial y^2}) - \frac{k_0}{M_0}(h - H) + W \tag{4}$$

Here $\sigma, k, k_0, M_0, H \approx const$, $W(x, y, t)$—is a function which defines the infiltration process, $h(x, y, t)$ and $W(x, y, t)$ are the power functions for any small ε.

Then,

$$h(x, y, t) = h_0(t) + \sum_{k=1}^{\infty} \varepsilon^k \cdot h_k(x, y, t)$$

$$W(x, y, t) = W_0(t) + \sum_{k=1}^{\infty} \varepsilon^k \cdot W_k(x, y, t) \tag{5}$$

Taking expression (5) into consideration, the following differential equation is obtained

$$\sigma \frac{\partial h_0(t)}{\partial t} + \frac{k_0}{M_0} \cdot h_0(t) = W_0 \tag{6}$$

$$\sigma \frac{\partial h_1}{\partial t} = k \cdot h_0(t) \cdot \Delta h_1 - \frac{k_0}{M_0} \cdot h_1 - W_1 \tag{7}$$

Δ—is the Laplace operator.

In equation (7) except one expression, all the others can be considered as second order linear partial differential equations. The first expression is an ordinary differential equation, which has the following solution:

$$h_0(t) = C_0 \cdot e^{\cdot a t} + \frac{1}{\sigma} \cdot \int_0^t W_0(\tau) \cdot e^{-a(t-\tau)} d\tau \tag{8}$$

where a $= \frac{k_0}{\sigma \cdot M_0}$, $C_0 = h_0 + \Delta + \frac{W_0(0)}{a}$, and $W_0(0) = W(x, y, t)$ for $t = 0$ it is an initial value of infiltration.

In practice, we can take $W = W(t)$. Then, Eq. (7) is homogeneous. Taking into consideration of an initial condition, considering the Laplace transformation and applying the method of separation by x and y, integrating, we have:

$$h_1(x, y, t) = \frac{h_0}{\pi} \int_0^x \int_0^y Ei(-u_1^2) - Ei(-u_2^2) dx_1 dy_1$$

where, $u_1^2 = \frac{r^2}{4a(t-\tau)}$, $u^2 = \frac{r^2}{4a\tau}$, $r^2 = x_1^2 + y_1^2$. $Ei(z)$— is a power function, for which from (Knill, 2009), we have:

$$Ei(-u_1^2) = C_1 + \ln\left|-u_1^2\right| + \sum_{k=1}^{\infty} \frac{(-u_1^2)^k}{k \cdot k!}$$

$$Ei(-u_2^2) = C_1 + \ln\left|-u_2^2\right| + \sum_{k=1}^{\infty} \frac{(-u_2^2)^k}{k \cdot k!}$$

After some routine transformations, we have:

$$h_1(x, y, t) = \frac{h_0(t)}{\pi} \left(\ln \frac{t - \tau}{t} + \sum_{k=1}^{\infty} \frac{x^2 + y^2}{k \cdot k! \cdot 4a} \left(\frac{1}{\Delta t} - \frac{1}{t} \right) \right) \tag{9}$$

As ε takes small values for practical calculations, we can neglect them. Finally, we have:

$$h(x, y, t) = h_0(t) + \varepsilon \cdot h_1(x, y, t) + \ldots\ldots \tag{10}$$

If we assume that the velocity of water in the drainage system $v(t) = v_0 = const$, then we have:

$$Q_4 = \frac{n \cdot D_0}{4} \cdot v_0, \tag{11}$$

where $v_0 = C\sqrt{R \cdot i}$ is the velocity of water in the drainage system $(m/san;)$, C—is the Chezy coefficient, which is defined by Manning-Pavlovsky equation (Zwillinger, 1997) as:

$$C = \frac{1}{n} \cdot R^{\frac{1}{6}}, \tag{12}$$

$R = 0.25 \cdot D_0$ is the hydraulic radius, m; n—is the stiffness coefficient of the drainage system. As $H_0 - h_0 \leq y \leq H_0$, then the amount of water, which is percolated from the field Ω_1, is defined as:

$$Q_1 = F_1 \cdot \int_0^{h_0} h(x, y, t) dy = Q_1(x, t) \tag{13}$$

In practice and for the simplification of the calculation, the hydrologic parameter of the medium is taken as $F_1 = const$. Some amount of water remains in the layer h0 (as humidity) and other part is going to the lower layer. Hence, the lower layer becomes enriched with water. If $h_2 \leq y \leq h_1 - \Delta$, then in the field Ω_2 for enriching the ground by water (depending on the type and structure of the soil and region), it is necessary to have:

$$Q_2 = F_2 \cdot v_2$$

and at the level $h(x, y, t)$, the amount of V_2 always remains fertile.
If $0.5 D_0 \leq y \leq |\pm h_2|$, then for Ω_3 we have $Q_3 = F_3 \cdot v_3$.

Under the condition $Q_4 \geq \sum_{i=1}^{3} Q_i$, varying variables of the function $h(x, y, t)$ we can get $h(x, y, t) \leq h(x, H_0 - h_0 - \Delta, t)$, where Q_4—is the amount of water delivered by the drainage system.

The solution of (1) can be found under initial and boundary conditions.

If $Q_4 < \sum_{i=1}^{3} Q_i$ and $h(x, y, t) \le h(x, H_0 - h_0 - \Delta, t)$, additional amount of water passing through the drainage system which can be explained in the field Ω_3 (field of percolation) along the axis x.

According to the above-mentioned theoretical results, using the package of the programs "Derive", we get the following numerical results:

$$C_0 = 161.5; 303.21; 486.37,$$
$$a = 1.25; 0.83; 0.62,$$
$$\sigma = 0.1; 0.15; 0.2,$$
$$W = 200; 600,$$
$$\Delta t = 0.5, y = 2.0,$$
$$k_0 = 0.5, t = 1 \div 14,$$

In the expression (9), we take $k = 1 \div 20$. In the Fig. 2, the graphical behaviour of the level of underground water depending on the amount of water passing through the drainage system is given. It is clear, as shown in the Fig. 2, that if the amount of water passing through the drainage system is increasing, then the level of underground water should go down.

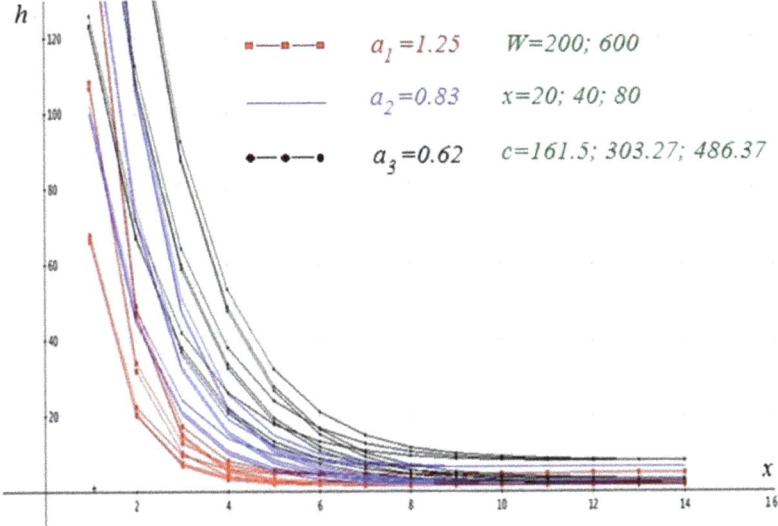

Fig. 2 Behaviour of the groundwater using the drainage system

4 Statistical Approach

Another method of solution can be based on statistical approach such as the statistical analysis for different specific regions. As mentioned above, amount of water passing through the drainage system depends on various random parameters and new parameters are derived during the observations, thus the regression models with increasing number of parameters can be applied (Hajiyev, 2009, 2011). The use of such models allows for a more accurate estimation of the function, describing the behaviour of the underground water.

5 Conclusion

In this chapter, the behaviour of the groundwater (temporal variation) is analysed. Under the given conditions, it was derived that for an isotropic and homogeneous case, the level of the priming relative to axis x is approximately stable. A change in other parameters leads to reduce the effectiveness of drainage system and, consequently, to lower the level of water. Hence, using some engineering processes, it is necessary to increase the velocity of water through the drainage system. If $t \geq 6$, then the stabilization of the water level is observed. This phenomenon can be explained as the power of the drainage system is not changing and the velocity of water is increasing, then it is derived that water must be delivered to the collector.

References

Аверьянов, С. Ф. (1978). Борьба с засолением орошаемых земель. М: Колос, 288 с.

El-Hafez, S. A., Mahmoud, M. A., & El-Bably, A. Z. (2020). *Development and recent information and data on irrigation technology and management*. Springer Nature.

Gnedenko, B., Belyaev, Y., & Solovyev, A. D. (2002). *Mathematical methods of reliability theory*. Academic Press.

Hajiyev, A. H. (2009). Nonlinear regression models with increasing numbers of unknown parameters. *Russian Academy of Sciences Mathematical, 79*, 3:339–341.

Hajiyev, A. H. (2011). *Regression models with increasing numbers of unknown parameters*. (pp. 1208–1211). Springer.

Hajiyev, A. H. (2014). A glimpse into ecological engineering based engineering management. *International Journal Management Science and Engineering Management, 9*, 4

Hajiyev, A, & Li, Z., (2019). A multi-criteria decision making method for urban flood resilience evaluation with hybrid. *International Journal of Disaster Risk Reduction, 36*, 101140.

Hajiyev, A. H., & Rustamov, Y. (2013). Statistical estimation of reliability in water pipe-line systems. *Melioration and Water Problems, 2*, 26–28

Knill, O. (2009). *Probability and stochastic processes with applications*. Overseas Press (India) Pvt.

Plaster, E. (2013). *Soil science and management*. Delmar Cengage Learning.

Rustamov, Y. (2010). Reliability of closed horizontal drainage system and method of its increasing. *Azerbaijan Agrar Science, 3–4*, 86–88 (in Azerbaijani language).

Samiha, O., Abd El-Hafeez, Z., & Tahany, N. (2020). *Deficit irrigation*. Springer.
Zwillinger, D. (1997). *Handbook of differential equations*. Academic Press.
Полубаринова, П. Я., Кочина, В. Г., & Пряжинская, В. Н. (1969). Математические методы в вопросах орошения. М: Наука,. 414 с.

Asaf Hajiyev is a Professor at the Azerbaijan National Academy of Sciences, Azerbaijan. He holds a Doctorate of Sciences from Bauman Moscow State Technical University, and has done a Ph.D. and post-doctoral research at Lomonosov Moscow State University, Russia. He has a vast research and teaching experience and serves as the Chair of many departments: Department of Probability and Statistics; Department of Controlled Queues, Institute of Control Systems at the Azerbaijan National Academy of Sciences; and Department of Theory of Probability and Mathematical Statistics, Baku State University. He is also a Senior Scientific Researcher, Department of Probability Statistics, Royal Institute Technology in Stockholm. Being a renowned researcher, he has served at several universities around the world, including China, Germany, Italy, Portugal, Sweden, Turkey, and USA. He serves on the editorial boards of many prestigious national and international academic journals. He has been an organizing member and the keynote speaker at numerous international conferences. He has been honored with many prestigious awards like: Azerbaijan Lenin Komsomol Prize Winner on Science and Engineering; Grand Prize at the International Conference "Management Science and Engineering Management", Macao; and Grand Prize at the International Conference "Management Science and Engineering Management" at Islamabad, Pakistan. He is the Honorary Academician of Academy of Sciences of Moldova, Foreign Member of the Mongolian National Academy of Sciences, Member of TWAS (The World Academy of Sciences), Honorary Professor of Chengdu University (China), and Elected Member of the International Statistical Institute. He has also the honor of holding the Office of the Vice-President of the Parliamentary Assembly of the Black Sea Economic Cooperation Organization (PABSEC). Since 2015, he serves as the Secretary General of PABSEC. He shares his talents and expertise on the boards of many international academic organizations and institutions. He has more than 135 peer-reviewed scientific publications to his credit, published in highly reputed journals.

Yasin Rustamov is an Associate Professor at the Azerbaijan National Academy of Sciences, Institute of Control Systems, Azerbaijan. He earned his Doctor of Engineering degree from Baku State University, Azerbaijan. Presently, he holds the position of Head of Laboratory 2.2 Stochastic Control and Applied Statistics. His areas of interest include: land reclamation and irrigated agriculture; development of closed drainage filters from local career materials for reclamation systems; land reclamation, remediation and land conservation; and theoretical and practical foundations of improving reliability of hydraulic land reclamation systems and devices. He has extensively published in highly reputed scientific journals.

Mohamed Mujithaba Mohamed Najim is working as the Vice Chancellor of the South Eastern University of Sri Lanka since 2015 and as a Professor at the Faculty of Science, University of Kelaniya, Sri Lanka. He earned his B.Sc. degree in Agriculture from University of Peradeniya, Sri Lanka in 1994 with a first-class honours pass. He completed his M. Eng. in *Irrigation Engineering and Management* from the Asian Institute of Technology, Thailand in 2000 and was honored with the "Hisamatsu Price" for the outstanding performance in the field of study Irrigation Engineering and Management. He completed his Ph.D. in Water Resources Engineering from the University Putra Malaysia, Malaysia in 2004. Dr. Najim has published extensively (more than 50 full papers out if which some are in Q1–Q4 journals and more than 70 short communications) on the issues associated with water resources in the national and international journals. He has also presented issues in water resources management and agriculture at various international conferences. Hydrological modeling and stream flow assessment, impact of climate change on agriculture and aquaculture, and agricultural water management and wastewater agriculture are his areas of interest. He has attempted to develop strategies to conserve water resources and promote environmental conservation addressing issues of climate change. He was honored by many prestigious awards such as Presidential Awards for Scientific Research, NRC Merit Award for Scientific Publication, Award for the popularization of Science, etc. Dr. Najim started his scientific career in 1995 as an academic at the University of Peradeniya, Sri Lanka. He served at the Sabaragamuwa University of Sri Lanka as an academic from 1997-1998, as an academic again at the University of Peradeniya, Sri Lanka from 1998 to 2007, as a Senior Academic and a Professor at University of Kelaniya, Sri Lanka since 2007, and as the Vice Chancellor of South Eastern University of Sri Lanka since 2015. He serves on the Editorial Boards of many international journals and a member of many professional organizations.

Chapter 13
Readiness of Entrepreneurs Towards Group Performance Development of OTOP Product: A Case Study in Northeastern Thailand

Seksak Chouichom

Abstract One Tambon (Village) One Product (OTOP) model of Thailand is a government project intended to boost the household income for Thai villagers. However, it should be noted that OTOP products, not only encompass general consumable goods, but also other non-consumable products which are the fruit of traditional and indigenous knowledge. In this research, thirty-five OTOP entrepreneurs who engaged in rice production and processing in Nakhon Phanom Province, northeastern Thailand, have been recruited as participants. The sample size of this study was determined by purposive sampling technique. An interview schedule with open-ended and close-ended questions was used to collect the data, which were subsequently interpreted through descriptive statistics. The responses were scored on a five-point Likert's scale ranging from 'very strongly agree (5)' to 'very strongly disagree (1)' to assess the level of readiness or willingness among OTOP rice entrepreneurs. The results of this study revealed that the majority of the respondents were male (51.43%). Most of them were aged between 41 and 60 (40.00%). Their highest education was secondary school (45.45%), and their monthly earnings were under 20,000 Thai Baht (62.85%). The participants were also active members of OTOP rice production along with 21 and 30 others in their communities (34.28%). Most of them have been in the business for 3 to 5 years (42.85%), and they run their business on their independent funding (60.00%). The majority of the respondents (57.14%) received the Food and Drug Administration Certification from Food and Drug Administration (FDA) of Thailand, and were awarded four stars of Thai OTOP Standard. Moreover, they handled their OTOP business and activity as a group cluster (62.85%). The respondents strongly agreed on both production issues (total mean = 3.95), and buying-selling issue (total mean = 3.80). Supporting issue was also a concern among the respondents (total mean = 4.11), followed by competition issue (total mean = 3.80). A large proportion of OTOP entrepreneurs felt that marketing for OTOP products was the most troubling issue (80.00%), especially for the distribution of new products that have been upgraded.

S. Chouichom (✉)
Department of Agricultural Extension and Communication, Faculty of Agriculture, Kasetsart University, 50 Ngam Wong Wan Rd., Lat Yao, Chatuchak, Bangkok 10900, Thailand
e-mail: seksak.c@ku.ac.th; seksak.ku@gmail.com

Keywords OTOP · Thai enterprise · Marketing · Thailand

1 Introduction

In 1997, many South-East Asian countries were hit by social and economic crisis, and a leading agricultural producing country such as Thailand was no exception. A survey result at the time showed that the number of people living below poverty line stood at 6.80 million people in 1994, before climbing to 9.90 million in 1997 (an increase by 15.9%). In addition to that, 1.40 million people had to go unemployed (Sukhothai Thamathirat University, 1997).

In 2011, the Thai Government launched One Tambon (Village) One Product (OTOP) project that capitalizes on local indigineous knowledge for product development and production. The main aims of OTOP project were to alleviate community poverty and to boost income for Thai farmers and entrepreneurs. This project was hoped to support and extend community economy and improve the rural standard of living through the use of indigenous knowledge and rural community resources (Claymome, 2007; Pongsakornrungsilp, 2008).

In fact, the OTOP project in Thailand was initially called OVOP (One Village One Product). Derived from the master model of Oyama village, Otita prefecture, Japan, the idea behind this project was to enhance human resource development, self-reliance, and creative thinking on local and global scale. Following its launch, the site has been visited by many government agents from various countries to cooperate and exchange knowledge (Hiramatsu, 1999, 2008).

In 2017, the OTOP project in Thailand was joined by 4473 groups of villagers who managed to earn over 6855 billion Baht of combined income (100 Thai Baht = 3.27 USD, September, 30, 2019) a year (CDD, 2019). However, when OTOP project is brought into discussion, most Thai people frequently associate it with tangible products such as local Thai silks, local vines, preserved food, fruit juices, just to name a few. Apart from that, the majority of Thai OTOP entrepreneurs remained unfamiliar with the main OVOP principles such as indigenous knowledge transfer, self-reliance, and human resource development (Chaweewan & Kornchakorn, 2012; Kurokawa, 2009; Watunyu, 2019).

What's more worrisome is that there was a great deal of obstacles that keep the OTOP entrepreneurs from upgrading their products to meet requirements for certification (Chouichom, 2019). Some of the ongoing OTOP activities such as rice production are still facing production issues, causing setbacks and weakening the morale of the group members.

In hope to make a contribution to the OTOP industry, the objectives of this study were to survey some socio-economic background of OTOP entrepreneurs who engaged in rice production, to assess their readiness, and to identify some obstacles that could prevent them from achieving their rice production goals.

2 Methodology: Study Area, Data Collection and Analysis

In September, 2018, thirty-five OTOP entrepreneurs who were active in rice production were recruited to participate in this study, which took place in Ban-Tong Sub-district, That-Phanom District, Nakhon Phanom province, Thailand using the purposive sampling technique following the pretest conducted one month earlier. The study employed a semi-structured and structured questionnaire. In order to complement both quantitative and qualitative data, more information was collected through focus group discussion by face-to-face interviews. This analysis used population-based survey to assess the readiness of rice OTOP entrepreneurs in group activity involving rice commodity production. Interviews were conducted on farm sites as well as in the farmers' households. The interview schedules were composed of open-ended and close-ended questions, some of which were capable of eliciting quantitative data. The modified interview schedule includes sixteen statements of willingness or readiness of group performance development regarding rice commodity production. The received responses were scored on a five-point Likert's scale ranging from "strongly agree (5)" to "strongly disagree (1)" (Likert, 1932). The data were then analyzed with the Statistical Package for the Social Science for Windows. Descriptive statistics was also applied to show the percentage, arithmetic mean, minimum and maximum of the gathered data.

3 Results and Discussion

3.1 Demographic Characteristics of OTOP Entrepreneurs

As shown in Table 1, the majority of respondents in this study were male (51.43%) and female (48.57%). Most of them were aged between 41 and 60 (40.00%). The respondents reported to have completed secondary school education (45.45%). Regarding their financial aspect, the participants' monthly income was below 20,000 Thai Baht (62.85%). The data obtained from this study also corresponds with that from Chouichom (2014a) who found that that rice farmers in northeastern Thailand were mostly male (56.84%) and aged between 41 and 50. Despite their primary school education, the farmers were reasonably literate showing no difficulty with reading or writing. However, when looking at their earning, the famers in Surin province appeared to have higher income than those in Nakhon Phanom province. This could be attributed to variables such as atmospheres, soil quality, social and economic circumstances. Respondents were part of their communities with 21–30 members (34.28%), and they have between 3 to 5-year experience in the business (42.85%).

It is also worth noting that the subjects conduct their OTOP business as a group cluster (62.85%), and typically self-finance their activities (60.00%). Likewise, Chaiwet (2007), who conducted a study in Nhong-Khai province, northeastern Thailand, had discovered that communal OTOP enterprise would consist of 11 to 20

Table 1 General profile of OTOP entrepreneur involved in rice production

Data	Frequency (persons)	Percentage
1. Gender		
Male	18	51.43
Female	17	48.57
Total	35	100.00
2. Age (Years)		
21–40	13	37.14
41–60	14	40.00
>60	8	22.86
Total	35	100.00
3. Educational level		
Elementary education	13	40.00
Secondary education	16	45.45
Bachelor's degree	3	10.24
Master's degree	2	4.31
Total	35	100.00
4. Group income (Thai Baht/month)		
<20,000	22	62.85
20,000–30,000	6	17.14
30,001–40,000	4	11.42
40,001–50,000	2	5.72
>50,000	1	2.87
Total	35	100.00
5. Group member (Person)		
1–10	8	22.85
11–20	12	34.28
21–30	13	37.14
>40	2	5.73
Total	35	100.00
6. Group Being (Years)		
<3	13	37.15
3–5	15	42.85
>5	7	20.00
Total	35	100.00
7. Financial sources/Support		
Own group	21	60.00
Commercial bank	1	2.87

(continued)

Table 1 (continued)

Data	Frequency (persons)	Percentage
Provincial fund	6	17.13
Encouraged economic project	6	17.13
Others	1	2.87
Total	35	100.00
8. The Food and drug administration certification (Thai FDA)		
Verified	20	57.14
Unverified	15	42.86
Total	35	100.00
9. Level of Thai OTOP standard (Stars)		
3 Stars	7	20.00
4 Stars	11	31.43
5 Stars	10	28.57
No getting	7	20.00
Total	35	100.00
10. Business management		
Oneself	12	34.28
Group	22	62.85
Others	1	2.87
Total	35	100.00

Source The author

with a start-up capital of 50,000 to 100,000 Thai Baht. Their experience in the business was between 1 and 3 years. The majority of respondents (57.14%) obtained the Food and Drug Administration (FDA) Certification from the FDA of Thailand, and was awarded Thai OTOP Standard of 4 stars (level 5 was the highest). Mukda (2014) stated in her research that the OTOP Standard should be assessed and granted by a related government organization such as Community Development Department (CDD). The number of stars should serve as an indicator of the product quality. For example, four and five-stars rated products can be considered trustworthy, value-added, marketing-needed, and also price-upgraded (CDD, 2019).

3.2 Level of Readiness of OTOP Entrepreneurs Towards Group Performance Development of Rice Production

As shown in Table 2, a strong harmony in readiness and willingness levels (mean score = 3.50–4.49) was present in almost every main issue related to rice production

Table 2 Readiness of OTOP entrepreneur towards group performance development

Details/issues	Readiness levels					Mean
	5 (%)	4 (%)	3 (%)	2 (%)	1 (%)	
1. Production issue						
1.1 Increased member trend	16 (45.71)	13 (37.14)	6 (17.15)	0 (0.00)	0 (0.00)	4.27 (Very strongly agree)
1.2 Hired labor trend	14 (40.00)	15 (42.85)	6 (17.15)	0 (0.00)	0 (0.00)	4.01 (Strongly agree)
1.3 Enough labor	16 (45.71)	13 (37.14)	6 (17.15)	0 (0.00)	0 (0.00)	4.30 (Very strongly agree)
1.4 Provided raw material	15 (42.85)	12 (34.28)	8 (22.87)	0 (0.00)	0 (0.00)	4.20 (Very strongly agree)
1.5 Raw material sources	13 (37.14)	11 (31.43)	11 (31.43)	0 (0.00)	0 (0.00)	4.09 (Strongly agree)
1.6 Applied machine and utilization	13 (37.14)	15 (42.85)	7 (20.01)	0 (0.00)	0 (0.00)	4.14 (Strongly agree)
1.7 Improved production processing	5 (14.28)	11 (31.44)	14 (40.00)	5 (14.28)	0 (0.00)	3.46 (Strongly agree)
1.8 Computer application for group management	6 (17.15)	8 (22.85)	15 (42.85)	6 (17.15)	0 (0.00)	3.73 (Strongly agree)
1.9 Group accounting	4 (11.42)	11 (31.43)	15 (42.85)	5 (14.30)	0 (0.00)	3.37 (Strongly agree)
Total						3.95 (Strongly agree)
2. Buying and selling issue						
2.1 Total sales trend	8 (22.85)	14 (40.00)	13 (37.15)	0 (0.00)	0 (0.00)	3.85 (Strongly agree)
2.2 Total customers trend	9 (25.71)	15 (42.85)	11 (31.44)	0 (0.00)	0 (0.00)	3.93 (Strongly agree)

(continued)

Table 2 (continued)

Details/issues	Readiness levels					Mean
	5 (%)	4 (%)	3 (%)	2 (%)	1 (%)	
2.3 Previous clients trend	9 (25.71)	16 (44.18)	10 (30.11)	0 (0.00)	0 (0.00)	3.95 (Strongly agree)
2.4 New clients trend	6 (17.14)	14 (40.00)	15 (42.86)	0 (0.00)	0 (0.00)	3.75 (Strongly agree)
2.5 Clients data-based management	5 (14.28)	13 (37.14)	14 (40.00)	3 (8.58)	0 (0.00)	3.56 (Strongly agree)
Total						3.80 (Strongly agree)
3. Supporting issue						
3.1 Supporting from clients	12 (34.28)	15 (42.86)	8 (22.86)	0 (0.00)	0 (0.00)	4.11 (Strongly agree)
3.2 Supporting from community official	13 (37.11)	16 (45.71)	6 (17.15)	0 (0.00)	0 (0.00)	4.19 (Strongly agree)
3.3 Supporting from district office	12 (34.28)	14 (40.00)	9 (25.72)	0 (0.00)	0 (0.00)	4.11 (Strongly agree)
3.4 Supporting from provincial office	11 (31.42)	15 (42.86)	9 (25.72)	0 (0.00)	0 (0.00)	4.06 (Strongly agree)
3.5 Supporting from Bangkok / central office	12 (34.28)	15 (42.86)	8 (22.86)	0 (0.00)	0 (0.00)	4.11 (Strongly agree)
Total						4.11 (Strongly agree)
4. Competition issue						
4.1 Cumulative indigenous	15 (42.86)	10 (28.57)	10 (28.57)	0 (0.00)	0 (0.00)	4.15 (Strongly agree)
4.2 Appropriated price	14 (40.00)	13 (37.14)	8 (22.86)	0 (0.00)	0 (0.00)	4.20 (Very strongly agree)
4.3 Sale promotion strategies	13 (37.14)	15 (42.86)	7 (20.00)	0 (0.00)	0 (0.00)	4.16 (Strongly agree)
4.4 Sale and distribution channel	7 (20.00)	15 (42.86)	13 (37.14)	0 (0.00)	0 (0.00)	3.86 (Strongly agree)

(continued)

Table 2 (continued)

Details/issues	Readiness levels					Mean
	5 (%)	4 (%)	3 (%)	2 (%)	1 (%)	
4.5 Product development	7 (20.00)	13 (37.14)	15 (42.86)	0 (0.00)	0 (0.00)	3.76 (Strongly agree)
4.6 Product quality control	3 (8.57)	14 (40.00)	13 (37.14)	5 (14.29)	0 (0.00)	3.42 (Strongly agree)
4.7 Product standard and certification	3 (8.57)	13 (37.14)	15 (42.86)	4 (11.43)	0 (0.00)	3.40 (Strongly agree)
4.8 Award reception	4 (11.43)	14 (40.00)	13 (37.14)	4 (11.43)	0 (0.00)	3.50 (Strongly agree)
Total						3.80 (Strongly agree)

development upon a closer look at the numbers, participants were highly confident that labour was not going to be an issue (mean $= 4.30$) and were very positive about having more members to join the group (mean $= 4.27$). These implicate that laborer and trend of member in the community can support and enhance rice production in the community. This could be a positive sign as Mukda (2014) had stated that increased productivity goes hand in hand with good quality laborers. Moreover, Stevenson (2002) asserted that labors and employees can play an important role in production efficiency and quality. However, a study by Arthitkawin and Wongwirat (2014) mentioned that the entrepreneurs in northern Thailand did not employ many people and laborers in their enterprise, which could manufacture OTOP products with good skills comparable to food machines. Thus, Chouichom (2019) found that the high participation in production and implementation stage was due to the inclination of group members to act like business owners.

In case of the buying and selling issue, entrepreneurs also had a strong agreement particularly on previous client trend (mean $= 3.95$) and total customers trend (mean $= 3.93$) which allowed the products to reach more consumers and generate more income in the future. A study by Sangkhasuk et al. (2017) indicated that the utilization of electronic database can improve the performance of production and inventory management and also they affirmed more that the use of electronic channels will allow customers to access the data more conveniently (64.75%). The easy information access will help a business to stand out and secure its competitive edge (Hogefroster, 2014). The utilization of information technology is also recognized by Kotler (1997) to help marketers identify prospective customers, and stay in touch with old customers.

As far as supporting issues are concerned, the entrepreneurs expressed their strong agreement level (mean = 4.19). This denotes that rural governing body has become involved in rice production development of OTOP groups. Srisurapol and Sripokangkul (2016) mentioned in their research that rural community organizations always lend their support to OTOP producers by providing them with information useful for product development, arranging necessary training and even financing OTOP product upgrades. To help OTOP producers stay afloat in today's economy, government organizations should support OTOP entrepreneurs in every matter until they are able to stand on their own feet (Chiarakul, 2014).

Regarding the competition issue, the respondents stated that reasonable price for rice products was the most important aspect (mean = 4.20). The participants also strongly agreed that both sale promotion strategies and cumulative indigenous were of great importance (mean = 4.16 and 4.15, respectively). This means that group member knowledge and sale promotion can support product development, making the price more competitive. Puttaphoompitak (2015) declared that Thai OTOP entrepreneurs should continue to develop locally-made products, so that they can compete the ASEAN Economic Community (AEC) market. Furthermore, she also gave advice that more entrepreneurs should utilize rural indigenous knowledge together with technological innovation to enhance OTOP products. Chouichom (2019) stated that one of problem of OTOP herbal products was the expensive production machines. According to Chiarakul (2014), the cost of raw materials for decorating products and souvenirs could at times be very even for basic products.

3.3 OTOP Entrepreneurs' Constraints from This Study

As illustrated in Table 3, the entrepreneurs considered marketing to be their biggest problem (80.00%). This is because the lack of product development could make OTOP products less attractive to customers. Moreover, without clear product standardization, consumers would hasitate to make their purchase. Chouichom (2014b) and Claymome (2007) stated that the main problems facing OTOP producers were lack of marketing knowledge and lack of physical storefront. In addition, Nuisuk (2006) raised a concern about poor quality of OTOP products, lack of marketing knowledge, and limited marketing channels. Chiarakul (2014) and Chouichom (2019) recommended that entrepreneurs should constantly adapt and innovate to make their products more functional and appealing to customers. They should also select effective distribution channels for their OTOP products. Benrit (2008) and Johanson (1996) emphasized the need of having electronic distribution channel by stating that it has become so fundamental that marketing OTOP products would not be possible without it (Wathinee, 2006). Suh and Kim (2014) conducted a study on Small and Mid-Size Enterprises (SMEs) in South Korea and found that entrepreneurs have to pay attention to technology and innovation, so that they can remain competitive in terms of price performance, product quality and after-sale services, all of which can have an effect on customer satisfaction.

Table 3 OTOP entrepreneurs' constraints in rice production from this study

Details	Frequency (Persons)	Percentage (100%)
1. Production	22.00	62.85
2. Modern machine	18.00	51.40
3. Raw material	16.00	45.70
4. Production skills	15.00	42.85
5. Marketing	28.00	80.00
6. Government supporting	13.00	37.10
7. New knowledge and technology	19.00	54.30
8. Packaging	23.00	65.70
9. Budget or fund	17.00	48.50
10. Public relations	20.00	57.10

Lastly, new or emerging entrepreneurs need to be assisted with funding, business knowledge and regulations in order to propel their business enterprises (Numprasertchai et al., 2018). Also, there are other scholars who proposed some other different factors such as quality of product and responsiveness to the markets (William et al., 2005; Chouichom and Liao, 2019).

4 Conclusion

This study examined the level of the readiness and willingness of OTOP entrepreneurs towards group performance development among rice farmers in Nakhon Panom Province, northeastern Thailand. Most OTOP entrepreneurs expressed positive and strong agreement in readiness in group performance development on all four main issues.

The results of this study showed that most respondents had very strong agreement on the aspect of labor (mean = 4.30) and increased member trend (mean = 4.27). The main obstacles for the OTOP entrepreneurs observed in this survey included lack of marketing knowledge for OTOP products, difficulty in improving packaging design, and inferior production system.

References

Arthitkawin, A., & Wongwirat, Kh. (2014). OTOP Participatory integrated product management toward 4 to 5 stars level: A case study of Laow Hua Kee Chinese Sausage. *Journal of Modern Management Science, 2*(7), 26–37.

Benrit, P. (2008). The management of one tambon one product: A case study of the rubber-leaved flowers group in Pattani. Prince of Songkla University, Thailand. *Journal of Humanities and Social Sciences, 4* (1), 7–26.

CDD. (2019). *One Tambon (Village) one product guideline for community enterprise 2019.* Community Development Department (CDD). Ministry of the Interior.

Chaiwet, S. (2007). *Management the business community group in Nong-Khai Province.* Faculty of Business Administration, Mahasarakham University master's thesis. (in Thai, with English abstract).

Chaweewan, D., & Kornchakorn, A. (2012). Similarity and difference of One Village One Product (OVOP) for rural development strategy in Japan and Thailand. *Japanese Studies Journal Special Issue: Regional Cooperation for Sustainable Future in Asia, 2*(10), 52–62.

Chiarakul, T. (2014). The problems and the adaptation of OTOP to AEC. *Executive Journal of the Executive Standard Department, Rahjapaht Institutional Office, 34*(1), 177–191. (in Thai, with English abstract).

Chouichom, S. (2014a). Some socio-economic factors affecting farmers' participation of agricultural extension education effort: A case study in northeastern Thailand. In M. Behnassi (Ed.), *Science, policy and politics of modern agricultural system: global context to local dynamics of sustainable agriculture.* (pp. 47–60). Springer.

Chouichom, S. (2014b). Preferences, expectations, and opinions of consumers towards decision making in fiber food products purchasing: A case study in Bangkok, Thailand. In *Burapaha university international conference 2014: Global warming and its impacts* (pp. 23–31). Burapaha University.

Chouichom, S. (2019). Science, technology and Innovation (STI) for OTOP upgrade program. In *Thailand institute of scientific and technological research (TISTR)*, Ministry of Science and Technology. (completed report in Thai).

Chouichom, S., & Liao, L. M., et al. (2019). Participation of female farmers groups in *kai* algal processing and production in Northern Thailand. In M. Behnassi (Ed.), *Human and environmental security in the era of global risks; perspectives from Africa, Asia and the Pacific Islands.* (pp. 265–274). Springer Science.

Claymome, Y. (2007). A study for sustainable local development through one town one product: An overview of OTOP in Thailand. In *International OVOP policy association (IOPA) Conference.* Zhejiang University.

Hiramatsu, M. (1999). Think globally——The "One Village, One Product" movement transcends generation and national borders. In *Keynote speech in international symposium in commemoration of international cooperation Day 1999*, at Sankei Kaikan.

Hiramatsu, M. (2008). *One Village, One product spreading throughout the world.* Oita Japan: Office: Oita OVOP International Exchange Promotion Committee.

Hogefroster, M. (2014). Future challenges for innovation in SMEs in the Baltic Sea region. *Procedia-Social and Behavioral Sciences, 110*, 241–250.

Johanson, B. (1996). The dynamics of entrepreneurial networks. In Reynolds, P., et al. (Eds.). *Frontiers of entrepreneurship research* (pp. 253–267).

Kotler, P. (1997). *Marketing management: Analysis, planning, implementation and control.* (9th ed.). Prentice-Hall.

Kurokawa, K. (2009). Effectiveness and limitations of the One Village One Product (OVOP) approach as a government-led development policy: Evidence from Thai OTOP, Studies in Regional Science. *The Journal of the Japan Section of the Regional Science Association International, 39*(4), 977–989.

Likert, R. (1932). A technique for the measurement of attitudes. *Archives of Psychology, 140*, 5–55.

Mukda, W. (2014). Guidelines management of product manufacturing group on One Tambon One Product in Tak Province. *Suan Dusit University Research Journal, 10*(1), 187–205. (in Thai, with English abstract).

Nuisuk, Ch. (2006). *The production and cost of herbal soap in One Tambon One Product (OTOP).* Faculty of Economics, Chulalongkorn University master's thesis. (in Thai, with English abstract).

Numprasertchai, H., Srinammuang, P., & Skuna, J. (2018). Critical success factors of Thai SMEs in the new product and service development sector. In *Integrated economy and society: Diversity, creativity, and technology* (pp. 389–396), 16–18 May 2018.

Pongsakornrungsilp, S. (2008). The important factors in consumers' purchasing decision making of One Tambon One Product goods in Nakhon Sri Thammatrat Province. *Songklanakarin: E-Journal of Social Sciences and Humanities, 14*(2), 307–320.

Puttaphoompitak, W. (2015). OTOP to AEC. *Western University Research Journal of Humanities and Social Science, 1*(1), 100–112. (in Thai, with English abstract).

Sangkhasuk, R., Naklungka, K., Ekphon, W., & Surasawasdee, W. (2017). Development of e-commerce channel among community enterprise network. *Phanakhorn Rajapahat Research Journal (Humanities and Social Sciences), 12*(1), 38–49. (in Thai, with English abstract).

Srisurapol, P., & Sripokangkul, S. (2016). Guidelines for the strengthening of vocation groups in Nachumsang Sub-District, Phu Wiang District, Khon Khean Province, Thailand. *Phimoldhamma Research Institute Journal, 4*(1), 85–98. (in Thai, with English abstract).

Stevenson, W. J. (2002). *Operations management* (9th edn.). McGraw Hill-Irwin Inc.

Sukhothai Thamathirat University. (1997). *Economic crisis effective in 1997*. Sukhothai Thamathirat University. (in Thai).

Suh, Y., & Kim, M. (2014). Internationally leading SMEs versus internationalized SMEs: Evidence of success factors from South Korea. *International Business Review, 23*, 115–129.

Wathinee, P. (2006). *OTOP project evaluation: A case study of metal ornamented wood product groups, Tambon San Kamphaeng, Ampkoe San Kamphaeng, Changwat, Chiang Mai*. Chiang Mai University master's thesis. (in Thai, with English abstract).

Watunyu, J. (2019). OTOP and OVOP: Analysis of placing a value on development. *Thai International Journal of East Asian Studies, 23*(1), 148–167.

William, G., James, M., & Susan, M. (2005). *Fundamentals of business: Starting a small business*. McGraw-Hill/Irwin.

Seksak Chouichom is a Lecturer at the Department of Agricultural Extension and Communication, Faculty of Agriculture, Kasetsart University, Bangkhen, Bangkok, Thailand. He has conducted research and published widely on Thai organic rice and socio-economic agriculture as a part of his Ph.D. research. He holds both B.Sc. and M.S. degrees in Agricultural Extension and Ph.D. in Agricultural Economics with a focus on rural farming and development systems from Kasetsart University, Thailand and Hiroshima University, Japan, respectively.

Postface

This contributed volume deals with the daunting challenges to food production and security for Asia, Middle East, and Africa. It focuses on the often-cited need for transformative change due to current and future resource constraints brought about by climate change, water scarcity, and soil degradation. The chapters analyse many of these dynamics within the context of individual countries which are reflective of the overall situation within the region. These dynamics are complicated because they deal with interacting systems with complex and changing characteristics. Food, water, agriculture, climate, biodiversity, and land/soils have economic, environmental, and social dimensions that require sustainable practices as a pre-requisite for their long-term viability to support human life and livelihoods.

This volume identifies and assesses the impacts of different risk dynamics – especially environmental/climate risks and resource scarcity – which are currently challenging agricultural production and food security in many vulnerable regions. The book contains 13 chapters which are mostly empirically based and provide context-specific analyses and recommendations based on a variety of case studies from Africa (Morocco, Tunisia, Egypt, Nigeria, and sub-Saharan countries), Middle East (Gulf countries), and Asia (Pakistan and Thailand).

Agriculture is critical because it provides inputs into the food production system, which nourishes humans and keeps people food secure. Yet agriculture does not operate in a vacuum and it is facing many challenges from climate change, water scarcity, and other resource constraints. However, agriculture is not only affected by such challenges but contributes as well in worsening their magnitude and implications (for instance, agriculture is affected by water scarcity and climate change but also contributes to water insecurity and GHG emissions through its environmental and climate footprint). Furthermore, agriculture, especially in parts of Asia, Middle East, and Africa, will face higher temperatures, changing rainfall patterns, and more floods and droughts, along with increasing food demands. These regions and their farmers need to find ways to adapt and adopt climate-smart agricultural technologies that will help them adjust to these changing weather patterns while lowering their carbon footprint.

Water scarcity has long been a concern because of population pressures, but climate change and the concomitant changing rainfall/snow patterns have greatly enhanced its severity. Flooding will be more pronounced in some areas, while droughts will be longer and more severe in others. Managing water availability and access under such conditions will be challenging given increased urbanization. Economic and social issues again come into play concerning water rights and the pricing of water. Water has historically been treated as a low-priced or free input in the past, but this has resulted in great waste that cannot be allowed in the future.

The problem of food security is equally complex. The confounding factors of economic, social, and political development come into play, along with constraints of local food supply. For many food-insecure people, the best way to increase food security is through local food production, but this pushes up against the constraints mentioned earlier, which are often more difficult where food insecurity is most prevalent. Many families do not have the means to purchase food and are not able to produce it. This calls for policy actions and political developments to supply the necessary foodstuffs without jeopardizing future economic and social development. There has been great progress in reducing food insecurity in the past twenty years, but these accomplishments are seriously threatened by future climate scenarios.

The food production and security challenges faced by Asia, Middle East, and Africa in the future decades differ by country, but there are many common elements. Through the insights shared in this volume, the editors aim to contribute to the growing academic literature pertaining to food security and food systems while enhancing political discussions and policy agendas on how to effectively address these current and projected challenges. Interested scholars, students, practitioners, and decision makers worldwide will find the publication a useful and instructive reading.

Lightning Source UK Ltd.
Milton Keynes UK
UKHW020620010822
406667UK00002B/18